Case Studies in Small Animal Point of Care Ultrasound

This illustrated practical guide covers the many facets of small animal point-of-care ultrasound (POCUS) using clinical cases commonly seen in practice. It details specific point-of-care ultrasound techniques, views, interpretations, pitfalls and knowledge gaps, highlighting the utility of initial and serial POCUS for both diagnostic and patient management purposes. Divided into seven sections, the book covers ultrasound of the pleural space and lungs, the heart, and abdomen, as well as the Caudal Vena Cava Collapsibility Index (CVC-CI), miscellaneous, and advanced cases.

Authored by leading authorities in the field, this guide recognizes the strengths of POCUS in screening and monitoring while also highlighting the limitations of using it as a sole diagnostic tool. Focusing on commonly asked questions and applications, the authors take a binary question approach to veterinary POCUS, which has been shown to decrease errors in human POCUS.

Case videos to accompany Chapters 4, 5, 7, 9, 11, 13 and 15 are available on the book's Companion Website at https://routledgetextbooks.com/textbooks/9780367547257/. These can be accessed directly on a smartphone using the QR codes in the relevant chapters.

This full-color handbook is the ideal reference for small animal and mixed practitioners, including general veterinary practitioners, emergency veterinarians, interns, residents, and veterinary students.

VETERINARY COLOR HANDBOOK SERIES

PUBLISHED TITLES

A Colour Handbook of Skin Diseases of the Dog and Cat UK Version, Second Edition, *by Patrick J. McKeever, Tim Nuttall, and Richard G. Harvey*

Urinary Stones in Small Animal Medicine: *A Colour Handbook,* *by Albrecht Hesse and Reto Neiger*

Small Animal Dental, Oral and Maxillofacial Disease: *A Colour Handbook,* *by Brook A. Niemiec*

Small Animal Emergency and Critical Care Medicine: *A Color Handbook,* *by Elizabeth A. Rozanski and John E. Rush*

Small Animal Fluid Therapy, Acid-base and Electrolyte Disorders: *A Color Handbook,* *by Elisa M. Mazzaferro*

Skin Diseases of the Dog and Cat, Third Edition, *by Tim Nuttall, Melissa Eisenschenk, Nicole A. Heinrich, Richard G. Harvey*

Small Animal Anesthesia and Pain Management: *A Color Handbook, Second Edition,* *edited by Jeff Ko*

Infectious Diseases of the Dog and Cat: *A Colour Handbook,* *edited by Scott Weese, Michelle Evason*

Equine Anesthesia and Pain Management: *A Color Handbook,* *edited by Michele Barletta, Jane Quandt, Rachel Reed*

Case Studies in Small Animal Point of Care Ultrasound: *A Color Handbook,* *edited by Erin Binagia, Søren Boysen, Tereza Stastny*

For more information about this series, please visit:
https://www.routledge.com/Veterinary-Color-Handbook-Series/book-series/CRCVETCOLHAN

Case Studies in Small Animal Point of Care Ultrasound

A COLOR HANDBOOK

EDITED BY
ERIN BINAGIA, SØREN BOYSEN, TEREZA STASTNY

CRC Press is an imprint of the
Taylor & Francis Group, an **informa** business

Cover image is courtesy of Erin Binagia.

First edition published 2025
by CRC Press
2385 NW Executive Center Drive, Suite 320, Boca Raton FL 33431

and by CRC Press
4 Park Square, Milton Park, Abingdon, Oxon, OX14 4RN

CRC Press is an imprint of Taylor & Francis Group, LLC

©2025 selection and editorial matter, Erin Binagia, Søren Boysen, Tereza Stastny; individual chapters, the contributors

Reasonable efforts have been made to publish reliable data and information, but the author and publisher cannot assume responsibility for the validity of all materials or the consequences of their use. The authors and publishers have attempted to trace the copyright holders of all material reproduced in this publication and apologize to copyright holders if permission to publish in this form has not been obtained. If any copyright material has not been acknowledged please write and let us know so we may rectify in any future reprint.

Except as permitted under U.S. Copyright Law, no part of this book may be reprinted, reproduced, transmitted, or utilized in any form by any electronic, mechanical, or other means, now known or hereafter invented, including photocopying, microfilming, and recording, or in any information storage or retrieval system, without written permission from the publishers.

For permission to photocopy or use material electronically from this work, access www.copyright.com or contact the Copyright Clearance Center, Inc. (CCC), 222 Rosewood Drive, Danvers, MA 01923, 978-750-8400. For works that are not available on CCC please contact mpkbookspermissions@tandf.co.uk

Trademark notice: Product or corporate names may be trademarks or registered trademarks and are used only for identification and explanation without intent to infringe.

Library of Congress Cataloging-in-Publication Data
Names: Binagia, Erin, editor. | Boysen, Soren, editor. | Stastny, Tereza, editor.
Title: Case studies in small animal point of care ultrasound : a color handbook / edited by Erin Binagia, Soren Boysen, Tereza Stastny.
Other titles: Veterinary color handbook series.
Description: First edition. | Boca Raton : CRC Press, 2024. | Series: Veterinary color handbook series | Includes bibliographical references and index. |
Identifiers: LCCN 2024024975 (print) | LCCN 2024024976 (ebook) | ISBN 9780367547257 (pbk) | ISBN 9781032566566 (hbk) | ISBN 9781003436690 (ebk)
Subjects: MESH: Pets–physiology | Ultrasonography–veterinary | Animal Diseases–diagnostic imaging | Point-of-Care Testing | Case Reports
Classification: LCC SF772.58 (print) | LCC SF772.58 (ebook) | NLM SF 772.58 | DDC 636.089/607543–dc23/eng/20240703
LC record available at https://lccn.loc.gov/2024024975
LC ebook record available at https://lccn.loc.gov/2024024976

ISBN: 9781032566566 (hbk)
ISBN: 9780367547257 (pbk)
ISBN: 9781003436690 (ebk)

DOI: 10.1201/9781003436690

Typeset in Janson Text LT Std
by Deanta Global Publishing Services, Chennai, India

Case videos to accompany Chapters 4, 5, 7, 9, 11, 13 and 15 are available on the book's Companion Website at https://routledgetextbooks.com/textbooks/9780367547257/. These can be accessed directly on a smartphone using the QR codes in the relevant chapters.

Printed and bound in Great Britain by
TJ Books Limited, Padstow, Cornwall

I dedicate this book to my brother John and to the rest of my family, who always support my projects, dreams, and adventures. I would also like to extend this dedication to my resident mate, friend, and co-author, Dr Erin Binagia, the best collaborator one could ask for; my professor, mentor, and co-author, Dr Søren Boysen, who inspired my initial love of critical care and veterinary POCUS; and lastly, to the next generation of veterinary POCUS aficionados!

T

I dedicate this book to my parents who have always supported me throughout my adventures and journey on this round football called earth. To my resident mates and mentors who helped kickstart the integration of FAST and POCUS into veterinary medicine, my co-editors (Erin and T), without whose convincing this project would have stalled before it even got started. The amazing authors who humbled me with their knowledge, dedication, and fantastic case contributions. To the numerous POCUS students, colleagues, and co-researchers who always make the educational journey much easier with their endless enthusiasm to question and learn, and finally, my family. My wife and two lovely daughters who endured long nights, weekends, missed football matches, and oaths promising this would be the last chapter and book I will ever write… at least until the next one…

Søren

I dedicate this book to my daughter, Eloise, who was born into this project and has been with me every step of the way. To my mother, Gail, mother-in-law, Carmen, and husband, Will, who have supported me and entertained my daughter so that I can pursue my dreams – without them, this book would have taken much longer to complete. To my resident mate and friend T– we have tackled many challenges together over the years, and I hope our bond will continue to bring good things to the veterinary field, to ourselves, and to others in this world. To Søren – thank you for entrusting us with your book idea and allowing us to bring it to life. You allowed two young, eager veterinarians to convince you to go forward together as a team and gave us the opportunity to add an exciting and useful textbook to the veterinary literature. Thank you to all the authors who contributed cases, images, and videos – without you, this book would not exist. Lastly, thank you to all the students, interns, and residents I've ever worked with – you give me the motivation to start new projects and continue learning.

Erin

CONTENTS

VII

Preface	xvi
About the Editors	xviii
List of Contributors	xx

Section 1	**Introduction to veterinary point-of-care ultrasound**	**1**
CHAPTER **1**	**Basic Ultrasound Physics** *Erin Binagia*	3
CHAPTER **2**	**Binary Question Approach to Point-of-Care Ultrasound** *Søren Boysen*	10
CHAPTER **3**	**Basics of Image Acquisition for Point-of-Care Ultrasound** *Tereza Stastny*	13

Section 2	**Pleural space and lung ultrasound (PLUS): Section Editor Søren Boysen**	**17**
CHAPTER **4**	**Introduction to Pleural Space and Lung Ultrasound (PLUS)** *Søren Boysen*	19
CHAPTER **5**	**Pleural and Lung Ultrasound (PLUS) Cases**	35
Case 5.1	**Dachshund with Respiratory Distress and Coughing** *Hugo Swanstein and Kris Gommeren*	36
Case 5.2	**Labrador Retriever with Mild Dyspnea and Profuse Bleeding from a Parasternal Penetrating Injury** *Simone Cutler and Daria Starybrat*	42

Case 5.3	**Mixed Breed Dog with Respiratory Distress Following Motor Vehicular Trauma** *Søren Boysen*	48
Case 5.4	**Havanese with Acute Respiratory Distress, Known Myxomatous Mitral Valve Disease, and 3-Week History of Coughing** *Chiara Debie and Christopher Kennedy*	53
Case 5.5	**Cane Corso with Acute Hemoptysis, Lethargy, and Anorexia** *Céline Pouzot-Nevoret*	59
Case 5.6	**English Setter Subjected to Prolonged Anesthesia and Abnormal Lung Function** *Chiara Di Franco and Angela Briganti*	64
Case 5.7	**Beagle with Acute Onset of Respiratory Distress and Anuria Following a Laparotomy** *Andrea Armenise and Anna De Nitto*	70
Case 5.8	**Mixed Breed K9 with Progressive Lethargy and Hypoxemia for 3 Weeks** *Serge Chalhoub*	75
Case 5.9	**Dogue de Bordeaux with Exercise Intolerance and a 5-Day History of Tachypnea** *Emilie Van Renterghem and Anne-Christine Merveille*	79
Case 5.10	**Abyssinian with Pyrexia, Coughing, and Prior History of Pyothorax** *Tove M Hultman and Ivayla Yozova*	87
Case 5.11	**English Springer Spaniel with Acute Onset of Respiratory Distress and Fever** *Julie Menard*	91

Case 5.12	**Non-Ambulatory Labrador with Tachypnoea and Moderate Respiratory Effort Following Motor Vehicular Trauma** *Marta Garcia Arce and Daria Starybrat*	97
Case 5.13	**Border Collie with Acute Onset of Respiratory Distress after Having a Leash Wrapped Around Her Neck** *Julia Delle Cave and Jo-Annie Letendre*	101
Case 5.14	**Toy Poodle with Hemolytic Anemia and Severe Respiratory Distress Following Prednisone and a Whole Blood Transfusion** *Andrea Armenise*	107
Case 5.15	**Mixed Breed K9 with Acute Lethargy and Anorexia of 3 Days' Duration** *Serge Chalhoub*	116
Case 5.16	**Labrador Retriever with Respiratory Distress and Superficial Abrasions Following Motor Vehicular Trauma** *Søren Boysen*	122
Case 5.17	**Chihuahua with Severe Dyspnea and Tremors Following Milbemycin Toxicity and Intra-Lipid Therapy** *Elizabeth Gribbin and Rachael Birkbeck*	127
Case 5.18	**West Highland White Terrier with Progressive Respiratory Distress, Dry Cough, and Exercise Intolerance of 2 Months' Duration** *Elodie Rizzoli and Géraldine Bolen*	132
Case 5.19	**Chocolate Lab with Respiratory Distress and Hind Limb Lameness Following Motor Vehicular Trauma** *Søren Boysen*	137
Case 5.20	**Bloodhound with Severe Myositis Secondary to a Snake Bite That Occurred 4 Days Previously** *Chiara Di Franco and Angela Briganti*	142

| Case 5.21 | **Stable Female Spayed Domestic Shorthair Following Motor Vehicle Trauma**
Søren Boysen | 147 |

Section 3 Veterinary cardiac point-of-care ultrasound: Section Editor Tereza Stastny — 153

| CHAPTER 6 | **Introduction to Veterinary Cardiac Point-of-Care Ultrasound**
Tereza Stastny | 155 |

CHAPTER 7	**Cardiac Point-of-Care Ultrasound Cases**	162
Case 7.1	**Chihuahua with a 3-Day History of Heart Murmur, Tachypnea, and a 24-Hour History of Severe Respiratory Distress** *Tereza Stastny*	163
Case 7.2	**Golden Retriever with a 2-Day History of Coughing and Exercise Intolerance Following an Episode of Vomiting** *Chiara Debie and Christopher Kennedy*	168
Case 7.3	**Acute Collapse in a Cavalier King Charles Spaniel** *Shari Raheb and Xiu Ting Yiew*	174
Case 7.4	**Golden Retriever with Collapse, Lethargy, Ataxia, and Mild Respiratory Distress** *Julien Guillaumin*	181
Case 7.5	**Heartworm-Positive German Shepherd Mix with Sudden Onset Lethargy, Weakness, and Dark Red–Colored Urine** *LM Bacek and Kendon Wu Kuo*	187
Case 7.6	**Main Coon with Acute Onset of Labored Breathing** *Liz-Valérie S Guieu and Charles T Talbot*	192

Case 7.7	**Great Dane with a 5-Day History of Weakness, Syncope, Difficulty Breathing, Occasional Coughing, and Abdominal Distension** *Priscilla Burnotte*	198
Case 7.8	**Boxer with Acute Vomiting, Cough with Terminal Wretch, Tachypnea, Weakness, Weight Loss, and Lethargy** *Dana Caldwell and Amanda Liggett*	206
Case 7.9	**Terrier with a 2-Week History of Cough, Exercise Intolerance, Progressive tachypnea, and Suspected Syncope** *Dave Beeston and Kris Gommeren*	213
Case 7.10	**Domestic Short Hair Cat Referred for Cardiovascular Collapse Following Suspected Trauma** *Steffi Jalava and Ivayla Yozova*	220
Section 4	**Vascular applications of veterinary point-of-care ultrasound: Section Editor Tereza Stastny**	**227**
CHAPTER **8a**	**Introduction to Ultrasound-Guided Vascular Access** *Tereza Stastny*	229
CHAPTER **8b**	**Introduction to Caudal Vena Cava Collapsibility Index** *Tereza Stastny*	234
CHAPTER **9**	**Vascular Point-of-Care Cases**	238
Case 9.1	**Borzoi with Acute Onset of Vomiting, Followed by Severe Dyspnea and Hyperthermia** *Laurentin Duriez and Kris Gommeren*	239
Case 9.2	**Golden Retriever with a 24-Hour History of Acute Vomiting, Diarrhea, and Severe Lethargy** *Aurélie Jourdan and Kris Gommeren*	246

Case 9.3	**Beauceron with Status Epilepticus Secondary to Heat Stroke, Fever (40°C [104°F]), and Severe Phlebitis** *Laurentin Duriez and Kris Gommeren*	253
Case 9.4	**Oliguric Domestic Shorthair on IV Fluids for Suspected Lily Intoxication that has Gained Weight while Hospitalized** *Tove M Hultman and Ivayla Yozova*	257

Section 5	**Abdominal point-of-care ultrasound: Section Editor Erin Binagia**	**265**
CHAPTER 10	**Introduction to Abdominal Point-of-Care Ultrasound** *Erin Binagia*	267
CHAPTER 11	**Abdominal Point-of-Care Cases**	273
Case 11.1	**Pekinese with a 7-Day History of Progressive Obtundation, Polyuria, Polydipsia, Dysuria, Stranguria, and Pollakiuria** *Georgina Hall and Erica Tinson*	274
Case 11.2	**Maltese with a 2-Day History of Lethargy, Vomiting, Inappetence, and Acute Abdominal Pain** *Andrea Armenise*	278
Case 11.3	**Shih Tzu with Acute Collapse and Abdominal Pain Following 10-Days of Inappetence, Vomiting, Polyuria, and Polydipsia** *Esther Gomez Soto, Thom Watton, and Laura Cole*	283
Case 11.4	**Domestic Shorthair with Polyuria, Polydipsia, Vomiting, and Hyporexia** *Xiu Ting Yiew*	288
Case 11.5	**Chihuahua with a 2-Day History of Lethargy, Anorexia, and Vaginal Discharge** *LM Bacek and Kendon Wu Kuo*	294

Case 11.6	**Samoyed with a 4-Day History of Vomiting after Passing Parts of a Rubber Ball** *Anais Allen-Deal and Daria Starybrat*	299
Case 11.7	**Labrador Retriever with Acute Collapse and a 24-Hour History of Lethargy and Anorexia** *Charles T Talbot and Liz-Valérie S Guieu*	305
Case 11.8	**Pit Bull Terrier with Dystocia** *Igor Yankin*	312
Case 11.9	**Golden Retriever with a 2-Day History of Lethargy and Hyporexia Following Ingestion of an Oxtail Treat** *Nadine Jones and Erica Tinson*	318
Case 11.10	**Mixed-Breed K9 with Acute Vomiting** *Erin Binagia*	324
Case 11.11	**Husky Mix with Obtundation, Lethargy, Vomiting, and Anorexia 48 Hours after Landing on the Tailgate Jumping into a Truck** *Serge Chalhoub*	329
Case 11.12	**Domestic Shorthair with a 2-Day History of Lethargy and Vomiting** *Annelies Valcke and Laura Cole*	333
Case 11.13	**Spaniel with a 3-Day History of Lethargy and a Single Episode of Vomiting** *Olivia X Walesby and Daria Starybrat*	339
Case 11.14	**Staffordshire Bull Terrier with 2 Days of Vomiting, Anorexia, and No Observed Urination for 48 Hours Despite Water Intake** *Alexandra Nectoux and Mark Kim*	343
Case 11.15	**Rottweiler with Frequent Regurgitation and Anorexia Following Gastrointestinal Foreign Body Removal** *Nadine Jones and Erica Tinson*	348

| Case 11.16 | Belgian Shepherd with 1 Week of Severe Exercise Intolerance Following Trauma that Occurred while Jumping onto a Boat | 354 |

Pauline Jaillon and Kris Gommeren

Section 6   Miscellaneous point-of-care ultrasound cases: Section Editor Erin Binagia   361

| CHAPTER 12 | Introduction to Miscellaneous Point-of-Care Ultrasound Cases | 363 |

Erin Binagia

| CHAPTER 13 | Miscellaneous Point-of-Care Ultrasound Cases | 364 |

| Case 13.1 | Cavalier King Charles with a Grand Mal Seizure 2 Hours Prior to Presentation and a 1-Month History of Abnormal Behavior | 365 |

William Glenn Lane

| Case 13.2 | Domestic Short Hair that Presents with Dyspnea an Hour after Recovering from Dental Treatment Performed under General Anesthesia | 370 |

Serge Chalhoub

Section 7   Systemic and advanced point-of-care ultrasound cases: Section Editor Erin Binagia   375

| CHAPTER 14 | Introduction to Systemic and Advanced Point-of-Care Ultrasound Cases | 377 |

Erin Binagia

| CHAPTER 15 | Systemic and Advanced Point-of-Care Ultrasound Cases | 378 |

| Case 15.1 | Golden Retriever with Acute Vomiting, Diarrhea, and Collapse 10 Minutes after Going Outside | 379 |

Erin Binagia

Case 15.2	**Labradoodle With Lethargy and Inappetence Progressing To Acute Hematochezia, Tenesmus, and Prayer Posturing** *Armi M Pigott*	386
Case 15.3	**Pit Bull with 2 Days of Progressive Vomiting, Lethargy, and Inappetence after Getting into the Trash 72 Hours Earlier** *Armi Pigott*	391
Case 15.4	**Mixed-Breed K9 with a 5-Day History of Anorexia, Vomiting, and Abdominal Pain** *Elizabeth Gribbin and Rachael Birkbeck*	397
Case 15.5	**Australian Shepherd That Is Unresponsive Following a Motor Vehicle Accident 30 Minutes Prior to Presentation** *Mark W Kim and Alexandra Nectoux*	403
Case 15.6	**Great Dane with a 3-Day History of Lethargy and Pigmenturia Following a Single Episode of Vomiting 24 Hours Previously** *Dave Beeston and Stefano Cortellini*	408

Appendix: Table of Contents by Diagnosis   413

Index   416

PREFACE

Veterinary point-of-care ultrasound (POCUS) is a rapidly expanding diagnostic field. However, many veterinary curricula have not yet incorporated POCUS, leaving a large gap in its clinical application to daily general and emergency practice. Without extensive training, learning veterinary POCUS is a challenging, evolving, and lifelong endeavor. This textbook titled *Case Studies in Small Animal Point of Care Ultrasound* uses case examples to help bridge the clinical skills and knowledge gap between veterinary curricula and the practicing veterinary clinicians who lack the confidence, training, and experience to apply POCUS on a daily basis. Through the book's 58 true-to-life cases, commentaries, images, videos, and referenced literature, readers explore many of the applications of POCUS encountered in clinical practice. Most of the case scenarios build from commonly encountered clinical presentations that can be diagnosed without extensive ultrasound training, but the book also includes more unusual cases to illustrate specific points and to emphasize the application of POCUS in the more advanced setting. As such, this book is intended to be more of an atlas and should appeal to sonographers with all levels of experience. The book will be useful for both novice sonographers looking to learn the basics of ultrasound through assessment of common cases as well as more experienced sonographers who want to challenge their knowledge and clinical reasoning through assessment of less common and more complicated cases.

The book is conveniently divided into broader applications of POCUS and includes case examples in pleural space and lung ultrasound (PLUS), abdominal POCUS, cardiac POCUS, vascular POCUS, miscellaneous POCUS, and advanced multi-organ assessments. Each section starts with an introduction on technical aspects for both image generation and interpretation, as well as tips and tricks on how to perform POCUS evaluation, which is designed to assist readers in how to interpret the subsequent cases. Introductory sections are included to explain the process of focusing the POCUS exam through binary questions, a review of basic ultrasound physics, and obtaining optimal images.

Clinical scenarios are arranged by POCUS systems. Within each section, the cases are in random order without a titled diagnosis so that the reader enters the case free of preconceived thoughts and must read through the case and apply clinical knowledge to answer questions and arrive at a diagnosis. For the busy practitioner who just wants to review ultrasound images for a specific case, a quick search is located at the end of the book. Each clinical scenario starts with a succinct summary of the patient's history, examination, and initial investigation. Still ultrasound images and/or cineloops are provided to query the reader's diagnostic, investigatory, and management plans. The case then proceeds with answers to each question and an explanation of how POCUS is clinically integrated into the case from the perspective of expert authors in the field. The answers provide a detailed discussion on each topic with further illustration where appropriate. Key still ultrasound images and cineloops are discussed, while important terminology and key take-home points are highlighted. At the end of each case, key teaching points are summarized and select references are included to provide additional reading. The authors give clear and concise advice on immediate and further sonographic application, differential diagnoses, tips and pitfalls, and the prioritization of ultrasound assessment.

Authors represent specialists and clinicians from multiple veterinary medical schools and other institutions around the world, incorporating a breadth of knowledge and experience.

ABOUT THE EDITORS

Erin Binagia received her DVM degree from Texas A&M in 2015. She then completed an emergency-focused rotating small animal internship at what is now Arizona Veterinary Emergency and Critical Care Center in Gilbert, Arizona, and stayed on as an emergency veterinarian the following year. She completed her residency training in emergency and critical care at Michigan State University in 2020. Since finishing her residency, she has moved back to Texas and has been working as a locum criticalist at multiple hospitals in Texas and Arizona and as an online consultant. She is also currently an adjunct professor at Texas A&M University. Her favorite part of veterinary medicine is teaching students, interns, and residents, and her clinical interests include anaphylaxis, toxicities, sepsis, tremor disorders, and tetanus. When not working, she usually enjoys running and soccer, but lately has been spending quality time with her dogs, cats, and her daughter, Eloise.

Søren R. Boysen obtained his DVM from the Western College of Veterinary Medicine (WCVM) in Canada, completed an internship at the Atlantic Veterinary College (AVC) in Canada, and a residency in small animal emergency and critical care at Tufts University in Massachusetts. He became a diplomate of the American College of Veterinary Emergency and Critical Care (ACVECC) in 2003. Former chief of veterinary ECC at the University of Montreal (2003–2008), he helped establish veterinary ECC in the province of Quebec. In January 2009, Søren accepted a position at the University of Calgary to help with curriculum development for the newly formed college of veterinary medicine, where he is currently employed as a full professor in small animal emergency and critical care. A recipient of numerous teaching, research, and speaker excellence awards, he has become an internationally recognized educator. A founding member of the veterinary emergency and critical care ultrasound (VECCUS) group, he continues the promotion and education of veterinary ECC and point-of-care ultrasound (POCUS) globally through his numerous contributions to research, book chapters, journal publications, and work with various ACVECC, ECVECC, VECCS, ECVECC, and VetCOT committees. With the help of many great colleagues from Tufts and around the world, he co-developed the small animal FAST exam in 1999, followed by abdominal POCUS and pleural space and lung (PLUS) exams, and continues to pioneer novel ultrasound training techniques and workshops for non-specialist practitioners. Along with POCUS, his research interests include veterinary education/simulation, hemorrhage, coagulation, resuscitation, and perfusion.

Dr. Tereza Stastny is a Diplomate of the American College of Veterinary Emergency and Critical Care. She earned her Doctor of Veterinary Medicine degree from the University of Calgary, completed a rotating small animal internship at Texas A&M University, followed by an emergency and critical care residency at Michigan State University. After residency, she served as a critical care specialist at the Arizona Veterinary Emergency & Critical Care Center in the Phoenix metropolitan area, before joining Penn Vet as an Assistant Professor of Emergency and Critical Care Medicine. Dr. Stastny's research focuses on optimizing mechanical ventilation for patients with acute respiratory failure by assessing respiratory mechanics and exploring new ventilation modes. Her clinical interests include respiratory physiology and mechanical ventilation, extracorporeal therapies, sepsis & SIRS, acid-base disturbances, and veterinary point-of-care ultrasound. Dr. Stastny is dedicated to teaching and lecturing, having assisted and co-led numerous POCUS workshops and lectures, underscoring her commitment to advancing ultrasound expertise in the veterinary field. Outside of veterinary medicine, Dr. Stastny enjoys trail running, traveling, hiking, and coaching youth soccer.

LIST OF CONTRIBUTORS

Anais Allen-Deal
Hospital for Small Animals
University of Edinburgh
Edinburgh, UK

Andrea Armenise
Ospedale Veterinario Santa Fara
Bari, Italy

LM Bacek
BluePearl Specialty + Emergency Pet Hospital
Tampa, FL

Dave Beeston
Clinical Sciences and Services
Royal Veterinary College
London, UK

Géraldine Bolen
Faculty of Veterinary Medicine
University of Liège
Liège, Belgium

Rachel Birkbeck
The Ralph Veterinary Referral Hospital
Furneux Pelham, UK

Angela Briganti
Department of Veterinary Sciences
University of Pisa
Pisa, Italy

Priscilla Burnotte
Faculty of Veterinary Medicine
University of Liege
Liege, Belgium

Dana Caldwell
Arizona Veterinary Emergency & Critical Care Center
Mesa, AZ

Serge Chalhoub
Faculty of Veterinary Medicine
University of Calgary
Calgary, AB, Canada

Laura Cole
Royal Veterinary College
Hatfield, UK

Stefano Cortellini
Royal Veterinary College
Hatfield, UK

Simone Cutler
Royal (Dick) School of Veterinary Studies
University of Edinburgh
Edinburgh, UK

Chiara Debie
Faculty of Veterinary Medicine
University of Liege
Liege, Belgium

Julia Delle Cave
Faculty of Veterinary Medicine
University of Montreal
Montreal, QC, Canada

Chiara Di Franco
University of Pisa
Pisa, Italy

List of contributors

Anna De Nitto
Ospedale Veterinario Santa Fara
Bari, Italy

Laurentin Duriez
Faculty of Veterinary Medicine
University of Liege
Liege, Belgium

Marta Garcia Arce
Hospital for Small Animals
University of Edinburgh
Edinburgh, UK

Esther Gomez Soto
The Ralph Veterinary Referral Centre
Marlow, UK

Kris Gommeren
Faculty of Veterinary Medicine
University of Liege
Liege, Belgium

Elizabeth Gribbin
Royal Veterinary College
London, UK

Liz-Valérie S Guieu
James L. Voss Veterinary Teaching Hospital
Colorado State University
Fort Collins, CO

Julien Guillaumin
College of Veterinary Medicine and
Biomedical Sciences
Colorado State University
Fort Collins, CO

Georgina Hall
Queen Mother Hospital for Animals
Royal Veterinary College
Hertfordshire, UK

Tove Hultman
Massey University
Palmerston North, New Zealand

Pauline Jaillon
Faculty of Veterinary Medicine
University of Liege
Liege, Belgium

Steffi Jalava
Veterinary Teaching Hospital
Massey University
Palmerston North, New Zealand

Nadine Jones
Department of Clinical Science and Services
Royal Veterinary College
Hatfield, UK

Aurélie Jourdan
Faculty of Veterinary Medicine
University of Liege
Liege, Belgium

Christopher Kennedy
Faculty of Veterinary Medicine
University of Liege
Liege, Belgium

Mark Kim
Vetagro Sup
Lyon, France

William Glenn Lane
Toronto, ON, Canada

Jo-Annie Letendre
Faculty of Veterinary Medicine
University of Montreal
Montreal, QC, Canada

Amanda Liggett
Desert Veterinary Medical Specialists
Gilbert, AZ

Julie Menard
Faculty of Veterinary Medicine
University of Calgary
Calgary, AB, Canada

Anne-Christine Merveille
University of Liège
Liège, Belgium

Alexandra Nectoux
VetAgro Sup
Lyon, France

Céline Pouzot-Nevoret
VetAgro Sup
Lyon, France

Armi M Pigott
Department of Clinical Sciences
Cornell University College of Veterinary Medicine
Ithaca, NY

Shari Raheb
Department of Clinical Studies
University of Guelph
Guelph, ON, Canada

Elodie Rizzoli
Faculty of Veterinary Medicine
University of Liège
Liège, Belgium

Daria Starybrat
Royal (Dick) School of Veterinary Studies
University of Edinburgh
Edinburgh, UK

Hugo Swanstein
Copenhagen University
Copenhagen, Denmark

Charles T Talbot
James L. Voss Veterinary Teaching Hospital
Colorado State University
Fort Collins, CO

Erica Tinson
Queen Mother Hospital for Animals
Royal Veterinary College
Hertfordshire, UK

Annelies Valcke
Faculty of Veterinary Medicine
University of Liege
Liege, Belgium

Emilie Van Renterghem
University of Liège
Liège, Belgium

Olivia X Walesby
Hospital for Small Animals
University of Edinburgh
Edinburgh, UK

Thom Watton
Royal Veterinary College
Hertfordshire, UK

Kendon Wu Kuo
Auburn University College of Veterinary Medicine
Auburn, AL

Igor Yankin
Texas A&M University
College Station, TX

Xiu Ting Yiew
Ontario Veterinary College
University of Guelph
Guelph, ON, Canada

Ivayla Yozova
School of Veterinary Science
Massey University
Palmerston North, New Zealand

SECTION 1

INTRODUCTION TO VETERINARY POINT-OF-CARE ULTRASOUND

CHAPTER 1
BASIC ULTRASOUND PHYSICS

Erin Binagia

This chapter briefly reviews key points of ultrasound physics that a veterinarian must understand to interpret ultrasound images. Please refer to other sources for a more in-depth review.[1-3] An ultrasound machine is comprised of a probe, a computer processor, and a screen. The probe creates, transmits, and receives high-frequency sound waves via piezoelectric crystals. The sound waves produced by the probe are aimed at a target tissue by the operator, and those sound waves either pass through or are reflected off the tissue. The reflected sound waves are received by the probe. The processor then takes these signals from the probe and turns them into a cross-sectional image of the body, which is then displayed on a screen.[1,2]

IMAGE DESCRIPTIONS:

Echogenicity refers to the degree of "whiteness, grayness, or blackness" of the image.

- **Hyperechoic**: target tissue appears white compared to surrounding tissues because all ultrasound waves are reflected back to the ultrasound probe. Examples include bone and air.
- **Hypoechoic**: target tissue is a darker shade of gray compared to surrounding tissues. A classic example is the liver which is hypoechoic relative to the spleen.
- **Isoechoic**: tissues are the same shades of gray (same echogenicity).
- **Anechoic**: target tissue is pure black because no ultrasound waves are returned to the receiver. All normal fluids are anechoic (**Figure 1.1**).

IMAGE ORIENTATION:

Longitudinal image (Figure 1.2): the probe should be directed toward the spine and the marker directed cranially. When the patient is in *left lateral recumbency*, the cranial view will be on the left side of the screen, the caudal view on the right, the ventral view at the top of the screen, and the dorsal view at the bottom.

Transverse image (Figure 1.3): the probe directed toward the spine and the marker directed dorsally. When the patient is in *left lateral recumbency*, the patient's right side will be on the left side of the screen and the left side on the right, the ventral view at the top of the screen, and the dorsal view at the bottom.

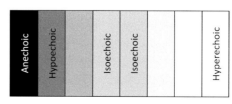

FIGURE 1.1 Echogenicity describes the different shades of white, gray, and black.

DOI: 10.1201/9781003436690-2

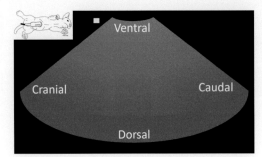

FIGURE 1.2 Longitudinal image orientation with patient in left lateral recumbency, probe directed toward spine, marker directed cranially. The square represents the reference icon which corresponds to the probe marker.

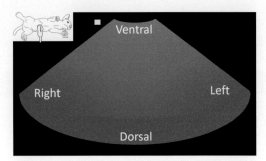

FIGURE 1.3 Transverse image orientation with patient in left lateral recumbency, probe directed toward spine, marker directed dorsally. The square represents the reference icon which corresponds to the probe marker.

ULTRASOUND PRINCIPLES THAT AFFECT AN IMAGE[1,2]:

Wavelength (m): the wavelength refers to the distance between successive crests of a sound wave, or the distance traveled during one cycle (**Table 1.1**).

Frequency (hertz, Hz): frequency is the number of times a wavelength is repeated (cycles) per second. Sound wave frequencies commonly used in diagnostic ultrasound are in the range of 2–15 MHz and higher. Higher-frequency sound waves provide superior detail but become attenuated (see definition below), leading to decreased penetration of tissues. Lower-frequency waves have less attenuation, allowing for deeper tissue penetration at the expense of detail (**Figure 1.4**).

Velocity (m/sec): velocity is the speed at which sound travels through tissues and is calculated as frequency (cycles/second) × wavelength (m). Sound passes slowly through air and quickly through bone. The average sound velocity through most tissues (blood, fat, organs, muscle) is 1540 m/s.[4]

Acoustic impedance: acoustic impedance is the reflection or transmission characteristics of a tissue and can be estimated by the density differences between tissues. It can be calculated as acoustic impedance (Z) = velocity (u) × tissue density (p). It is the small differences in acoustic impedance between the body's tissues that are ideal for imaging (**Table 1.2**).

Attenuation: attenuation is the process of sound losing energy as it passes through tissues, resulting in weaker returning sound waves. Factors that affect attenuation are *absorption*, *scatter*, *reflection*, and *refraction* (**Figure 1.5**).

- **Absorption**: absorption is the process of sound waves that enter the body and are then absorbed by tissues and converted to heat. These sound waves do not return to the probe and do not contribute to an image.
- **Scatter**: this refers to the process of sound waves that are scattered by tissues with irregular surfaces and either do not return to the probe or return in an altered path, decreasing the image resolution (detail). For the best image detail, the ideal angle of ultrasound reflection is 90°. Linear probes use this principle to create a highly detailed image.
- **Reflection**: this describes the process of all ultrasound waves contacting a tissue that it cannot penetrate (such as bone, foreign material, or air), thus reflecting all the sound back to the probe. When a sound

Table 1.1 **Artifacts to understand when performing POCUS to help diagnose or avoid pitfalls**

BASIC POCUS ARTIFACTS

ARTIFACT	APPEARANCE ON ULTRASOUND	WHY IT OCCURS	CLINICAL EXAMPLE(S)	HOW TO INTERPRET/FIX/USE
Distal acoustic shadowing	Area of intense whiteness at the interface, followed by blackness deeper to the interface	Imaging strong reflectors (bone, foreign material, or air)	Ribs Feces in colon SI foreign body Urinary bladder stones	Find an acoustic window that avoids bone and gas Diagnose foreign material if seen within the small intestine Diagnose stones in urinary bladder
Edge shadowing	A thin hypoechoic to anechoic area lateral and distal to the edge of a fluid-filled structure with a curved surface	Imaging fluid-filled structures causing refraction	Urinary bladder Gallbladder	Will appear as if there is a free fluid next to the bladder wall and misdiagnose a rupture – change the ultrasound angle to 90° to the wall
Distal acoustic enhancement	Tissues distal to a fluid-filled structure appear brighter than solid tissues at the same depth	Sound passing through fluid is attenuated less	Urinary bladder Gallbladder Cyst	The presence of this artifact can help determine if an object is fluid filled
Mirror artifact	Will see a mirror image distal to the curved reflective surface (e.g., GB visualized on both sides of the diaphragm)	Imaging a structure close to a curved, strong reflector, sound beam is reflected and strikes adjacent tissues	Imaging the diaphragm	Can misdiagnose a diaphragmatic hernia if unfamiliar with this artifact
A-lines or reverberation	Regularly spaced, parallel hyperechoic lines that extend all the way to the bottom of the screen in an endless loop	Sound entering strong reflective layers and bouncing back and forth before returning to the probe	Lung imaging	Normal artifact that occurs when imaging the lung
B-lines or lung rockets	Narrow, vertical lines originating at the pulmonary pleural line, extending distally to the bottom of the screen. Eliminates A-lines, moves synchronously with breathing	Created by strong impedance of air adjacent to small amounts of water	Any lung pathology	Interpreted as "something within the lungs" and can be edema of any cause, mucous, pus, neoplasia, etc.
Comet-tail or ring-down artifact	Appears similar to A-lines	Reverberation created by strong reflectors with high AI (metallic FBs) or low AI relative to adjacent structures (gas)	Metallic FBs Implants Needles Stylets Gas in bowel	Presence of this artifact during US-guided procedures helps identify the location of the needle

(*Continued*)

Table 1.1 (Continued) **Artifacts to understand when performing POCUS to help diagnose or avoid pitfalls**

		BASIC POCUS ARTIFACTS		
ARTIFACT	APPEARANCE ON ULTRASOUND	WHY IT OCCURS	CLINICAL EXAMPLE(S)	HOW TO INTERPRET/FIX/USE
Slice-thickness artifact	Hyperechoic artifact with round surface that appears within the lumen of a fluid-filled structure being imaged	Part of the beam is just outside a fluid-filled structure	Gallbladder Urinary bladder	Misinterpreted as sludge in GB or sediment in UB. True sediment will have a flat surface and will be stirred by repositioning patient or ballottement
Side-lobe artifact	Will see a weaker image within the distal wall of the UB, will look like sediment	Smaller US beams lateral to the main beam bounce off a curved, reflective surface and are misinterpreted as coming from the main beam by the processor	Imaging the urinary bladder	Can be misinterpreted as sediment within the UB Change the probe, drop the focal point, or lower the gain settings – the artifact will change or disappear True sediment will be stirred by repositioning patient or ballottement

Note: See references for image examples and more detailed information.[1,2]

Abbreviations: SI: small intestine; US: ultrasound; UB: urinary bladder; GB: gallbladder; FB: foreign body; AI: acoustic impedance.

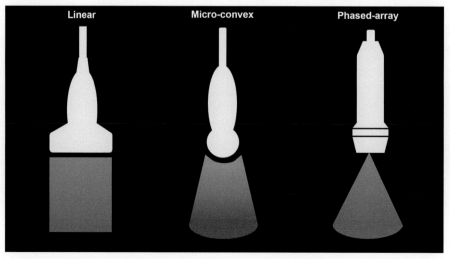

FIGURE 1.4 **Images of basic probe options and the corresponding image shape.**

Basic Ultrasound Physics

Table 1.2 **Chart describing the three basic types of ultrasound probes**

ULTRASOUND PROBE CHOICES

PROBE	PROBE HEAD	FREQUENCY	IMAGE SHAPE	USES
Linear	Large, flat bar shaped	High*	Rectangle (wide)	Best image detail, used for abdominal imaging (e.g., intestinal layers)[1,2]
Curvilinear or micro-convex	Convex, smaller radius	Low+ to mid	Pie shape, near field is curved	Very versatile, can be used to answer most POCUS questions[1,5]
Phased array	Narrow, flat head	Low+	Pie shape, near field is pointed	Avoids ribs, creates a wide field of view at deeper depths. Best for cardiac ultrasound[1,2]

Note: See references for more details.[1,2]
*Higher frequency (i.e., 8–13 MHz) probe means superior detail, but less tissue penetration.
+ Lower frequency (i.e., 1–5 MHz) probes mean deeper penetration, but less detail.

beam encounters a soft tissue–bone or soft tissue–gas interface, all sound is reflected and thus unable to image structures deep to that interface. This phenomenon then produces a *distal acoustic shadow*.
- **Refraction**: as a sound wave passes from one medium to another, the sound wave velocity changes. If the medium interface is at an oblique angle, the beam will bend (or refract) and produce an artifact of improper location for the imaged structure. Refraction and reflection contribute to a thin, poorly echogenic area lateral and distal to a curved structure (e.g., gallbladder), a phenomenon termed *edge shadow*.

IMAGE OPTIMIZATION[1,2]:

- **Depth**: adjust the depth so that the area of interest fills most of the screen. Note that the depth of sound wave penetration is inversely proportional to the frequency used. Lowering the frequency will allow better penetration of tissues but at the same time decrease the image detail.
- **Gain**: gain refers to image brightness. The main gain knob will adjust the overall brightness of the image, while the sliders (time gain compensation [TGC]) will adjust the brightness along individual bands across the image. The goal is to have a consistent, "just right" brightness from top to bottom. An image with too much gain (too bright) or not enough (too dark) will create an image that lacks good contrast, making it difficult to visualize details. When adjusting the gain, use the anechoic fluid within the urinary bladder for reference, as it should appear pure black.
- **Frequency**: for superior image detail, choose a higher-frequency probe such as a linear array. For deeper tissue penetration (bigger dogs, deeper organs), choose a lower-frequency probe, such as a convex probe. If the probe cannot be changed (many emergency rooms only have a micro-convex probe), choose the highest

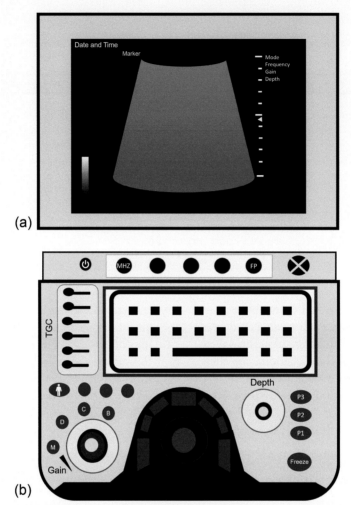

FIGURE 1.5 (a, b) Image of where settings are displayed on the screen (a) and where the most common image optimization knobs are usually located (b) on an ultrasound machine. MHz represents the frequency knob. FP represents the focus knob, which allows adjustment of the number and position of the focal point(s). To adjust the focal point, enter B-mode and press the knob to switch between number and position, rotate to adjust. The focal point(s) is(are) displayed as a triangle along the depth scale on the screen. The time gain compensation sliders will adjust the gain along the corresponding bands at different depths across the screen. To adjust, slide left or right. To adjust gain and depth, rotate the knob left or right. The depth scale (cm) is located to the right of the image. The selected settings are also displayed to the right of the image. Print keys (P1–P3) are used for image storage. Other keys to mention include those labeled B, M, D, and C. The B key enters the traditional 2D B-mode which is used for almost all point-of-care ultrasound (POCUS) exams. The M key enters the M-mode which is used in some echocardiogram evaluations. The D key represents Doppler and the C key represents color mode which are rarely used to interpret POCUS questions and are not discussed in this chapter. See references for more information on these alternative modes.[1,2]

frequency (MHz) possible that will penetrate the desired tissue depth.
- **Focal position and number**: the focus position represents where the beam narrows to give a more detailed image at that set depth. Set the focal position at the area of interest. Choose only one focal point (otherwise the frame rate will be reduced, creating a rougher image).

REFERENCES

1. Fulton RM. Chapter 1: Focused-Basic Ultrasound Principles and Artifacts. In: Lisciandro G, ed. *Focused Ultrasound Techniques for the Small Animal Practitioner*. 1st ed. John Wiley & Sons; 2014, pp. 1–16.
2. Mattoon JS, Nyland TG. Chapter 1: Fundamentals of Diagnostic Ultrasound. In: Mattoon JS, Nyland TG, eds. *Small Animal Diagnostic Ultrasound*. 3rd ed. Saunders; 2015, pp. 1–49.
3. Kremkau FW. *Sonography Principles and Instruments*. 10th ed. Elsevier Saunders; 2021.
4. Coltrera MD. Ultrasound Physics in a Nutshell. *Otolaryngol Clin North Am* Dec 2010;43(6):1149–1159. doi:10.1016/j.otc.2010.08.004.
5. Gommeren K, Boysen SR. Chapter 189: Point-of-Care Ultrasound in the ICU. In: Silverstein DC, Hopper K, eds. *Small Animal Emergency and Critical Care*. 3rd ed. Elsevier; 2023, pp. 1076–1092.

CHAPTER 2

BINARY QUESTION APPROACH TO POINT-OF-CARE ULTRASOUND

Søren Boysen

INTRODUCTION

Point-of-care ultrasound (POCUS) is defined as the acquisition, interpretation, and immediate clinical integration of ultrasonographic imaging performed patient-side by an attending clinician, regardless of skill level, with the goal of answering a focused question or series of questions, rather than assessing all structures of an organ(s) by a specialist.[1] Thus, it is designed to quickly narrow differential diagnoses and rule out life-threatening conditions through the assessment of a limited number of organs or body regions in patients with well-defined clinical symptoms, driven by Bayes' theorem and the pretest probability of a problem being present.[2,3] Although research regarding its clinical application is active and ongoing, with new applications identified regularly, the skills employed should be determined by an individual sonographer's training, experience, and proficiency. It provides clinically significant information that is otherwise unattainable on physical examination alone, and is therefore complimentary to triage, physical examination, and other point-of-care diagnostic and clinical tests or findings; however, it does not replace them.

CLINICALLY DRIVEN BINARY QUESTIONS

When applying POCUS, a clinically driven, often binary approach allows a rapid and thorough investigation of prioritized questions to further direct diagnostics and therapies.[4–6] Not having a clinically driven question to answer can lead to "fishing expeditions" where the transducer is randomly placed on the patient in the hope of finding something without knowing what to look for or where to locate it. With POCUS, the binary approach should be fluid in nature, allowing multiple complimentary binary questions to be rapidly and chronologically answered based on clinically driven results. For example, if a cat presents in respiratory distress with rapid shallow breathing and muffled heart/lung sounds ventrally, pleural effusion would be a high-priority question to rapidly assess (is pleural effusion present, yes/no?). If pleural effusion is present, the underlying cause should be considered, which would lead to the subsequent question: is the left atrial to aortic ratio enlarged (yes/no)? All POCUS findings must be interpreted considering the entire clinical picture, which may lead to additional and more comprehensive POCUS evaluations as results become available and/or patient stabilization is initiated.

Finally, the question asked, even if binary, must be thoroughly answered to ensure that the correct diagnosis is reached. For example, if the objective is to rule out a small volume of pleural effusion, placing the transducer only at the level of the costochondral junction over the heart is likely to miss the effusion: a thorough evaluation of the entire ventral pleural space with the patient in a standing or sternal position will increase the chances of identifying fluid.[7]

CLINICAL INTEGRATION OF POCUS

With the development of less expensive and smaller portable ultrasound units, combined with evidence that non-radiologists and non-cardiologists can perform POCUS, it is currently used in nearly all aspects of veterinary medical care, from triage, diagnosis, monitoring, and screening, to procedural guidance (**Figure 2.1**). Therefore, POCUS may comprise single-targeted ultrasound examinations (e.g., urinary volume estimation, renal pelvic dilation, or pleural effusion) or multiple-targeted ultrasound examinations (e.g., underlying cause of shock, respiratory distress, and trauma) depending on the specific question or series of questions to be answered.[8,9] Some clinical POCUS questions, by their nature, require an assessment of organs from more than one domain (e.g., when using POCUS to assist with non-specific physical examination findings such as shock).

POCUS has been dubbed an extension of the physical examination[10,11]; therefore, like the physical examination, POCUS is applied differently depending on the question to

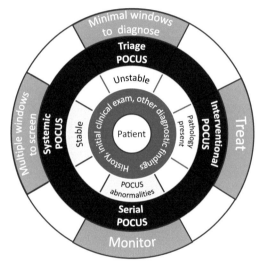

FIGURE 2.1 POCUS is patient centered, targeted, and dynamic, designed to answer a focused clinically driven question or series of questions. It is also integrative and has four general applications which vary depending on the targeted objectives of the scan, the pretest probability of a problem being present (based on history, initial clinical exam, and other diagnostic findings), and the clinical setting encountered: (1) triage POCUS uses the minimum number of windows possible to identify the most immediate life-threatening and critical conditions; (2) serial POCUS is applied to monitor the progression or resolution of any pathology, and response to therapy; (3) systemic or multiorgan POCUS uses multiple windows in more stable patients (with or without specialist assistance) to detect asymptomatic conditions and new developments and/or to ensure sonographically detectable problems have not arisen prior to undertaking procedures, anesthesia, or discharge/service transfer; and (4) therapeutic POCUS is used to reduce complications of interventions where applicable. These applications may involve single-targeted ultrasound examinations (e.g., urinary volume estimation, renal pelvic dilation, or pleural effusion) or multiple-targeted ultrasound examinations (e.g., underlying cause of shock, respiratory distress, and trauma) depending on the specific question or series of questions to be answered. POCUS = point-of-care ultrasound. Source: Dr Søren Boysen, with permission.

ask and answer, as well as the clinical situation encountered (**Figure 2.1**). For example, extensive physical and POCUS examinations are not immediately recommended in unstable patients where triage and immediate stabilization take precedence. Like a triage exam being followed by a more comprehensive physical exam, once the patient is stable and life-threatening conditions have been dealt with, more comprehensive total POCUS evaluations can be completed. Therefore, it is essential to interpret POCUS findings within the clinical context of a particular case. POCUS should be individualized, patient centered, targeted, and dynamic in its cognitive application (**Figure 2.1**).

Although standardized POCUS protocols are important and recommended in many situations, patient position, species, breed, history, physiology, and pathology can impact sonographic findings. Therefore, no single POCUS protocol will work for every patient and to ensure the greatest chance of detecting underlying pathology the POCUS technique should be adapted considering the specific question asked, the clinical situation encountered, and the proficiency of the operator.

REFERENCES

1. Díaz-Gómez JL, Mayo PH, Koenig SJ. Point-of-care ultrasonography. *N Engl J Med* 2021; 385(17): 1593–1602.
2. Shokoohi H, Duggan NM, Adhikari S, Selame LA, Amini R, Blaivas M. Point-of-care ultrasound stewardship. *J Am Coll Emerg Phys Open* 2020; 1(6): 1326–1331.
3. Tanael M. Users' guide to point-of-care ultrasonography. *J Am Coll Emerg Phys Open* 2020; 1(6): 1777.
4. Bruijns SR, Engelbrecht D, Lubinga W, Wells M, Wallis LA. Penetrating the acoustic shadows: Emergency ultrasound in South African emergency departments. *S Afr Med J* 2008 Dec; 98(12): 932–934.
5. Abu-Zidan FM. Point-of-care ultrasound in critically ill patients: Where do we stand? *J Emerg Trauma Shock* 2012; 5(1): 70–71. doi: 10.4103/0974-2700.93120.
6. Atkinson P, Bowra J, Lambert M, Lamprecht H, Noble V, Jarman B. International federation for emergency medicine point of care ultrasound curriculum. *CJEM* 2015 Mar; 17(2): 161–170.
7. Boysen S, Chalhoub S, Romero A. Veterinary point-of-care ultrasound transducer orientation for detection of pleural effusion in dog cadavers by novice sonographers: A pilot study. *Ultrasound J* 2020 Oct; 12(Suppl 1): 45. doi: 10.1186/s13089-020-00191-6.
8. Choi WJ, Ha YR, Oh JH, Cho YS, Lee WW, Sohn YD, Cho GC, Koh CY, Do HH, Jeong WJ, Ryoo SM, Kwon JH, Kim HM, Kim SJ, Park CY, Lee JH, Lee JH, Lee DH, Park SY, Kang BS. Clinical guidance for point-of-care ultrasound in the emergency and critical care areas after implementing insurance coverage in Korea. *J Korean Med Sci* 2020 Feb 24; 35(7): e54.
9. Lau YH, See KC. Point-of-care ultrasound for critically-ill patients: A mini-review of key diagnostic features and protocols. *World J Crit Care Med* 2022 Mar 9; 11(2): 70–84.
10. López Palmero S, López Zúñiga MA, Rodríguez Martínez V, Reyes Parrilla R, Alguacil Muñoz AM, Sánchez-Yebra Romera W, Martín Rico P, Poquet Catalá I, Jiménez Guardiola C, Del Pozo Pérez A, Lobato Cano R, Lazo Torres AM, López Martínez G, Díez García LF, Parrón Carreño T. Point-of-care ultrasound (POCUS) as an extension of the physical examination in patients with bacteremia or candidemia. *J Clin Med* 2022 Jun 23; 11(13): 3636.
11. Lee L, DeCara JM. Point-of-care ultrasound. *Curr Cardiol Rep* 2020 Sep 17; 22(11): 149. doi: 10.1007/s11886-020-01394-y.

CHAPTER 3
BASICS OF IMAGE ACQUISITION FOR POINT-OF-CARE ULTRASOUND

Tereza Stastny

Sonographic skills necessary to obtain quality POCUS images are easily mastered by both general and emergency practitioners. This chapter will describe patient positioning, preparation, coupling agents, and transducer manipulations.

PATIENT POSITIONING

Patients can be scanned in lateral recumbency (right or left), sternal recumbency, and/or standing position. Dorsal recumbency should be avoided, as it may cause unstable patients to decompensate or even arrest due to compression of the caudal vena cava by intra-abdominal organs and a subsequent reduction in preload.[1,2] The machine should always be brought to the patient, not the patient to the machine. For this reason, a portable ultrasound machine that can be transported cage-side is recommended. Alternatively, when a portable machine is not available, the ultrasound machine should be stationed at the triage table. Although the order of sites scanned during each POCUS is not important, clinicians should develop a consistent and systematic approach to scanning their patients.

PREPARATION AND COUPLING AGENTS

Shaving patient fur is typically not required to obtain quality POCUS images. Fur is parted, and an acoustic coupling agent is applied over the scanning site. Certain Northern dog breeds with thick undercoats may require shaving to improve the sonographic image quality. Alcohol will usually suffice as the sole acoustic coupling agent. Alcohol-based hand gels may also be used. Ultrasound conducting gel applied following the application of alcohol can serve to enhance the acoustic coupling and should be smoothed against the skin when applied to eliminate any air bubbles which interfere with image acquisition. Alternatively, ultrasound conducting gel may also be used as the sole acoustic coupling agent.

TRANSDUCER ORIENTATION

Each transducer has an indicator marker on one side of the transducer that corresponds to a symbol displayed on one side of the ultrasound screen (**Figure 3.1**). This indicates transducer orientation and allows the transducer to be manipulated such that the object of interest can be moved to the desired location on the ultrasound's screen.

DOI: 10.1201/9781003436690-4

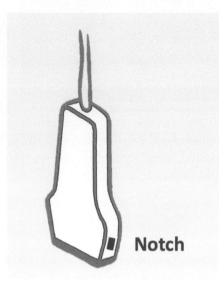

FIGURE 3.1 **Illustration of an indicator marker (in this case a notch) depicted to the right side of the transducer head.**

LONGITUDINAL (LONG) OR TRANSVERSE (SHORT) AXIS

The transducer is orientated in either a longitudinal or transverse axis to the structure(s) of interest. The longitudinal axis is obtained by placing the transducer in the "long" view relative to the body or organ being evaluated. Typically, this view will be obtained by having the notch on the transducer facing cranial, or towards the head, but may vary depending on the organ being assessed. The transverse axis is obtained by placing the transducer in the "short" view relative to the body or organ being evaluated. The short axis is acquired by rotating the transducer 90 degrees from the long axis, with the marker of the transducer facing dorsal, or towards the spine. This may vary depending on the organ being assessed.

MANIPULATIONS

The part of the transducer that contacts the sonographic surface is termed the "head" and where the cable attaches to the transducer is termed the "tail" (**Figure 3.2**). Note that the terms probe and transducer are used interchangeably. The "angle of insonation" refers to the angle of the ultrasound beam relative to the organ of interest. Five main transducer manipulations aid in the acquisition of sonographic images: sliding, sweeping, rotating, rocking, and fanning.[3,4]

Sliding denotes moving the entire transducer in a longitudinal axis, at a constant angle, towards the target. This is often used to follow a specific structure, such as a vessel or to find a better imaging window (**Figure 3.3**).[3,4]

Sweeping denotes moving the entire transducer in a transverse axis, at a constant angle, towards the target. This manipulation is also used to follow a specific structure such as a vessel or to find a better imaging window (**Figure 3.4**).[3,4]

Rotating involves keeping the transducer head in one location perpendicular to the

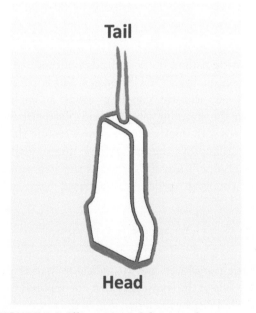

FIGURE 3.2 **Illustration of the transducer "head" that contacts the sonographic surface and the transducer "tail" where the cable attaches.**

Basics of Image Acquisition for Point-of-Care Ultrasound

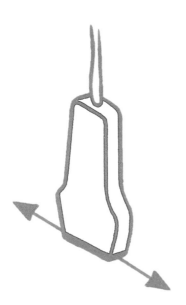

FIGURE 3.3 **Illustration of a transducer "sliding" along its longitudinal axis.**

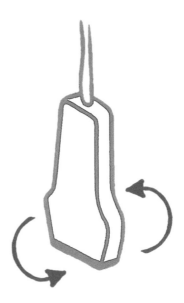

FIGURE 3.5 **Illustration of a transducer head "rotating" along its central axis counterclockwise.**

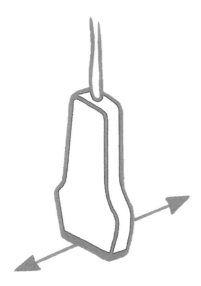

FIGURE 3.4 **Illustration of a transducer "sweeping" along its transverse axis.**

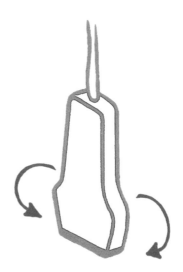

FIGURE 3.6 **Illustration of a transducer head "rocking" side-to-side along its long axis.**

sonographic surface and turning the transducer along its central axis clockwise or counterclockwise. Rotation is used to switch between the long and short axis of specific structures (**Figure 3.5**).[3,4]

Rocking involves keeping the transducer head stationary and moving the tail of the transducer side-to-side along the transducer's long axis (**Figure 3.6**). Rocking helps to center the area of interest and is also referred to as "in-plane" motion because the image is kept in-plane throughout the manipulation.[3,4]

Fanning involves keeping the transducer head stationary and moving the tail of the transducer side-to-side along the transducer's short axis. Fanning changes the angle of insonation allowing visualization of multiple cross-sectional images of the structure of interest (**Figure 3.7**).[3,4]

When rotating, rocking, and fanning, the head of the transducer is manipulated around the point of contact. With sliding and sweeping, the head of the transducer departs from the initial point of contact on the sonographic surface. It is important to note that small manipulations of 1-2 degrees or 1-2 millimeters in distance have a large impact on the plane in which the structure of interest is being scanned.[3,4]

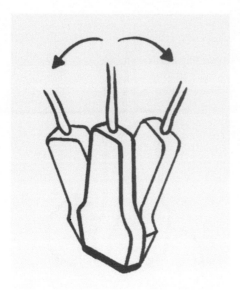

FIGURE 3.7 **Illustration of a transducer head "fanning" side-to-side along its short axis.**

REFERENCES

1. Goya S, Wada T, Shimada K, et al. Effects of postural change on transesophageal echocardiography views and parameters in healthy dogs. J Vet Med Sci. 2017;79(2):380-386. doi:10.1292/jvms.16-0323
2. Nakao S., Come P. C., Miller M. J., Momomura S., Sahagian P., Ransil B. J., Grossman W.1986. Effects of supine and lateral positions on cardiac output and intracardiac pressures: an experimental study. Circulation 73: 579–585. doi: 10.1161/01.CIR.73.3.579
3. Moore CL, Copel JA. Point-of-care ultrasonography. N Engl J Med 2011; 364(8): 749-757.
4. Søren BO. Chapter 18: Emergency Point-of-Care Ultrasound. In: Grimm KA, Lamont LA, Tranquilli WJ, et al. Lumb & Jones' Veterinary Anesthesia and Analgesia. 6th ed. 2020. In publishing process.

SECTION 2

PLEURAL SPACE AND LUNG ULTRASOUND (PLUS)

Section Editor Søren Boysen

CHAPTER 4

INTRODUCTION TO PLEURAL SPACE AND LUNG ULTRASOUND (PLUS)

Søren Boysen

INTRODUCTION

Veterinary pleural space and lung ultrasound (PLUS) has made significant advancements from the 2008 landmark publication that applied thoracic focused assessment with sonography for trauma (TFAST) to identify pleural effusion and pneumothorax in dogs suffering trauma.[1] Today PLUS is widely used in the clinical setting to evaluate numerous multi-species pleural space and lung pathologies across many aspects of veterinary medicine.[2-14] This chapter will briefly review key concepts to interpret normal and abnormal PLUS findings, and how to obtain images. See additional references for more detailed information.[8,15]

Patient preparation, transducer selection and machine settings

Although placing animals in lateral recumbency (particularly right) has been advocated for thoracic FAST exams,[1] research suggests that restraining cats and dogs in any position to perform diagnostic procedures can increase their anxiety and stress.[16] In patients with cardiovascular and/or respiratory compromise, resistance to restraint may lead to increased work of breathing and oxygen consumption, and ultimately patient decompensation. PLUS should therefore be performed in the position least likely to increase patient anxiety and discomfort, which is most often a standing or sternal position.[8,15] It is essential to realize that localization of some sonographically identifiable pathology will change with patient positioning; fluid falls and gas rises.[8,11] Where the transducer is situated on the patient, and the ultrasound beams directed, should be modified to ensure the location of positionally-dependent pathology is assessed.[16] If required, patients initially assessed in lateral recumbency can be gently rolled into a sternal position, or even encouraged to stand to assess PLUS windows not initially accessible with the patient in lateral recumbency. However, it is not obligatory to assess all PLUS windows in all patients to answer a focused question or to arrive at a diagnosis, particularly if the patient is unstable. Do not discontinue stabilization efforts or compromise patient safety/comfort to complete a PLUS exam; bring the machine to the patient, and if necessary complete the PLUS evaluation when the patient is more stable.

To maintain efficiency, fur is not clipped but parted to provide access to the skin, and alcohol is used as the coupling agent, with or without gel. Alcohol may cause radiographic artifact. It also poses a fire hazard if electrocautery, defibrillators, or laser devices are used following its application. Gel can be added to the transducer head in addition to applying alcohol to the patient if image quality is suboptimal. Alternatively, fur can be clipped, and ultrasound gel applied as the sole coupling agent at the transducer-skin interface, particularly if image quality remains suboptimal. Sedation is rarely required as PLUS is non-invasive and requires minimal restraint. In some situations, sedation may decrease anxiety and work of breathing, which may improve image acquisition in certain settings, particularly for assessment of the lungs.

DOI: 10.1201/9781003436690-6

A micro-convex/curvilinear transducer is used for most PLUS exams, with a frequency between 5 and 7.5 MHz. Low frequency settings may make it easier to identify lung sliding. By contrast, a linear transducer with higher frequency (10-13 MHz) may provide greater detail for assessment of the pleural line and some lung surface pathology. In general, the depth is adjusted according to body condition, so the field of view in the ultrasound image extends to approximately 4-6 cm and the pleural line is visible at the junction between the proximal and middle thirds of the field of view. Adjust the gain to maximize appearance of the thoracic wall, while also allowing any far field image artifact to be visible. It may be helpful to turn off artifact reduction settings (e.g. turn off speckle reduction), fan the transducer so ultrasound beams are reflected away from the transducer by the lung surface, sit the transducer over a single rib (e.g. "dead BAT" sign, **Figure 4.1**), decrease the

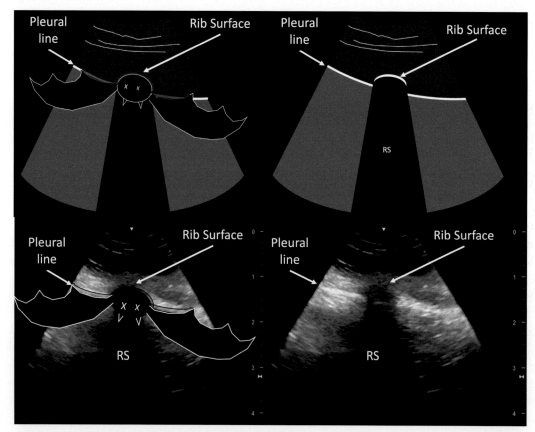

FIGURE 4.1 Schematic and still images of the "dead BAT" sign (also referred to the "one eyed gator" in veterinary medicine). When the transducer is situated over a single rib the proximal surface of the rib and pleural line to either side of the rib shadow (RS) form an imaginary image likened to a dead bat, where the body of the bat represents the rib surface and the underside of the two wings represent the pleural line located on either side of the rib shadow. The "dead BAT" sign is used to identify the pleural line, while situating the transducer over a single rib enhances visible lung sliding because less ultrasound waves are reflected to the transducer footprint, making the pleural line appear less hyperechoic, thicker and "grainier." Source: Dr. Søren Boysen, with permission.

depth setting (e.g. to 2 cm), decrease the gain setting, and/or using a combination of these techniques to better identify lung sliding. If the depth and gain are decreased to visualize lung sliding, they should be re-adjusted (increased) to look for B-lines, the curtain sign, and lung consolidations.[15] See cases for machine settings when looking for specific PLUS pathology.

NORMAL PLEURAL SPACE AND LUNG ULTRASOUND (PLUS) FINDINGS

More than 90% of ultrasound beams are reflected from soft tissue-to-air interfaces, rendering anything below the air interface sonographically invisible (**Figure 4.2**). Therefore, because the chest wall forms a soft tissue-to-air interface with air filled lung, ultrasound can only be used to interpret the lung surface (visceral pleura) and will fail to detect pathology situated (>1–3 mm) deep to the lung surface.[17] Fortunately, many pulmonary diseases that cause lung pathology in people will result in sonographically visible lung surface changes

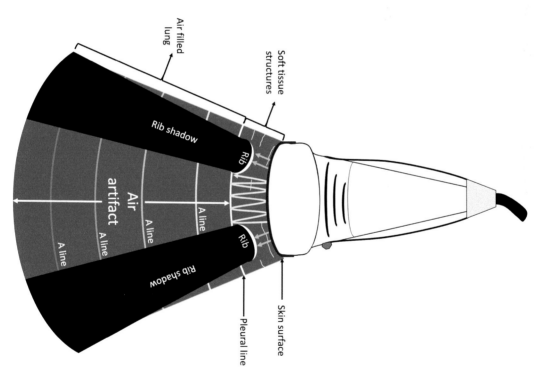

FIGURE 4.2 Schematic ultrasound image of a healthy animal depicting a soft tissue-to-air interface at the pleural line, where ultrasound beams (yellow arrow) are reflected between two highly reflective surfaces; 1) the lung surface at the pleural line and 2) the transducer. Each time the ultrasound beam is reflected from the lung surface and returns to the transducer an A-line is formed. If the ultrasound beams are not reflected between the transducer and soft-tissue air interface (can be deflected away from the transducer by fanning/rocking) then A-lines will not be visible. When the lung is air filled the lung tissue is not visible, and only artifact can be seen below its surface (air artifact). Ultrasound beams are also attenuated by bone creating rib shadows below the proximal bone surface (orange arrows). Source: Dr. Søren Boysen, with permission.

(B-lines and/or consolidations),[18] which is probably similar in small animals. When the lung is against the chest wall artifact will be sonographically visible arising from the pleural line, which is normally composed of the parietal and visceral pleura (see below). The pleural line is essential to identify as most PLUS pathology is diagnosed through changes in its sonographic appearance.[15]

To assist sonographers in locating the pleural line with confidence, the "BAT sign" (sometimes referred to as the "gator sign" in veterinary medicine) should be identified (**Figure 4.3**), which is composed of two ribs (wings of the bat) and the pleural line (body of the bat).[15] When the transducer is placed transverse (perpendicular) to the ribs over the lungs and intercostal space(s), the ultrasound beam transmits through the soft tissue structures of the thoracic wall (skin, subcutaneous tissue, muscle layers) and encounters either bone or air, two key enemies of ultrasound. This forms the image classically referred to as the "BAT sign". The bat pneumonic helps sonographers identify the

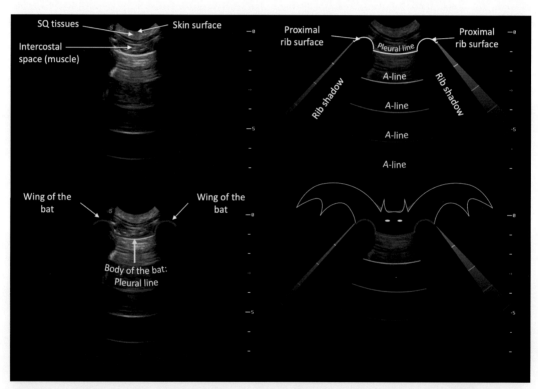

FIGURE 4.3 Labeled and unlabelled still and overlaid schematic images of the "BAT sign" and A-lines. The proximal rib surfaces of two adjacent ribs form the wings of the bat while the pleural line forms the body of the bat when the transducer is situated perpendicular to the ribs. The gain is adjusted based on the appearance of the thoracic wall (region above the pleural line) while also allowing A-lines to be visualized in the far field of the image. The depth is at 7 cm to allow far field A-lines and B-lines (not shown) to be appreciated. The depth can be decreased to put the pleural line at the junction between the proximal and middle thirds of the ultrasound image, which will make lung sliding easier to identify in real time. SQ = subcutaneous tissues. Source: Dr. Søren Boysen, with permission.

structures and transducer orientation needed to interpret thoracic wall anatomy: **b**one and **a**ir are encountered when the transducer is **t**ransverse to the ribs. With experience the pleural line can be located without having to identify the "BAT sign", and by turning the transducer parallel to the ribs within the intercostal space a larger lung surface area can be assessed without interference from the ribs.

The ratio of aerated to non-aerated lung is > 85% in healthy animals and scanning normal aerated lung regions should result in visible lung sliding (aka, the glide sign, **Video 4.1**) as the visceral pleura (lung surface) slides along the parietal pleura (inner lining of the thorax) at the hyperechoic pleural line during active respiratory expansion and relaxation of lung, diaphragm, and thoracic wall.[17,19-22] In addition, A-lines (**Figures 4.2, 4.3**), depending on the angle the beam strikes the pleurae,[16] and an occasional B-line (**Figure 4.4**), because of micro atelectasis and interlobar septa,[23] can normally be identified in healthy animals.

Depending on the quantity identified at a single transducer window and over an entire hemithorax, B-lines may be a normal finding or may be representative of alveolar-interstitial pathology (see increased B-lines below). If an entire hemithorax is scanned in dogs or cats most windows will have lung sliding and an absence of B-lines visible (**Figure 4.4**a). However, in up to 50% of cats and dogs it is possible to identify 1 or 2 B-lines at one or two windows over a hemithorax (**Figure 4.4**b, **Figure 4.4**c).[4,5,24-27] Although rare, up to 3 B-lines have been reported in a single window of the hemithorax of cats and dogs (**Figure 4.4d**).[24-26] More than 3 B-lines in a single window or finding more than 2 positive sites, regardless of the number of B-lines, on a hemithorax of dogs and cats should prompt consideration of lung pathology (see below). A recent publication in cats with no clinical signs and normal thoracic radiographs identified 2/67 cats with a single region of coalescing B-lines over a single hemithorax.[26] Due to the variation in number of B-lines reported with different lung protocols in healthy animals, and the small number of studies in healthy dogs and cats reported to date, the clinical context must be considered when deciding if identified B-lines represent a normal finding or if lung pathology is likely.

The most caudal lateral lung margin should also be easy to locate in heathy animals. It is identified by the vertical edge artifact referred to as the abdominal curtain sign, where aerated lung overlies and obscures underlying soft tissues of the abdomen along the costophrenic recess (**Figure 4.5**).[8] The abdominal curtain sign will move caudally and cranially with the inspiration and expiration respectively (**Figure 4.5**, **Video 4.2**). The region caudal to the abdominal curtain sign will vary in appearance depending on what soft tissue abdominal structures the aerated lung overlies (e.g. stomach, liver, intestines, kidneys etc.).[15]

Finally, it is possible to identify Z-lines and an occasional I-line when scanning the thorax of cats and dogs (**Video 4.3**). Z-lines are a normal artifact finding in healthy dogs and cats.[15,27] They are important to recognize as they are often confused for B-lines but can be differentiated from the latter as they arise at or above the parietal pleura (thoracic wall side of the pleural line), not the lung surface. Therefore, they do not move with lung sliding and they do not erase A-lines. They are ill-defined and disappear after 2-5 cm. Their significance is unknown (not associated with known pathology) and they are present in >

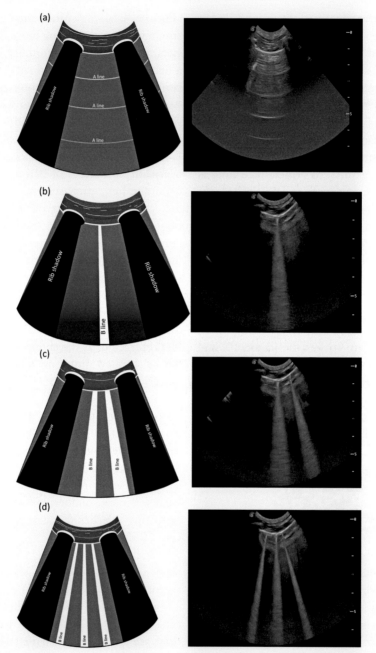

FIGURE 4.4 Schematic and still images of B-lines that can be found in healthy cats and dogs. A) Only A-lines are visible. This is the most common finding when scanning the lung surface of healthy cats and dogs. B) A single B-line is seen without A-lines. Up to 50% of cats and dogs will have 1 or 2 B-lines identified at one or two windows over a hemithorax. A-lines may or may not be visible in healthy animals depending on the angle the ultrasound beams strike the pleura C) Two B-lines are visible, which is less common that finding a single B-line, but still reported in healthy cats and dogs D) Three B-lines are seen, which although rare, has been reported in a single window of the hemithorax of cats and dogs. Lung sliding would be seen in all of the above examples but can't be appreciated in still images. Source: Dr. Søren Boysen, with permission.

Introduction to Pleural Space and Lung Ultrasound (PLUS)

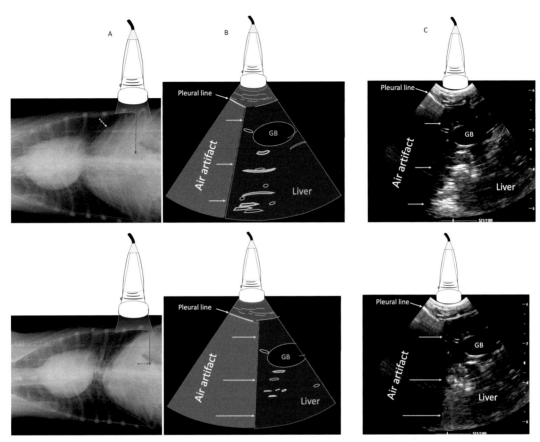

FIGURE 4.5 Radiograph, schematic and still images of the location and sonographic appearance of a normal abdominal curtain sign (ACS) artifact and its movement caudally during inspiration when the transducer is orientated perpendicular to the ribs with the indicator marker directed cranially. (A) Thoracic radiograph of a dog showing the transducer location to identify the ACS. Visible soft tissue structures are shown in green within the ultrasound beam (caudal to the air-filled lung shown in pink). A sharp vertical edge artifact (curtain sign, red dotted arrow) will occur at the demarcation created where air filled lung overlies soft tissue structures of the abdomen. Caudal to the red dotted arrow, the soft tissues of the abdomen are visible while air induced artifact is visible cranial to the red dotted arrow due to aerated lung. The dotted white arrow shows the diaphragm curving away from the chest wall which is not visible due to overlying air-filled lung. The blue dotted arrow shows the distance the vertical edge artifact that is the curtain sign moves caudally with inspiration. (B) Schematic image showing an ultrasonographic image with the transducer positioned as shown in A. The vertical edge artifact is shown as the dotted red arrow and the yellow horizontal arrows, with visible abdominal structures to the right of the red arrow and air artifact to the left of the red arrow. (C) Still ultrasound image showing the normal vertical edge artifact (yellow arrows) of the ACS. GB = gall bladder. Source: Dr. Søren Boysen, with permission.

80% of healthy cats and dogs. They can also be identified in patients with pneumothorax. I-lines meet the same criteria as B-lines with the exception they only extend a short distance from the pleural line.[15] The significance of I-lines is uncertain, although it has been hypothesized they may represent an incomplete B-line.

SONOGRAPHICALLY-DEFINED PLEURA AND LUNG BORDERS

Defining and identifying sonographic pleura and lung borders orientates operators to where they are on the hemithorax, may increase the sensitivity and specificity of detecting pathology, and decreases the chances of mistaking structures outside the thoracic cavity for PLUS pathology (e.g., gas in the stomach for B-lines).[8,11,15] When scanning the lateral surface of the lungs transthoracically, 5 distinct pleura and lung ultrasound borders can be identified (**Figure 4.6**).[15] Although not a border, the caudal lung surface which is not visible transthoracically can be identified on midline via the subxiphoid window. The 5 PLUS borders are not absolute and are defined by the location lung/and or pleura can no longer be sonographically identified because normal structures prevent their assessment when using a transthoracic approach to assess the lateral lung surfaces (**Table 4.1**). If the PLUS borders

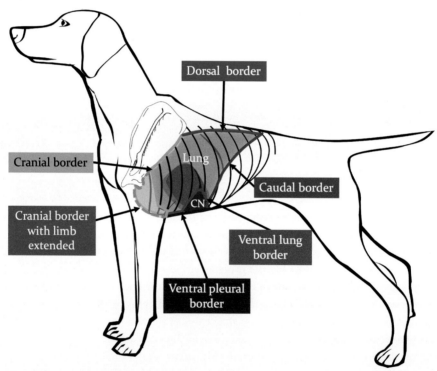

FIGURE 4.6 The five transthoracic sonographically-defined pleura and lung borders that can be identified when scanning the lateral surface of the lung of healthy patients. The caudal border is the abdominal curtain sign (red outline). The dorsal border is comprised of the hypaxial/sublumbar muscles (blue outline). The cranial border is defined by what is sonographically visible due to the musculature and presence of the thoracic limb (green outline). With the limb extended cranially and the transducer tucked into the axilla area the cranial ventral border can be significantly expanded (green dotted outline and green shaded lung region). The ventral pleural border (purple outline) is comprised of the interface between the ventral surfaces of the heart or lung and the sternal/pectoral muscles. The ventral lung border is shown in pink and deviates from the ventral pleural border at the cardiac notch (CN). Everything between the transthoracically scannable lung borders is lung surface (shaded in pink). The caudal- lung surface can also be assessed via the subxiphoid window (not shown). Source: Dr. Søren Boysen, with permission.

Table 4.1 **The five transthoracic sonographically-defined pleura and lung borders that can be identified when scanning the lateral surface of the lung and key factors that alter their location and/or appearance**

SONOGRAPHICALLY DEFINED TRANSTHORACIC PLEURAL AND LUNG BORDERS	LOCATION AND IDENTIFICATION IN HEALTHY ANIMALS	KEY FACTORS THAT ALTER THE LOCATION AND/OR APPEARANCE OF THE SONOGRAPHIC BORDER
Caudal lung border	Defined by the abdominal curtain sign, which is normally identified when the transducer (orientated perpendicular to the ribs) is slid off the caudal lung surface onto the costophrenic recess and underlying soft tissues of the abdomen.	This border may shift cranially with pathology that increases abdominal pressure and/or decreases tidal volumes/lung expansion, may shift caudally with increased tidal volumes and or air trapping (e.g. asthma), and will change in appearance with pleural space pathology including pleural effusion, pneumothorax, and other pleural space occupying lesions.
Dorsal pleura and lung borders	Defined by the loss of lung sliding when the transducer is swept (transducer orientation perpendicular to the ribs) dorsally along the hemothorax onto the hypaxial/sublumbar muscles.	Space occupying pleural pathology that accumulates dorsally (e.g. pneumothorax) will shift the sonographically detectable dorsal lung border ventrally. By contrast, the dorsal pleural border is fixed at the hypaxial/sublumbar muscles. Displacement of the lung ventrally results in a loss of lung sliding directly ventral to the sublumbar/hypaxial muscles and the detection of lung sliding at the site the lung recontacts the chest wall (e.g. the lung point in cases of pneumothorax); the more severe the dorsal space occupying pathology the more ventrally the dorsal lung border is displaced (Fig. 4.7).
Cranial lung border	Defined by the flexor muscles of the shoulder dorsally and the thoracic inlet ventrally when the transducer is slid (transducer orientation perpendicular to the ribs) cranially along the thorax until the lung is no longer visible.	Extending the limb cranially and sliding the transducer into the axilla region expands the cranial border allowing more cranial ventral lung surface area to be scanned.
Ventral pleural border	Defined by identification of the pectoral/sternal muscles when the transducer is slid (transducer orientation parallel to the ribs) ventrally along the hemithorax until normal thoracic structures (e.g. lung or heart) are seen interfacing with the pectoral/sternal muscles.	The ventral pleural border, like the dorsal pleural border is relatively fixed by anatomic structures, although the normal thoracic structures interfacing with the border (e.g., heart and lung) may be displaced dorsally due to pericardial and/or pleural space occupying pathology that accumulates ventrally.
Ventral lung border	Defined by the ventral lung margins which deviate dorsally from the ventral pleura border because of the cardiac notch.	Space occupying pleural and pericardial pathology (e.g., pleural effusion, diaphragmatic hernia, significant pericardial effusion, etc.) that accumulates ventrally will shift the ventral lung border dorsally.

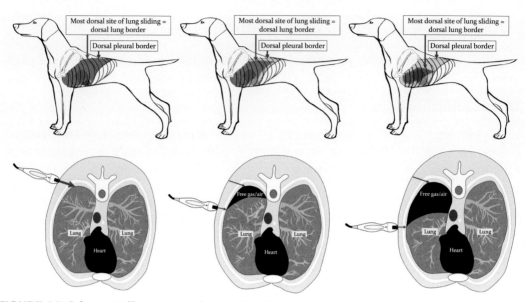

FIGURE 4.7 Schematic illustrations of a standing dog and correlating computed tomography describing how the lateral sonographically defined PLUS borders change with pathology when scanned transthoracically. (Upper and lower left images) Normal dog standing with the dorsal pleural (red arrow) and lung border (blue arrow) superimposed. Lung sliding should be visible over all lateral lung surface regions at and below the dorsal pleural and lung border of a healthy patient. (Middle upper and lower images) With the accumulation of pleural space pathology dorsally (e.g., free gas/air), the location of the pleural border is unchanged (red arrow); however, the dorsal lung border is displaced ventrally (blue arrow/transducer). Lung sliding will be lost at the dorsal pleural border and present at and below where the new dorsal lung border is formed. The return of lung sliding from a region without lung sliding (dorsal pleural border) to a region with lung sliding (new dorsal lung border) is referred to as the lung point. (Right upper and lower images) A more severe example of pleural space pathology creating greater displacement of the dorsal lung border and lung point ventrally (blue arrow/transducer). Note that with severe pathology, if the lung fails to recontact the chest wall, lung sliding will not be visible anywhere on that hemithorax (not shown, see case examples). Although not described in these images, the caudal lung border will also change with pleural space–occupying pathology as the lung is displaced cranially away from the costophrenic recess, which gives rise to abnormal curtain signs (see case examples). Source: Dr. Søren Boysen, with permission.

can be identified the lung surface can easily be assessed as it lies between the 5 PLUS borders of healthy animals. The importance of identifying sonographic borders as opposed to fixed external thoracic landmarks is highlighted by the fact that patients are not uniform in size or conformation, have varying fat content, and different underlying thoracic and abdominal pathology which can alter defined scannable lung regions and pleural borders (**Table 4.1**).

Interpreting abnormal pleural space and lung pathology

Interpretation PLUS findings requires an understating of normal thoracic anatomy and respiratory physiology, knowing how ultrasound interreacts with air at the lung surface, knowledge of where pathology will accumulate and distribute within the lung and pleural space, and consideration of history, clinical exam, and other diagnostic findings.

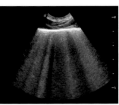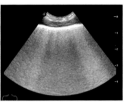

FIGURE 4.8 Left schematic and still image). Increased B-lines, which are starting to coalesce are visible. This is consistent with decreased peripherally aerated lung in cats and dogs, and should prompt consideration of interstitial and alveolar pathology. Right schematic and still image) B-lines can become so numerous they begin to coalesce and form a solid "white sheet" across the entire ultrasound window, which is referred to as "white lung". White lung should also prompt consideration interstitial and alveolar pathology. When B-lines coalesce and create "white lung" A-lines will not be visible regardless of the how the ultrasound beams interact with the lung surface. Source: Dr. Søren Boysen, with permission.

Furthermore, the operator should consider how anatomic species and breed differences, the mechanics of breathing (lung and thoracic wall recoil), the effects of gravity, and different patient positioning affect the location of pleural space and lung pathology, and modify the sonographic windows and structures evaluated accordingly. To assist with interpreting pathology, many of the cases in this book use PLUS profiles to map normal and abnormal sonographic pleural and lung ultrasound findings over both hemithoraces, considering concurrent POCUS and other clinical findings (see individual cases).

In general, the identification of PLUS pathology can be divided into diseases of the pleura, pleural space occupying conditions, and alveolar interstitial diseases. Diseases of the pleura are often interpreted with other PLUS findings. Pleural space diseases are often divided into 3 categories, 1) pneumothorax, 2) pleural effusion and 3) other pleural space occupying conditions (e.g., diaphragmatic hernia and masses, see individual cases). Lung pathology is sonographically identified through two key PLUS findings, increased B-lines and consolidations.[17-23] As the ratio of aerated lung at the lung periphery decreases (e.g., increased lung density due to increased tissue and/or fluid content or decreased air with atelectasis) the number of B-lines increase until the ratio of aerated to non-aerated lung falls below 5-10% (**Figure 4.8**).[28] At this point there is so little air in the periphery of the lung the ultrasound beams are no longer reflected from the lung surface and the lung parenchyma itself becomes visible as lung consolidation (**Figure 4.9**).

PLEURAL AND LUNG ULTRASOUND TECHNIQUE

Numerous pleura, pleural space, and lung ultrasound protocols have been described in human and veterinary medicine, all of which assess different percentages of pleura and lung surface area transthoracically, with or without inclusion of the subxiphoid window, the pericardiodiphragmatic window (**Figure 4.10**), and the abdominal curtain sign (**Figure 4.5**).[1,4-8,11,15]

Given there is a lack of veterinary literature to determine the level of agreement between different canine and feline lung and pleural space ultrasound protocols, and reference standards such as computed tomography, the decision regarding which protocol to use is currently at the discretion of the operator. In human medicine, evidence suggests lung ultrasound

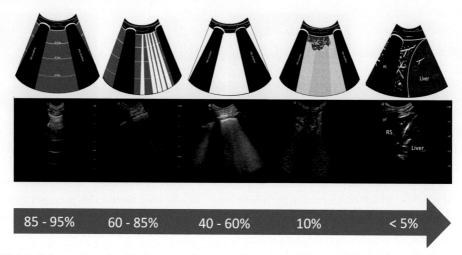

FIGURE 4.9 Schematic and still ultrasound images depicting different ratios of aerated to non-aerated lung. A) When the lung contains > 85% air (normal aerated lung), it is not possible to see lung tissue with ultrasound, although artifacts such as A lines (depicted) may be visible. B) When the lung contains < 85% but >10% air at the periphery (60-85% for illustrative purposes in this example), an increased number of B-lines become visible. The lung tissue remains non-visible, however, B-line artifact, which represents underlying lung pathology that reaches lung surface becomes apparent (5 B-lines in this example). Although not shown in this image A-lines are often visible between B-lines, depending on the angle the beams interact with the lung surface. C) Coalescing B-lines which suggests there a lot less than 85% air, but still ≥ 10% aerated lung at the lung periphery (40-60% for illustrative purposes in this example). A-lines tend not to be visible through "white lung" regardless of the angle the beams interact with the lung surface. D) A region of lung that contains < 5-10% air below the lung surface is depicted. When this situation is encountered the lung becomes sonographically visible as lung consolidation. Ultrasound beams will traverse consolidated lung tissue until they encounter a region of lung below the area of consolidation which contains more than 10% air. This creates a new soft tissue to air interface that reflects the ultrasound beam back to the transducer, preventing lung below the distal border of consolidation from being sonographically visible. D) In this example the ultrasound beam does not encounter a region of aerated lung below the consolidation with > 10% air, which allows the beams to continue to pass through the consolidated lung tissue until the opposite pleura is visible. Ultrasound beams will continue to travel through regions of lung consolidation until they either encounter more aerated lung (> 10% air) or they reach the distal surface of the lung. Source: Dr. Søren Boysen, with permission.

protocols that evaluate a larger lung surface area (12-28 sites vs 4-6 sites) have a higher sensitivity and specificity for detecting pathology but take longer to perform.[29,30] The clinician must therefore balance the risk of missing pathology against the time required to perform sonographic examination. Including the subxiphoid window and abdominal curtain sign may identify pathology otherwise missed with traditional transthoracic protocols and assessment of the abdominal curtain sign may allow earlier identification of pleural effusion in humans. The lack of research comparing veterinary protocols makes translation of results between studies, and establishment of cut-off values to detect pleura and lung pathology, challenging. Based on available veterinary literature, it appears that conditions resulting in marked dyspnea are likely to be diffuse in nature and PLUS should be limited to the minimal number of windows

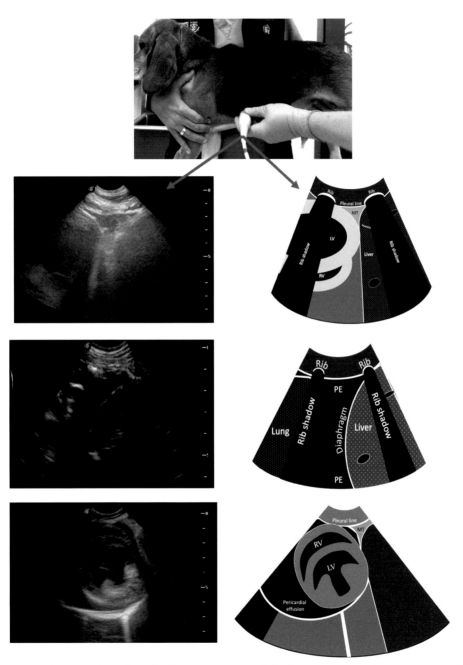

FIGURE 4.10 Photograph, still, and schematic images of the location, normal appearance, and abnormal findings identified at the pericardio-diaphragmatic (PD) window. (Top) Photograph of a dog demonstrating the location of the transducer to find the PD window. (2nd row) Still (left) and schematic images (right) showing the normal appearance of the PD window when the heart and diaphragm are visible within the same image. (3rd row) Schematic (left) and still image (right) of pleural effusion. (4th row) Schematic (left) and still image (right) of pericardial effusion which is easily differentiated from pleural effusion at this location. MT = mediastinal triangle, LV = left ventricle, RV = right ventricle. Source: Dr. Søren Boysen, with permission.

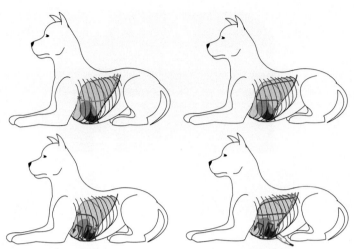

FIGURE 4.11 Schematic images of pleura and lung ultrasound (PLUS) to assess for pneumothorax, lung surface pathology, and pleural effusion using sonographically defined landmarks and lung borders with patients in a sternal position. (A) Abbreviated pleural space protocol to rapidly assess the abdominal curtain sign (ACS) and most gravity-independent sites for pneumothorax in respiratory compromised patients. In the sternal or standing patient, with the transducer situated perpendicular to the ribs (transducer indicator marker directed cranially), slide caudally from the starting point (just behind the front limb, blue Asterix) until the ACS is identified (purple dotted line), followed by sweeping the transducer dorsally along the ACS until the hypaxial muscles are encountered (loss of the pleural line; blue dotted line & green curved arrow). The transducer is then swept ventrally off the hypaxial muscles until the pleural line reappears (the most gravity-independent site; short green arrow). If the patient is in lateral recumbency (not shown) the widest non-gravity region of the thorax and ACS are assessed. (B) Abbreviated lung scanning protocol to rapidly assess multiple lung surface regions in the dorsal, middle, and ventral thirds of the thorax with the transducer orientated perpendicular to the ribs using an S-shaped pattern (red arrow) in respiratory compromised patients. Lung pathology tends to be less effected by position than pleural space pathology, and the same lung ultrasound protocol can often be used regardless of patient position, however, given some lung pathology can concentrate locally or regionally (e.g., aspiration pneumonia and positional atelectasis) different regions of the lung can be emphasized. In larger animals additional horizontal regions between the cranial and caudal borders can be assessed. At the pericardio-diaphragmatic (PD) window (yellow asterix) the transducer can be turned parallel to the ribs or maintained in a perpendicular orientation to scan the ventral lung surfaces. Time permitting, the subxiphoid site can also be assessed (not shown). (C) Abbreviated pleural space and lung ultrasound protocol to rapidly assess the most gravity-dependent pleural sites for pleural effusion and ventral lung pathology in respiratory compromised patients. Identifying the PD window (red arrow) with the transducer situated perpendicular to the ribs helps differentiate pleural and pericardial effusion. In the sternal or standing patient the transducer can be turned parallel to the ribs at the PD window (with the transducer indicator marker directed dorsally) and slid ventrally until the sternal muscles fill 1/3 to 1/2 of the ultrasound image. This allows the most ventral pleural regions to be assessed for the presence of smaller volume pleural effusion. With the transducer maintained in parallel orientation to the ribs it can be slid both dorsally and ventrally within each intercostal space (blue "zig zag" arrow) to assess both the ventral lung regions and the most ventral pleural space regions, respectively, from the ACS caudally to the thoracic inlet cranially. (D) Combined comprehensive PLUS protocol often used in more stable patients to assess the pleura, pleural space, and lungs bilaterally. The subxiphoid site is also included (red highlighted transducer) to assess the caudal lung and pleural space regions not otherwise accessible transthoracically. If the patient is in lateral recumbency, the widest gravity-dependent region of the thorax should be assessed for pleural effusion (not shown). Source: This work is a derivative of "Pleural and lung ultrasound protocol (PLUS)" by McMurray & Boysen, in Frontiers in Veterinary Science, is licensed under CC BY 2.0 by Søren Boysen

and shortest time possible to identify pathology. Once stable, or in less dyspneic animals, more comprehensive scanning protocols can be used, particularly when very localized or focal pleura and lung pathology is suspected (e.g., isolated lung nodules). Depending on the clinical setting encountered, the author currently uses the Calgary PLUS protocol (**Figure 4.11**) to scan the dorsal, middle, and ventral thirds of the lateral lung surfaces bilaterally, scanning between and assessing the 5 PLUS borders, incorporating the PD window, the abdominal curtain sign, and the caudal lung surface from the subxiphoid site, focusing on different sites depending on the pleural space and lung pathologies suspected.

REFERENCES

1. Lisciandro GR, Lagutchik MS, Mann KA, Voges AK, Fosgate GT, et al. Evaluation of a thoracic focused assessment with sonography for trauma (TFAST) protocol to detect pneumothorax and concurrent thoracic injury in 145 traumatized dogs. J Vet Emerg Crit Care. (2008) 18:258.
2. Łobaczewski A, Czopowicz M, Moroz A, Mickiewicz M, Stabińska M, Petelicka H, Frymus T, Szaluś-Jordanow O. Lung Ultrasound for Imaging of B-Lines in Dogs and Cats-A Prospective Study Investigating Agreement between Three Types of Transducers and the Accuracy in Diagnosing Cardiogenic Pulmonary Edema, Pneumonia and Lung Neoplasia. Animals (Basel). 2021 Nov 16;11(11):3279.
3. McMurray J, Boysen S, Chalhoub S. Focused assessment with sonography in nontraumatized dogs and cats in the emergency and critical care setting. J Vet Emerg Crit Care (San Antonio). 2016 Jan-Feb;26(1):64-73.
4. Rademacher N, Pariaut R, Pate J, Saelinger C, Kearney MT, Gaschen L. Transthoracic lung ultrasound in normal dogs and dogs with cardiogenic pulmonary edema: a pilot study. Vet Radiol Ultrasound. 2014;55(4):447-452.
5. Vezzosi T, Mannucci T, Pistoresi A, Toma F, Tognetti R, Zini E, Domenech O, Auriemma E, Citi S. Assessment of Lung Ultrasound B-Lines in Dogs with Different Stages of Chronic Valvular Heart Disease. J Vet Intern Med. 2017 May;31(3):700-704.
6. Armenise A, Boysen RS, Rudloff E, Neri L, Spattini G, Storti E. Veterinary-focused assessment with sonography for trauma-airway, breathing, circulation, disability and exposure: a prospective observational study in 64 canine trauma patients. J Small Anim Pract. 2019;60(3):173-182.
7. Murphy SD, Ward JL, Viall AK, et al. Utility of point-of-care lung ultrasound for monitoring cardiogenic pulmonary edema in dogs. J Vet Intern Med. 2021; 35:68–77.
8. Boysen S, McMurray J, Gommeren K. Abnormal Curtain Signs Identified With a Novel Lung Ultrasound Protocol in Six Dogs With Pneumothorax. Front Vet Sci. 2019;6:291
9. Cole L, Pivetta M, Humm K. Diagnostic accuracy of a lung ultrasound protocol (Vet BLUE) for detection of pleural fluid, pneumothorax and lung pathology in dogs and cats. J Small Anim Pract. 2021;62(3):178-186.
10. Walters AM, O'Brien MA, Selmic LE, Hartman S, McMichael M, O'Brien RT. Evaluation of the agreement between focused assessment with sonography for trauma (AFAST/TFAST) and computed tomography in dogs and cats with recent trauma. J Vet Emerg Crit Care (San Antonio). 2018;28(5):429-435.
11. Boysen S, Chalhoub S, Romero A. Veterinary point-of-care ultrasound transducer orientation for detection of pleural effusion in dog cadavers by novice sonographers: a pilot study. Ultrasound J. 2020 Oct; 12(Suppl 1): 45. doi: 10.1186/s13089-020-00191-6
12. Dicker SA, Lisciandro GR, Newell SM, Johnson JA. Diagnosis of pulmonary contusions with point-of-care lung ultrasonography and thoracic radiography compared to thoracic computed tomography in dogs with motor vehicle trauma: 29 cases (2017-2018). J Vet Emerg Crit Care (San Antonio). 2020 Nov;30(6):638-646.

13. Ward JL, Lisciandro GR, Ware WA, Miles KG, Viall AK, DeFrancesco TC. Lung ultrasonography findings in dogs with various underlying causes of cough. J Am Vet Med Assoc. 2019 Sep 1;255(5):574-583.
14. Fernandes Rodrigues N, Giraud L, Bolen G, et al. Comparison of lung ultrasound, chest radiographs, C-reactive protein, and clinical findings in dogs treated for aspiration pneumonia. J Vet Intern Med. 2022; 36(2):743-752.
15. Boysen, S., Chalhoub, S., and Gommeren, K. (2022). PLUS image interpretation: normal findings. In: The Essentials of Veterinary Point-of-Care Ultrasound: Pleural Space and Lung, 36–62. Zaragoza, Spain: Groupo Asis Biomedia.
16. Lloyd JKF. Minimising stress for patients in the veterinary hospital: Why it is important and what can be done about it. Vet Sci 2017; 4(2): 22.
17. Soldati, G., Sher, S., and Testa, A. (2011). Lung and ultrasound: time to "reflect". Eur. Rev. Med. Pharmacol. Sci. 15 (2): 223–227.
18. Lichtenstein DA, Lascols N, Mezière G, Gepner A. Ultrasound diagnosis of alveolar consolidation in the critically ill. Intensive Care Med, 2004, 30(2):276–281.
19. Miller A. Practical approach to lung ultrasound. BJA Education, 2016;16(2):39–45.
20. Picano E, Pellikka PA. Ultrasound of extravascular lung water: a new standard for pulmonary congestion. Eur Heart J. 2016, 37(27):2097–2104.
21. Lichtenstein DA. Current misconceptions in lung ultrasound: a short guide for experts. Chest, 2019, 156(1):21–25.
22. Lichtenstein DA, Mezière GA. Relevance of lung ultrasound in the diagnosis of acute respiratory failure: the BLUE protocol. Chest, 2008, 134(1):117–125.
23. Goffi A, Kruisselbrink R, Volpicelli G. The sound of air: point-of-care lung ultrasound in perioperative medicine. Can J Anaesth 2018; 65(4): 399-416.
24. Lisciandro GR, Fosgate GT, Fulton RM. Frequency and number of ultrasound lung rockets (B-lines) using a regionally based lung ultrasound examination named vet BLUE (veterinary bedside lung ultrasound exam) in dogs with radiographically normal lung findings. Vet Radiol Ultrasound 2014; 55(3): 315-322.
25. Rigot M, Boysen S, Masseau I, et al. Evaluation of B-lines with two point-of-care lung ultrasound protocols in cats with radiographically normal lungs. Vet Emerg Crit Care. 2022;32:S2–S37.
26. Lisciandro GR, Fulton RM, Fosgate GT, et al. Frequency and number of B-lines using a regionally based lung ultrasound examination in cats with radiographically normal lungs compared to cats with left-sided congestive heart failure. J Vet Emerg Crit Care 2017; 27(5): 499-505.
27. Martins A, Gouveia D, Cardoso A, et al. Incidence of Z, I and B lines detected with point of care ultrasound in healthy shelter dogs. J Vet Emerg Crit Care 2019; 29: S17–S39.
28. Cereda M, Xin Y, Goffi A, et al. Imaging the injured lung: mechanisms of action and clinical use. Anesthesiology 2019; 131(3): 716–749.
29. Buessler A, Chouihed T, Duarte K, et al. Accuracy of several lung ultrasound methods for the diagnosis of acute heart failure in the ED: A multicenter prospective study. Chest 2020; 157(1): 99-110.
30. Kok B, Schuit F, Lieveld A, et al. Comparing lung ultrasound: extensive versus short in COVID-19 (CLUES): a multicentre, observational study at the emergency department. BMJ Open 2021; 11(9): e048795.

CHAPTER 5
PLEURAL AND LUNG ULTRASOUND (PLUS) CASES

CHAPTER 5 – CASE 1
DACHSHUND WITH RESPIRATORY DISTRESS AND COUGHING

Hugo Swanstein and Kris Gommeren

HISTORY, TRIAGE, AND STABILIZATION

A 7-year-old female intact Dachshund presented with tachypnoea and respiratory distress after coughing for a week. She had vomited once over the previous 24 h.

Triage exam findings (vitals):
- Respiratory rate: 60 breaths per minute, increased effort and increased vesicular breath sounds
- Heart rate: 112 beats per minute, mildly decreased heart sounds on the right
- Mucous membranes: pink, <2 sec capillary refill time.
- Femoral and dorsal pedal arterial pulses: strong and regular
- Temperature: 39.5°C (103.1°F)

POCUS exam(s) to perform:
- Triage-performed POCUS:
  - PLUS
  - Subxiphoid view of abdominal POCUS

Abnormal POCUS exam results:
See **Figure 5.1.1** and **Video 5.1.1**.

Additional point-of-care diagnostics and initial management:
During POCUS, an IV catheter was placed and the dog received flow by oxygen, butorphanol (0.15 mg/kg SC), and ampicillin (10 mg/kg IV QID), and was placed in an oxygen cage. Venous blood gas analysis showed a metabolic acidosis (pH 7.29, pCO_2 35.4 mmHg, HCO_3 17.4 mmol/L, lactate 3.5 mmol/L, base excess −9.1 mmol/L). C-reactive protein (CRP) concentration was 130 mg/L

QUESTIONS AND ANSWERS

1. What are the differentials to rule in or out with POCUS based on history and physical exam?
2. What are the sonographic findings?
3. What is the sonographic diagnosis?
4. If necessary, what additional sonographic examination or findings would help rule in or out the differential diagnoses?

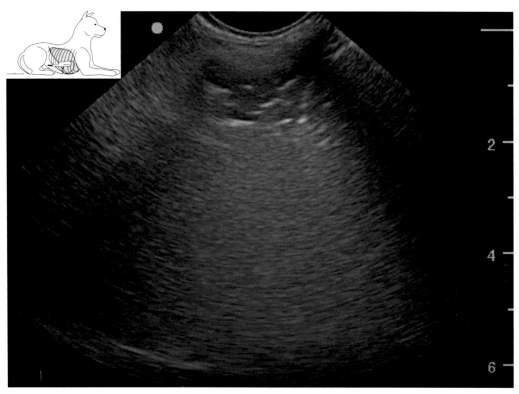

FIGURE 5.1.1 PLUS exam of the right ventral thoracic region at the heart base, with the dog in sternal recumbency, using a microconvex probe perpendicular to the ribs with the depth at 7 cm and the indicator marker directed cranially. Fur was not clipped and alcohol was used as the coupling agent. Video 5.1.1: video obtained from the same window as Figure 5.1.1.

1. **Differential diagnosis to rule in or rule out with POCUS**:
   - The most likely differential diagnoses based on history and physical exam findings are aspiration pneumonia or pneumonia (bacterial, viral, fungal, parasitic, protozoal). The extended differential diagnoses of expiratory dyspnea include cardiogenic pulmonary edema, chronic bronchitis, and non-cardiogenic pulmonary edema; however, these are not typically associated with hyperthermia. PLUS can identify changes compatible with aspiration pneumonia[3] as reported in the veterinary literature.[1,2]
2. **Describe your sonographic findings:**
   - Partial lung (non-translobar) consolidation (shred sign) of the right cranio-ventral hemithorax (**Figure 5.1.2**, **Video 5.1.2**).
   - Coalescent B-lines are seen surrounding the consolidated region (**Figure 5.1.2**, **Video 5.1.2**).
   - A-lines were visible over the entire left hemithorax (not shown).
   - A-lines were also seen in the dorsal and dorso-caudal areas of the right hemithorax (not shown).
3. **What is your sonographic diagnosis?**
   - Focal unilateral partial lung consolidation (shred sign) with dynamic air bronchograms and adjacent coalescent B-lines, localized predominantly to the right middle lung lobe region.
4. **What additional sonographic examination or finding would help you rule in or rule out your differential diagnoses (if necessary)?**

- Air bronchograms are sonographically visible when air persists within the smaller airways of the consolidated lung.
- They can be classified as dynamic or static, which helps differentiate underlying causes.
- Static air bronchograms can occur with underlying lung pathology but are also commonly seen with atelectasis.
- In contrast, dynamic air bronchograms have only been associated with lung pathology and are not see in cases with atelectasis.

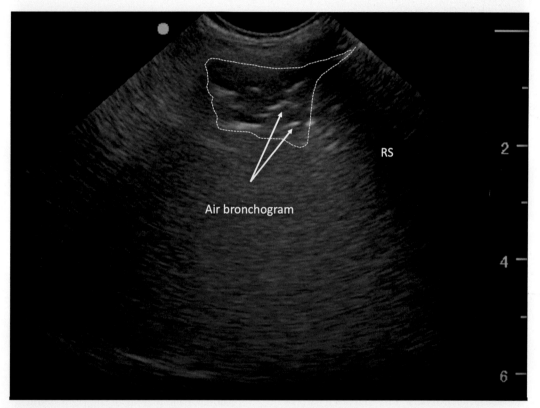

FIGURE 5.1.2 (Figure 5.1.1 labelled) A shred sign (dotted white line) is a type of partial lung consolidation, with the transition from visible consolidated lung to non-visible more aerated lung highlighted by a hyperechoic, irregular, serrated, and blurry transition. Shred signs often contain punctate to linear hyperechoic structures (air bronchograms) or hypoechoic fluid (fluid bronchograms).

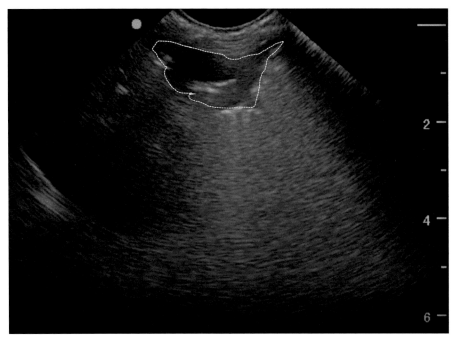

FIGURE 5.1.3 Day 5: the image was acquired in the same manner as described in Figure 5.1.1. The consolidation has reduced in size (white dotted line), with a smaller shred sign. The dog was doing much better clinically. **Video 5.1.3** Obtained at the windows from Figures 5.1.1 and 5.1.2 as a split screen video showing the first and serial PLUS findings. The heart can be seen pulsating within the cineloop.

CLINICAL INTEGRATION

Suspected diagnosis based on ultrasound findings:
Aspiration pneumonia is the most likely diagnosis. **Video 5.1.2 (Video 5.1.1 labelled)**.

Sonographic interventions, monitoring, and outcome:
- During the 6 days of hospitalization, serial PLUS was performed daily to track the progression/resolution of aspiration pneumonia (**Figure 5.1.3**, **Video 5.1.3**).
- The dog was monitored closely, continued on oxygen therapy, and received a ten-day course of ampicillin.
- The dog was discharged on the sixth day, based on clinical resolution, significantly reduced C-reactive protein concentration, and improved PLUS findings.
- Figure 5.1.3 and Video 5.1.3 show the progression/resolution of the aspiration pneumonia.

Important concepts regarding the sonographic diagnosis of aspiration pneumonia:
- Aspiration pneumonia typically affects the gravity-dependent lung lobes.
- In healthy dogs this is usually the right middle lung lobe, but in a critical care or peri-anesthetic setting the distribution of lesions can be atypical.
- A "classic PLUS profile" includes spared normal-looking lung with A-lines and lung sliding in the dorsal regions, transitioning to B-lines, and finally areas of lung

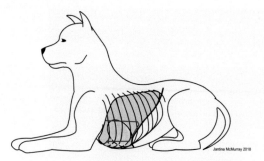

FIGURE 5.1.4 Overview of a modified PLUS approach in a sternal patient. The red arrow shows the initial probe placement at the cranial PLUS border and probe movements to reach the ventral portion of the thorax. The red arrow outlines the focused scanning to find signs of aspiration pneumonia or pleural effusion without having to do the entire PLUS protocol, which is particularly important in patients with marked respiratory distress. The yellow arrows indicate how to identify smaller amounts of pleural effusion or consolidations by turning the probe in parallel to the ribs. Once stable, a more comprehensive scan of the lungs can be performed, along with other POCUS examinations.

consolidation, particularly shred signs, as the probe is moved ventrally and cranially.
- B-lines occur because of decreased aerated lung (<85%), with the number of B-lines correlating with the degree of non-aerated lung. Thus, coalescent B-lines indicate a more severe loss of lung aeration at the lung periphery.
- A shred sign is a type of partial lung consolidation, with the transition from visible consolidated lung to non-visible more aerated lung highlighted by a hyperechoic, irregular, serrated, and blurry transition. Shred signs often contain punctate to linear hyperechoic structures (air bronchograms) or hypoechoic fluid structures (fluid bronchograms).
- Lung consolidation occurs when the aerated lung ratio of the lung surface is reduced below 10% of normal. Healthy lung consists of roughly 85%–95% air, impeding ultrasonography beyond the first 1–3 mm of the normal lung.
- The human literature has reported 8% of acquired pneumonias are not visible with POCUS.[4] C-reactive protein together with sonographic lung consolidation is reported to have better correlation with clinical findings than thoracic radiographs.[1]

Techniques and tips to identify the appropriate ultrasound images/views:
- When aspiration pneumonia is the working differential diagnosis, start scanning at the PLUS starting point (just caudal to the forelimb), slide the probe caudally to the curtain sign, descend to the ventral part of the thorax, and slide the probe cranially (**Figure 5.1.4**).
- If the scan is negative, follow up with a complete S-shaped scan to cover more of the lung surface to identify atypical lesions.
- Turning the transducer parallel to the ribs in the ventral regions can be helpful in detecting small ventral consolidations and parapneumonic effusion (**Figure 5.1.4**).[5]
- Placing the front legs of the patient in sternal recumbency on a rolled towel and extending the limb cranially to access the axillary region help gain access to the ventral lung regions without causing respiratory compromise.
- In the standing patient the limb can also be extended cranially to allow more ventral lung area and the axilla to be assessed.
- The cineloop presented in **Video 5.1.1** could be improved by increasing the gain and decreasing the depth, allowing for more contrast, and maximizing visualization of the consolidated region.

Take-home messages
- Lung consolidations, shred signs, increased B-lines, and a thickened pleura in the cranial ventral lung regions, along with increased C-reactive protein should always raise suspicion of aspiration pneumonia.
- Clinical evolution and C-reactive protein concentration are the best indications of antibiotic discontinuation; however, disappearing consolidations on PLUS can be a reassuring finding as well.

REFERENCES

1. Fernandes Rodrigues N, Giraud L, Bolen G, et al. Comparison of lung ultrasound, chest radiographs, C-reactive protein, and clinical findings in dogs treated for aspiration pneumonia. *J Vet Intern Med* 2022;36(2):743–752. doi:10.1111/jvim.16379
2. Fernandes Rodrigues N, Giraud L, Bolen G, et al. Antimicrobial discontinuation in dogs with acute aspiration pneumonia based on clinical improvement and normalization of C-reactive protein concentration. *J Vet Intern Med* 2022:1–7. doi:10.1111/jvim.16405
3. Ward JL, Lisciandro GR, Ware WA, Miles KG, Viall AK, DeFrancesco TC. Lung ultrasonography findings in dogs with various underlying causes of cough. *J Am Vet Med Assoc* 2019;255(5):574–583. doi:10.2460/javma.255.5.574
4. Reissig A, Copetti R, Mathis G, et al. Lung ultrasound in the diagnosis and follow-up of community-acquired pneumonia: A prospective, multicenter, diagnostic accuracy study. *Chest* 2012;142(4):965–972. doi:10.1378/chest.12-0364
5. Graziano N, Burnotte P, Gommeren K. A retrospective study on the incidence of parapneumonic effusion in 130 dogs with a clinical diagnosis of pneumonia. Unpublished Article. University of Liege, Belgium, 2022. EVECC Congress Proceedings 2022.

CHAPTER 5 – CASE 2

LABRADOR RETRIEVER WITH MILD DYSPNEA AND PROFUSE BLEEDING FROM A PARASTERNAL PENETRATING INJURY

Simone Cutler and Daria Starybrat

HISTORY, TRIAGE, AND STABILIZATION

An 18-month female spayed Labrador Retriever presented with acute lethargy and profuse bleeding from a left parasternal penetrating injury sustained on a walk 24 h prior to presentation. The dog was previously in good health.

Triage exam findings (vitals):
- Mentation: quiet, alert, and responsive
- Gait: ambulatory
- Respiratory rate: 36 breaths per minute, shallow breaths with mild increase in expiratory effort
- Bilaterally reduced ventral lung sounds.
- Heart rate: 140 beats per minute, regular rhythm, no murmur
- Mucous membranes: pale pink, tacky, 1 sec capillary refill time.
- Femoral and dorsal pedal arterial pulses: synchronous and strong.
- Temperature: 38.2°C (100.8°F).

POCUS exam(s) to perform:
- PLUS
- Cardiac POCUS
- Abdominal POCUS

Additional point-of-care diagnostics and initial management:
- Packed cell volume (PCV) 38%, total solids 4.8 g/dL, blood lactate 1.8 mmol/L, blood glucose 145 mg/dL (8.1 mmol/L).
- Manual platelet count: 150×10^9/L.
- Prothrombin time (PT): 12 sec (11–17), activated partial thromboplastin clotting time (aPTT): 94 sec (72–102).
- Supplemental oxygen.
- Methadone 0.2 mg/kg IV.

Abnormal POCUS exam results:
See **Figures 5.2.1** and **5.2.2** as well as **Videos 5.2.1** and **5.2.2**.

QUESTIONS AND ANSWERS

1. What are the differentials to rule in or out with POCUS based on history and physical exam?
2. What are the sonographic findings?

DOI: 10.1201/9781003436690-9

Mild Dyspnea and Profuse Bleeding from a Parasternal Penetrating Injury 43

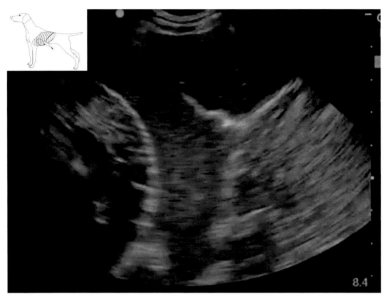

FIGURE 5.2.1 **PLUS with the dog standing and the probe directed cranially at the left caudal and ventral hemithorax. Curvilinear probe oriented perpendicular to the ribs with the frequency set at 8 MHz and the depth set at 8.4 cm. The fur was not clipped but parted and alcohol was used as the coupling agent. Video 5.2.1 Left caudal and ventral hemithorax of PLUS with the dog standing, as shown in Figure 5.2.1.**

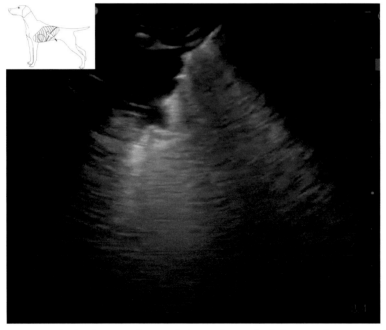

FIGURE 5.2.2 **PLUS with the dog standing and the probe at the left seventh intercostal space, at two-thirds the height of the hemithorax. Curvilinear probe oriented perpendicular to the ribs with the frequency set at 8 MHz and the depth set at 8.4 cm. The fur was not clipped but parted and alcohol was used as the coupling agent. Video 5.2.2 Left seventh intercostal space at two-thirds the height of the left hemithorax of PLUS with the dog standing, as shown in Figure 5.2.2.**

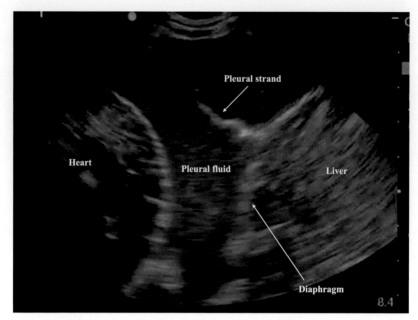

FIGURE 5.2.3 Pleural fluid is represented as a variably hypoechoic region leading to pleural surface separation. Linear stranding structure (middle right), most consistent with fibrin attached to the pleural surface. Video 5.2.3 (Video 5.2.1 labelled) Pleural effusion fills the costophrenic recess and separates the heart from the diaphragm.

3. What is the sonographic diagnosis?
4. If necessary, what additional sonographic examination or findings would help rule in or out the differential diagnoses?

1. **Differential diagnosis to rule in or rule out with POCUS:**
   - Pulmonary contusions, pneumothorax, hemothorax, chylothorax, and collapsed lung are all potential causes of respiratory distress in patients with suspected penetrating thoracic injuries.
2. **Describe your sonographic findings:**
   - Pleural fluid is represented as a variably hypoechoic region leading to pleural surface separation (**Figure 5.2.3, Video 5.2.3**).
   - B-lines (**Figure 5.2.4** and **Video 5.2.4**).
   - Shred signs (superficial lung consolidation with partial aeration) are visible where lung separated by pleural effusion recontacts the chest wall (similar to a lung point seen with a pneumothorax; **Figure 5.2.4, Video 5.2.4**).
3. **What is your sonographic diagnosis?**
   - Pleural effusion, partial lung consolidation with associated B-lines.
4. **What additional sonographic examination or finding would help you rule in or rule out your differential diagnoses?**
   - Absence of the following sonographic findings in the regions of the lung unaffected by pleural effusion or B-lines would raise suspicion of a concurrent pneumothorax: lung sliding (glide sign), lung pulse, and synchronous curtain sign (e.g. pneumothorax is often associated with asynchronous and double curtain signs).[1,2]
   - Ultrasound-guided diagnostic thoracocentesis with fluid analysis is valuable in classifying pleural effusion.[3]

Mild Dyspnea and Profuse Bleeding from a Parasternal Penetrating Injury 45

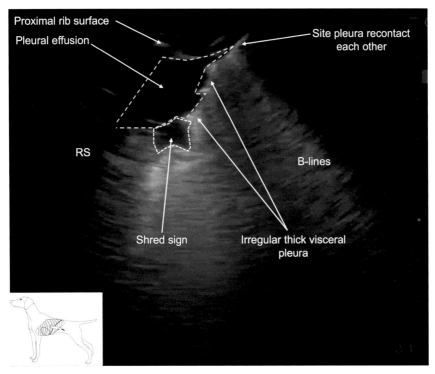

FIGURE 5.2.4 Pleural effusion (outlined by yellow dotted lines) can be seen separating the visceral and parietal pleura. Below the effusion the visceral pleura is irregular, and a shred sign can be seen (white dotted outline). The lung is visible recontacting the chest wall to the right of the image (similar to the lung point in cases with a pneumothorax). RS: rib shadow. Video 5.2.4 (Video 5.2.2 labelled) Lung sliding is absent to the left due to separation of the lung from the chest wall by pleural effusion. Underlying B-lines are visible. The return of lung sliding can be seen where the lung recontacts the chest and fluid is no longer separating the pleura. A shred sign (superficial lung consolidation with partial aeration) transitioning to a region where the lung recontacts the chest wall (similar to a lung point seen with a pneumothorax) and B-lines are visible.

CLINICAL INTEGRATION

Suspected diagnosis based on ultrasound findings:
Pleural effusion and lung contusions.

Sonographic interventions, monitoring, and outcome:
- The right hemithorax was assessed similarly and an equivalent volume of pleural effusion was identified with an absence of B-lines.
- A glide sign (lung sliding) and a synchronous curtain sign were identified at the caudo-dorsal aspects of both sides of the lung, and in the context of normal lung auscultation, a clinically significant volume pneumothorax was considered unlikely.
- Cardiac and abdominal POCUS were unremarkable.
- Ultrasound-guided thoracocentesis and pleural fluid analysis: consistent with hemorrhagic effusion.
- The dog responded well to supportive management and was eventually discharged.

Important concepts regarding pleural effusion/hemothorax:
- Pleural effusion is defined as fluid accumulation within the pleural space (between the visceral pleura overlying the lung and the parietal pleura lining the thoracic wall).[2]
- However, depending on the location of the pleural effusion, the separation of the visceral and parietal pleura may not be appreciated and instead fluid-filled cavities separating the diaphragm from the thoracic wall, the heart from the diaphragm, or the heart from the sternal structures are seen.
- The nature of the pleural fluid can affect the sonographic appearance with transudates and modified transudates having a more uniform anechoic to hypoechoic appearance relative to the surrounding lung when compared to highly cellular effusions.
- Mobile, semi-fixed stranding structures can sometimes be seen within the pleural space and can represent fibrin (suspected in this case), inflammatory precipitates, granulomatous material, or neoplastic aggregates.[3]
- Lung sliding will be absent secondary to pleural effusion due to separation of the visceral and parietal pleural surfaces and a return of lung sliding (similar to the lung point of a pneumothorax) can be seen where the pleural surfaces regain contact at the margins of the pleural effusion.
- The shred sign seen in **Figure 5.2.4**, located dorsally, is consistent with focal lung consolidation and likely represents pulmonary contusions. Shred signs related to atelectasis secondary to the compressive effects of recumbency or pleural effusion would instead be located within the ventral thorax and would resolve once the affected lung is reinflated.[1,2]

Techniques and tips to identify pleural effusion/hemothorax:
- Positioning the probe parallel to the ribs in the most ventral third of the thorax and scanning its entire length, from cranial to caudal, will maximize the ability to detect pleural effusion.
- The apparent volume/depth of pleural effusion will be affected by patient recumbency, and consistent patient positioning and probe location/orientation on the thorax (longitudinal or transverse) are recommended for serial monitoring.
- Ensure depth settings are adequate to include the full extent of the pleural effusion, reducing the depth and adjusting the focus point to assess the site where the lung recontacts the chest wall (adjacent pleural effusing associated lung point).

Take-home messages
- Presence of anechoic or hypoechoic effusion between thoracic structures, particularly within the costophrenic recess and/or displacement of the lung, heart, or diaphragm due to fluid accumulation within the thoracic cavity may be indicative of a hemothorax, especially in patients following trauma.
- Thoracocentesis and fluid analysis are indicated to confirm the diagnosis.

REFERENCES

1. Boysen SR, Lisciandro GR. The Use of Ultrasound for Dogs and Cats in the Emergency Room: AFAST and TFAST. *Veterinary Clinics of North America: Small Animal Practice*. 2013 Jul;43(4):773–97.
2. Lisciandro GR, Gambino JM. Diagnostic

Imaging: Point-of-Care Ultrasound. *Veterinary Clinics of North America. Small Animal Practice.* 2021;51(6):i–i.
3. Brogi E, Gargani L, Bignami E, Barbariol F, Marra A, Forfori F, Vetrugno L. Thoracic Ultrasound for Pleural Effusion in the Intensive Care Unit: A Narrative Review from Diagnosis to Treatment. *Critical Care.* 2017 Dec 28;21(1):325.

CHAPTER 5 – CASE 3

MIXED BREED DOG WITH RESPIRATORY DISTRESS FOLLOWING MOTOR VEHICULAR TRAUMA

Søren Boysen

HISTORY, TRIAGE, AND STABILIZATION

A 3-year-old male neutered mixed breed dog presented with respiratory distress and superficial abrasions following motor vehicular trauma.

Triage exam findings (vitals):
- Respiratory rate: 54 breaths per minute, increased effort, lung difficult to auscult
- Heart rate: 132 beats per minute
- Mucous membranes: pink, <2 sec capillary refill time
- Femoral and dorsal pedal arterial pulses: strong and regular
  Temperature: 38.6°C (101.5°F)

POCUS exam(s) to perform:
- Triage-performed PLUS

Abnormal POCUS exam results:
See **Figure 5.3.1** and **Video 5.3.1**.

Additional point-of-care diagnostics and initial management:
- Supplemental oxygen and hydromorphone (0.1 mg/kg IV).

QUESTIONS AND ANSWERS

1. What are the differentials to rule in or out with POCUS based on history and physical exam?
2. What are the sonographic findings?
3. What is the sonographic diagnosis?
4. If necessary, what additional sonographic examination or findings would help rule in or out the differential diagnoses?

Respiratory Distress Following Motor Vehicular Trauma

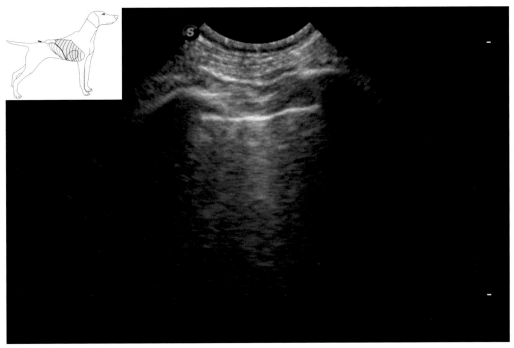

FIGURE 5.3.1 PLUS at the most caudal dorsal border in a standing patient using a microconvex probe, indicator marker cranial, depth set at 5 cm. Fur was not clipped and alcohol was used as the coupling agent. Video 5.3.1 Video obtained from the same window shown in Figure 5.3.1.

1. **Differential diagnosis to rule in or rule out with POCUS:**
   - Pulmonary contusions, hemothorax, diaphragmatic hernia, rib fractures, and pneumothorax are all likely causes of dyspnea in trauma patients.
2. **Describe your sonographic findings:**
   - Absence of lung sliding with Z-lines and occasional A-lines (**Figure 5.3.2**, **Video 5.3.2**, and **Video 5.3.3**).
3. **What is your sonographic diagnosis?**
   - Absence of lung sliding, B-lines, consolidation, pleural effusion, and a lung pulse with the presence of A-lines and Z-lines at the probe location.
4. **What additional sonographic examination or finding would help you rule in or rule out your differential diagnoses (if necessary)?**
   - Pneumothorax should be suspected when all the following sonographic findings are *absent* at a specific probe location: (1) lung sliding (aka the glide sign), (2) a lung pulse, (3) B-lines, (4) lung consolidations, and (5) pleural effusion.[1–3]
   - Pneumothorax can be sonographically confirmed when asynchronous and double curtain signs are identified and/or the lung point is identified.[1,4]
   - Once the dog is stable, and following thoracocentesis, the lung should be assessed to document the extent and distribution of any underlying lung pathology.

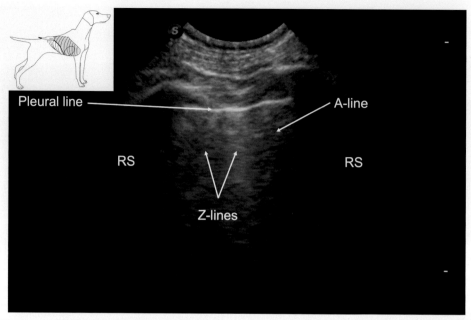

FIGURE 5.3.2 The hyperechoic pleural line appears smooth and regular, which although not possible to assess in a still image, has an absence of lung sliding (aka the glide sign). Z-lines appear as vertical, ill-defined hyperechoic lines that arise from, or superficial to, the parietal surface, disappear after a few centimeters, accentuate A-lines, and can be seen in healthy animals and animals with a pneumothorax. They are independent of lung sliding in patients where the lung is in contact with the chest wall. Although not well appreciated in this image, A-lines can be seen in patients with and without a pneumothorax and therefore do not help determine if a patient does or does not have a pneumothorax. Video 5.3.2 (Video 5.3.1 labelled) The absence of lung sliding (aka the glide sign) can only be appreciated in real time or with a cineloop as it requires the identification or absence of a horizontal "shimmer" along the pleural line which only occurs when the lung is in contact with the chest wall (lung sliding is lost in cases of a pneumothorax when the lung is not in contact with the chest wall). Video 5.3.3 Side-by-side videos using similar machine settings to demonstrate the presence (right video) and absence (left video) of lung sliding.

CLINICAL INTEGRATION

Suspected diagnosis based on ultrasound findings:
Pneumothorax.

Sonographic interventions, monitoring, and outcome:
- Abnormal curtain signs (asynchronous and double) were noted on both sides of the chest, confirming the presence of a bilateral pneumothorax.
- Bilateral thoracocentesis was performed.
- Repeat PLUS following thoracocentesis showed only a mild pneumothorax and the presence of B-lines consistent with concurrent pulmonary contusions.
- Cardiovascular and abdominal POCUS were performed following thoracocentesis, when the patient was more stable, and were found to be unremarkable (not shown).
- Serial PLUS was performed to monitor the degree/recurrence of a pneumothorax, and

to monitor the progression/resolution of increased B-lines over time.
- The dog was treated conservatively, monitored closely, and eventually discharged.

Important concepts regarding the sonographic diagnosis of absent lung sliding and pneumothorax:

- A pneumothorax should be suspected when the absence of lung sliding is noted; however, lung sliding can be very difficult to identify in patients where the lung is in contact with the chest wall, particularly if the patient has a very rapid shallow breathing pattern.[4,5]
- There will also be an absence of a visible lung pulse, B-lines, lung consolidations, and pleural effusion at the probe location over the region of the pneumothorax.[1-4]
- The sonographer has to carefully assess if lung sliding is visible across the entire pleural line, suggesting that the pleura are in contact and ruling out a pneumothorax at that probe location, or if lung sliding suddenly appears and disappears within the ultrasound window, suggesting a lung point, which confirms a pneumothorax (see lung point case).[1,3]
- When looking for lung sliding, focus only on the pleural line. Ignore all other structures and movement in the ultrasound image (e.g., the intercostal muscles will contract, the ribs will shift). Look only for the presence of a to-and-fro "shimmer" along the pleural line (the pleural line may shift vertically and/or in and out of the plane; however, this may occur in any animal, and it is the horizontal shimmer along the pleural line that must be carefully assessed to determine if lung sliding is present or absent).

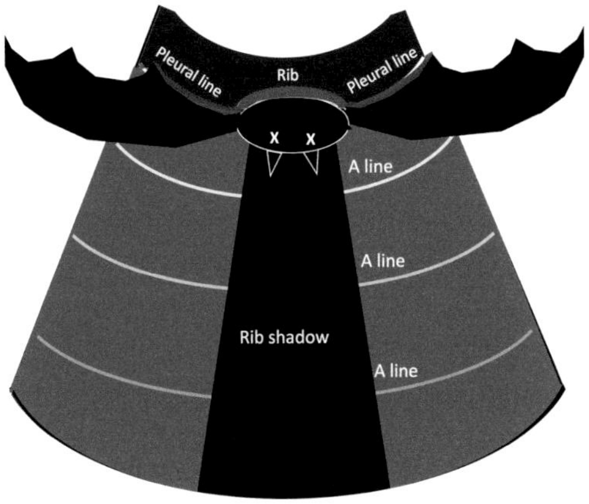

Techniques and tips to identify the appropriate ultrasound images/views:

- Search for a pneumothorax at the most sensitive sites for air to accumulate, which will vary with patient positioning: the caudo-dorsal site in the standing or sternal patient and the widest part of the chest in the laterally recumbent patient.[3]
- There are several tricks to make lung sliding easier to identify by making the pleural line appear "grainier"[3]:
  - Change the angle that the ultrasound beams strike the pleural line by fanning the transducer; the ultrasound beams strike the pleural line obliquely and not perpendicular.
  - Lung sliding is easier to see when A-lines are not visible and the angle of insonation is deferred from 90°.
    – Change the angle that the ultrasound beam strikes the pleural line by placing the transducer over a single

FIGURE 5.3.3 Situating the probe over a rib, such that a rib is centered in the ultrasound image, creates the "dead bat sign," where the wings of the upside-down bat represent the pleural line and the proximal rib surface the body of the bat. When the "dead bat sign" is obtained the pleural line often appears grainier to either side of the rib, making it easier to appreciate lung sliding and the fact that the pleural are in contact.

rib head referred to as the "dead bat sign" (**Figure 5.3.3**).
- Decrease the depth (e.g., to 2 cm) and frequency (e.g., lower frequency) settings to make the pleural line larger and "grainier," which makes lung sliding easier to appreciate.
- Decrease the gain setting on the ultrasound machine to make the pleural line "grainier."
- Turn off filters such as harmonics which "smooth" out structures, and make the pleural line less "grainy"
• Note: if the depth and gain are decreased, they should be increased again (e.g., return depth to 4–6 cm) to look for other pathology and allow easier identification of the curtain sign, which is more challenging to assess at low-depth settings.

Take-home messages
- Lung sliding can be difficult to identify in patients with rapid shallow respiratory patterns.
- If a sonographic diagnosis is uncertain, a search for abnormal curtain signs and lung points that confirm a pneumothorax should be undertaken provided the patient is sufficiently stable to do so.
- By contrast, when the patient is in obvious respiratory distress, there is an absence of lung sliding, and the history and clinical findings are supportive of a pneumothorax (e.g., absence of breath sounds in a trauma patient), thoracocentesis is immediately indicated.

REFERENCES

1. Boysen SR. Lung ultrasonography for pneumothorax in dogs and cats. *Vet Clin North Am Small Anim Pract* 2021 Nov;51(6):1153–1167.
2. Hwang TS, Yoon YM, Jung DI, Yeon SC, Lee HC. Usefulness of transthoracic lung ultrasound for the diagnosis of mild pneumothorax. *J Vet Sci* 2018;19(5):660–666.
3. Boysen S, Chalhoub S, Gommeren K. Clinical applications of PLUS: Is there pneumothorax, yes/no? In: Boysen S, Chalhoub C, Gommeren K, Eds. *The Essentials of Veterinary Point-of-Care Ultrasound: Pleural Space and Lung* Edra publishing, Laval, Quebec; 2022:63–90.
4. Boysen S, McMurray J, Gommeren K. Abnormal curtain signs identified with a novel lung ultrasound protocol in six dogs with pneumothorax. *Front Vet Sci* 2019;6:291.
5. Walters AM, O'Brien MA, Selmic LE, et al. Evaluation of the agreement between focused assessment with sonography for trauma (AFAST/TFAST) and computed tomography in dogs and cats with recent trauma. *J Vet Emerg Crit Care (San Antonio)* 2018;28(5):429–435.

CHAPTER 5 – CASE 4

HAVANESE WITH ACUTE RESPIRATORY DISTRESS, KNOWN MYXOMATOUS MITRAL VALVE DISEASE, AND 3-WEEK HISTORY OF COUGHING

Chiara Debie and Christopher Kennedy

HISTORY, TRIAGE, AND STABILIZATION

An 11-year-old neutered male Havanese presented with acute respiratory distress. The dog had known myxomatous mitral valve disease (MMVD) (ACVIM stage B2) and hyperadrenocorticism. It had a 3-week history of occasional coughing, which had increased in frequency over the previous 2 days.

Triage exam findings (vitals):
- Mentation: alert
- Respiratory rate: 80 breaths per minute and severe expiratory dyspnea
- Heart rate: 140 beats per minute
- Mucous membranes: cyanotic
- Cardiopulmonary auscultation: left apical systolic heart murmur, bilateral crackles dorsally

POCUS exam(s) to perform:
- PLUS
- Cardiac POCUS

Abnormal POCUS exam results:
See **Figures 5.4.1, 5.4.2,** and **5.4.3** as well as Videos **5.4.1** and **5.4.2**.

Additional point-of-care diagnostics and initial management:
- Oxygen supplementation (oxygen cage) and butorphanol 0.3 mg/kg IM.

QUESTIONS AND ANSWERS

1. What are the differentials to rule in or out with POCUS based on history and physical exam?
2. What are the sonographic findings?
3. What is the sonographic diagnosis?
4. If necessary, what additional sonographic examination or findings would help rule in or out the differential diagnoses?

DOI: 10.1201/9781003436690-11

CHAPTER 5 – CASE 4

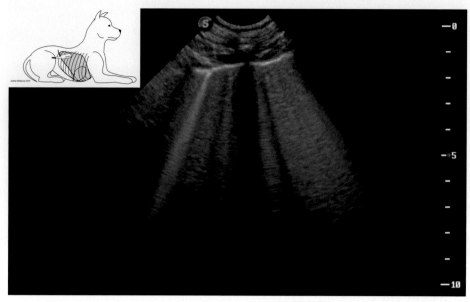

FIGURE 5.4.1 PLUS exam at the caudo-dorsal region of the right hemithorax with the patient in sternal recumbency, using a microconvex probe oriented perpendicular to the ribs, marker directed cranially, with the depth set at 10 cm. Similar findings were noted over multiple lung regions bilaterally. Fur was not clipped, only parted, and alcohol was used as the coupling agent. Video 5.4.1: Video obtained from the same PLUS window as Figure 5.4.1.

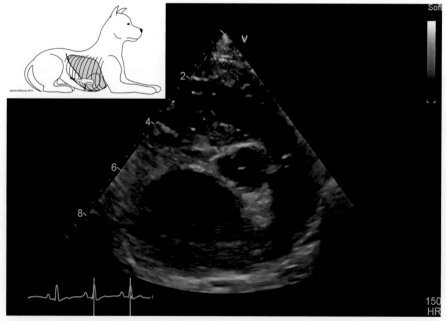

FIGURE 5.4.2 Cardiac POCUS at the right parasternal short-axis window over the heart base (papillary and basilar views) with the dog in sternal recumbency, using a phased-array probe, depth at 8 cm. Cardiac mode was used (image indicator to the right of the ultrasound screen) with the probe indicator marker directed cranially. Fur was not clipped, only parted, and ultrasound gel was used as the coupling agent.

1. **Differential diagnosis to rule in or rule out with POCUS:**
   - Left-sided congestive heart failure with cardiogenic pulmonary edema, non-cardiogenic pulmonary edema, pulmonary contusions, bronchopneumonia, pleural effusion, neoplasia, fibrosis (difficult to rule in or out specifically with ultrasound), pulmonary thromboembolism (difficult to rule in or out specifically with ultrasound).
2. **Describe your sonographic findings:**
   - On PLUS, multiple to coalescent B-lines are present (**Figures 5.4.1** and **5.4.4**, **Video 5.4.1**).
   - The pleural line is thin and smooth, which is suggestive of a non-inflammatory process (such as cardiogenic pulmonary edema).
   - On cardiac POCUS, the left atrial (LA) to aortic (Ao) ratio (LA:Ao ratio) is 2.5 (see **Figures 5.4.2** and **5.4.5**).
   - On the long-axis four-chamber view, the LA is subjectively very enlarged (see **Figures 5.4.3** and **5.4.6**, **Video 5.4.2**).
3. **What is your sonographic diagnosis?**
   - Increased B-lines (greater than three per view on multiple views and coalescent in some places) suggests a decreased aerated lung at the lung periphery, most likely increased lung water (interstitial syndrome, pulmonary edema). An enlarged LA is consistent with the previous diagnosis of MMVD.
4. **What additional sonographic examination or finding would help you rule in or rule out your differential diagnoses?**
   - Formal echocardiography.

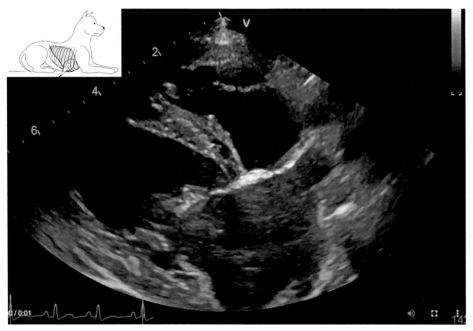

FIGURE 5.4.3 Cardiac POCUS at the right parasternal long-axis window with the dog in sternal recumbency using a phased-array probe, depth at 8 cm. Cardiac mode was used (image indicator to the right of the ultrasound screen) with the probe indicator marker directed cranially and dorsally (in line with the long axis of the heart). Fur was not clipped, only parted, and ultrasound gel was used as the coupling agent. Video 5.4.2: Video obtained from the same probe location as Figure 5.4.3.

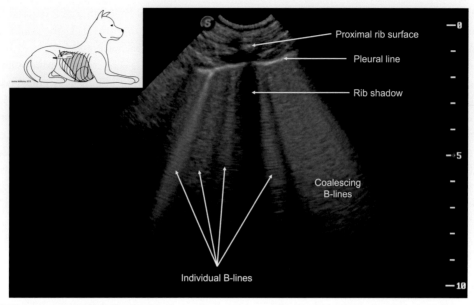

FIGURE 5.4.4 Multiple individual B-lines that start to coalesce are visible. Coalescent B-lines are B-lines that merge together and form a wider column than the smallest B-line visible within the same sonographic window. The pleural line is smooth, thin, and regular in appearance.

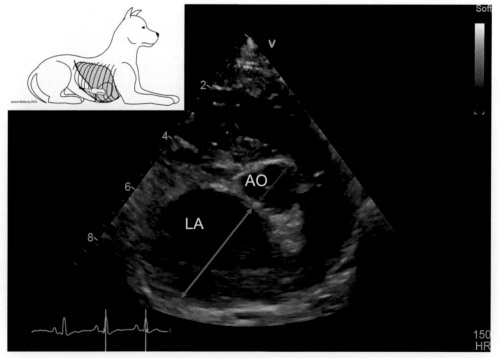

FIGURE 5.4.5 The left atrial width (blue arrow) to the aortic width (red arrow) results in an LA:Ao ratio of 2.5. An LA:Ao ratio ≥1.6 is indicative for left atrium enlargement. Ao: aorta; LA: left atrium.

CLINICAL INTEGRATION

Suspected diagnosis based on ultrasound findings:
Left-sided congestive heart failure with pulmonary edema, secondary to decompensated MMVD.

Sonographic interventions, monitoring, and outcome:
No further POCUS interventions are necessary. PLUS may be used to monitor the evolution of B-lines during hospitalization. Thoracic radiographs confirmed the suspected diagnosis: cardiogenic pulmonary edema, congested pulmonary veins, and LA enlargement. Echocardiography supported decompensated MMVD (ACVIM stage C). Treatment with furosemide and pimobendane successfully stabilized the dog.

Important concepts regarding the sonographic diagnosis of cardiogenic pulmonary edema:
- One to three B-lines per window can be seen physiologically.
- Greater than three B-lines in any single window, coalescent B-lines, and/or multiple B-lines in multiple locations (including both hemithoraces) support cardiogenic pulmonary edema.[1,3]
- B-lines should be differentiated from E-, Z-, and I-lines.[1]
- Theoretically, the pleural line in cardiogenic pulmonary edema should be thin and continuous. Inflammatory processes, e.g., bronchopneumonia and non-cardiogenic pulmonary edema, may be present with a thickened and discontinuous pleural line (based on preliminary studies in human medicine).[1,2]

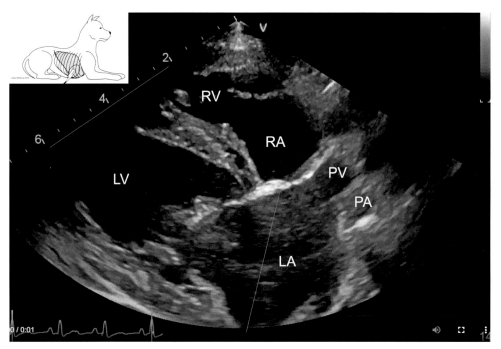

FIGURE 5.4.6 Subjective left atrial enlargement on a long-axis four-chamber view. Blue lines represent subjective estimation of the left atrial diameter. LV: left ventricle; RV: right ventricle; LA: left atrium; RA: right atrium; PV: pulmonary vein; PA: pulmonary artery.

- Lung consolidations or shred signs are not expected with cardiogenic pulmonary edema.

Techniques and tips to identify the appropriate ultrasound images/views:

- A few B-lines can be found in normal patients.[1,3]
- Z-lines and I-lines are not pathological, but are often present in healthy patients.[1]
- Consider the distribution of B-lines when formulating a diagnosis.[1]
- Pleural thickness should be used with caution as there is very limited evidence of pleural line assessment in veterinary medicine.[2]
- Left atrial enlargement alone cannot diagnose congestive heart failure.[6,7] The LA:Ao measurement is subject to marked inter-operator measurement error. Commonly, patients with respiratory distress due to left congestive heart failure have LA:Ao >2:1; such high ratios are unlikely to occur due to operator measurement error.
- Left atrial enlargement can also be identified via the right parasternal long-axis view (**Figure 5.4.3**).[6,7]
- Integration with clinical history and exam findings is essential.

Take-home messages

- The sonographic diagnosis of cardiogenic pulmonary edema should be made in combination with the history, clinical exam, PLUS, and cardiac POCUS.
- Thoracic radiographs may be useful to confirm or refute this diagnosis.

REFERENCES

1. Boysen S, Gommeren K, Chalhoub S. Clinical applications of PLUS: Are there increased B-Lines, yes/no? In: Boysen, Gommeren, Chalhoub eds. *The Essentials of Veterinary Point-of-Care Ultrasound: Pleural Space and Lung.* Eds Edra publishing, Laval, Quebec; 2022, 112–137.
2. Fischer, E. A., Minami, T., Ma, I. W. Y., & Yasukawa, K. (2021). Lung ultrasound for pleural line abnormalities, confluent B-lines, and consolidation. *Journal of Ultrasound in Medicine*, 41(8), 2097–2107. https://doi.org/10.1002/jum.15894
3. Lisciandro, G. R., Fulton, R. M., Fosgate, G. T., & Mann, K. A. (2017). Frequency and number of B-lines using a regionally based lung ultrasound examination in cats with radiographically normal lungs compared to cats with left-sided congestive heart failure. *Journal of Veterinary Emergency and Critical Care*, 27(5), 499–505. https://doi.org/10.1111/vec.12637
4. Picano, E., & Pellikka, P. A. (2016). Ultrasound of extravascular lung water: A new standard for pulmonary congestion. *European Heart Journal*, 37(27), 2097–2104. https://doi.org/10.1093/eurheartj/ehw164
5. Rademacher, N., Pariaut, R., Pate, J., Saelinger, C., Kearney, M. T., & Gaschen, L. (2014). Transthoracic lung ultrasound in normal dogs and dogs with cardiogenic pulmonary edema: A pilot study. *Veterinary Radiology and Ultrasound*, 55(4), 447–452. https://doi.org/10.1111/vru.12151
6. DeFrancesco, T. C., & Ward, J. L. (2021). Focused canine cardiac ultrasound. *Veterinary Clinics of North America-Small Animal Practice*, 51(6), 1203–1216.
7. Keene, B. W., Atkins, C. E., Bonagura, J. D., et al. (2019). ACVIM consensus guidelines for the diagnosis and treatment of myxomatous mitral valve disease in dogs. *Journal of Veterinary Internal Medicine*, 33(3), 1127–1140.

CHAPTER 5 – CASE 5

CANE CORSO WITH ACUTE HEMOPTYSIS, LETHARGY, AND ANOREXIA

Céline Pouzot-Nevoret

HISTORY, TRIAGE, AND STABILIZATION

A 9-year-old spayed female Cane Corso presented with acute hemoptysis, lethargy, and anorexia.

Triage exam findings (vitals):
- Mentation: decreased
- Respiratory rate: 68 breaths per minute, increased breath sounds, moderate hemoptysis
- Heart rate: 128 breaths per minute
- Mucous membranes: pale, 3 sec capillary refill time
- Femoral arterial pulses: weak
- Temperature: 39.5°C (103.1°F)

POCUS exam(s) to perform:
- Based on the vague clinical signs, a five-point abdominal POCUS and a sliding "S"-shaped PLUS exam were immediately performed
- Abdominal POCUS was unremarkable

Abnormal POCUS exam results:
See **Figure 5.5.1** and **Video 5.5.1**.

Additional point-of-care diagnostics and initial management:
Additional point-of-care diagnostics included a complete blood count (CBC) that showed moderate regenerative anemia (hemoglobin: 7.3 g/dL [13–20]; reticulocytes: 380 K/µL [10–110]), leukocytosis (white blood cell [WBC] count: 29.6 10⁹/L [5–16]), and severe thrombocytopenia (platelets: 26 10⁹/L [148–484]). Coagulation times were increased (prothrombin time [PT]: 20 sec [11–17 sec], activated partial prothrombin time [aPTT]: 110 sec [72–102 sec]).

The patient was administered nasal oxygen, an IV fluid bolus (Ringer's lactate 10 mL/kg), and slow IV tranexamic acid (10 mg/kg).

QUESTIONS AND ANSWERS

1. What are the differentials to rule in or out with POCUS based on history and physical exam?
2. What are the sonographic findings?
3. What is the sonographic diagnosis?
4. If necessary, what additional sonographic examination or findings would help rule in or out the differential diagnoses?

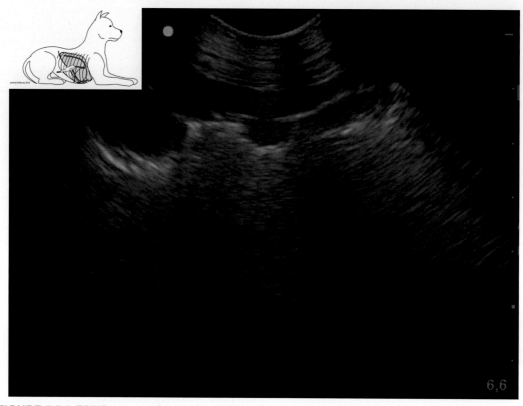

FIGURE 5.5.1 PLUS exam at the mid-caudal region of the right hemithorax with the dog in sternal, using a microconvex probe, perpendicular to the ribs with the depth set at 6.6 cm. Fur was not clipped, only parted, and alcohol was used as the coupling agent. Video 5.5.1 obtained from the same window as Figure 5.5.1.

1. **Differential diagnosis to rule in or rule out with POCUS:**
   - Hemothorax (pleural effusion), hemoabdomen (abdominal effusion), pericardial effusion with tamponade, and pulmonary hemorrhage are all likely causes of acute signs of shock associated with hemorrhagic diathesis.
2. **Describe your sonographic findings:**
   - Nodule signs (**Figure 5.5.2**).
   - The pleural line is less distinct and lung sliding is lost where the proximal nodule contacts the parietal lining of the chest wall (**Video 5.5.1**).
   - Also note that the nodules have a multifocal distribution over the lung surface, with heterogeneous sizes and characteristics, moving synchronously with respiration (**Video 5.5.1**).
3. **What is your sonographic diagnosis?**
   - Pulmonary nodules (nodule sign).
4. **What additional sonographic examination or finding would help you rule in or rule out your differential diagnoses (if necessary)?**
   - No additional sonographic examination is necessary.
   - Thoracic radiographs confirmed bilateral multifocal nodules (**Figure 5.5.3**).

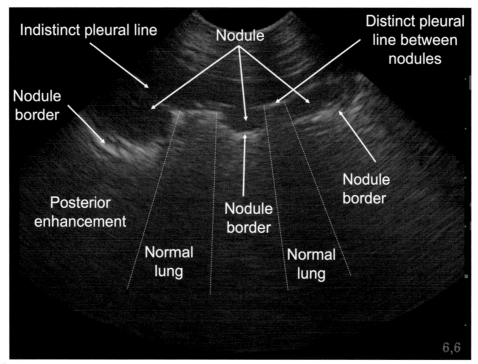

FIGURE 5.5.2 Three hypoechoic and homogeneous regions with well-delineated semi-circular deep boundaries and posterior enhancement are seen. Note the regions of lung between nodules lack B-lines. There is no air or fluid bronchograms within the nodules.

CLINICAL INTEGRATION

Suspected diagnosis based on ultrasound findings:
Metastatic neoplasia. Less likely fungal disease, angiostrongylosis, or multiple lung abscesses.

Sonographic interventions, monitoring, and outcome:
- The left hemithorax showed similar findings.
- Ultrasound-guided fine needle aspiration of a nodule confirmed a neoplastic process.
- Further coagulation testing (ROTEM®, fibrinogen, and D-dimers) confirmed disseminated intravascular coagulation.
- The dog was euthanized for humane reasons.

Important concepts regarding the sonographic diagnosis of metastatic neoplasia using nodule sign:
- A nodule sign is one of three partial lung consolidations that are identifiable with PLUS, together with shred and wedge signs.[1] Lung consolidation generally occurs when the ratio of aerated to non-aerated lung falls below 10% at the lung periphery.
- PLUS has been used to identify pulmonary nodules in dogs secondary to metastatic neoplasia,[2] angiostrongylosis,[3] and fungal disease.[4] Ultrasound criteria alone are not sufficient to discriminate the cause of nodule signs.
- Because nodules contain no air, the ultrasound beam can traverse consolidated

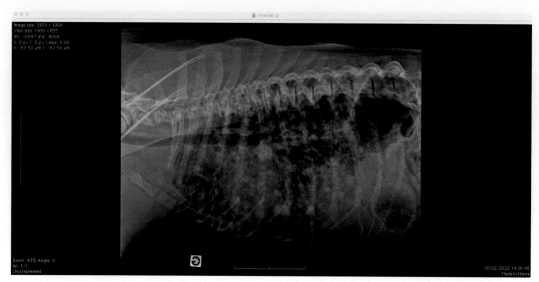

FIGURE 5.5.3A Thoracic radiographs showing multiple diffuse bilateral nodules. (Imaging service, VetAgro Sup, with permission)

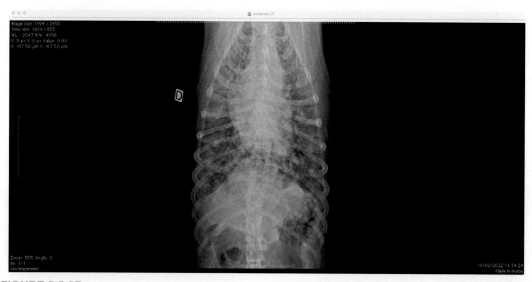

FIGURE 5.5.3B

nodules allowing them to become sonographically visible. At the nodule's distal borders, the ultrasound beam encounters lung tissue that is more than 10% aerated, creating a new tissue-to-air interface. This interface is well delineated, often having a semi-circular or circular deep boundary. Posterior enhancement and/or ring down artifact is inconstantly present.

- PLUS relies on the distribution and localization of pathology; however, as metastatic disease is non-specific and can be localized or multifocal, a large lung surface area may need to be assessed to identify nodule signs.
- In contrast to shred signs, which often have B-lines adjacent to them, the lung tissue adjacent to nodules is often normal in appearance.

- PLUS is a surface imaging technique. Pulmonary metastasis localized deep within lung parenchyma and/or surrounded by normal aerated lung is not sonographically visible, highlighting the importance of thoracic radiographs and/or computed tomography to compliment ultrasound for the identification of metastatic neoplasia.

Techniques and tips to identify nodule sign:
- As nodules are round structures, they can be identified with the probe oriented parallel or perpendicular to the ribs.
- Nodule signs can be diagnosed on still images.
- Nodule signs move synchronously with the respiratory cycle.
- The glide sign (aka lung sliding) will be visible to either side of nodule signs but is often attenuated to absent where the nodule contacts the thoracic wall.
- Based on their size, nodules can move out of the acoustic window during the respiratory cycle (**Video 5.5.2**).
- Keep the probe immobile when assessing lung sliding, nodules, and other lung pathology.
- Using the freeze and track function to slow the cineloop can be useful to visualize small and unique nodules.

Take-home messages
- Metastatic neoplasia can be diagnosed with PLUS through identification of nodule signs.
- Multifocal nodules should prompt a suspicion of metastatic neoplasia; however, angiostrongylosis, abscesses, and fungal disease cannot be ruled out on ultrasound alone.
- If identified, ultrasound-guided fine needle aspiration of a nodule can be performed.
- Small nodules (<1 mm) can be detected with PLUS if they reach the lung surface.
- Nodules must reach the lung surface to be visualized with PLUS, regardless of the size of the nodule.

REFERENCES

1. Boysen S, Chalhoub S, Gommeren K. Clinical applications of PLUS: Is there lung consolidation, yes/no? In: Boysen S, Chalhoub C, Gommeren K, editors. *The Essentials of Veterinary Point-of-Care Ultrasound: Pleural Space and Lung*. Edra publishing, Laval, Quebec; 2022, pp. 139–180.
2. Pacholec C, Lisciandro GR, Masseau I, et al. Lung ultrasound nodule sign for detection of pulmonary nodule lesions in dogs: Comparison to thoracic radiography using computed tomography as the criterion standard. *Vet J* 2021; 275:105727. DOI: 10.1016/j.tvjl.2021.105727
3. Venco L, Colaneri G, Formaggini L, De Franco M, Rishniw M. Utility of thoracic ultrasonography in a rapid diagnosis of angiostrongylosis in young dogs presenting with respiratory distress. *Vet J* 2021; 271:105649. DOI: 10.1016/j.tvjl.2021.105649
4. Ward J, Lisciandro G, Ware W, et al. Lung ultrasonography findings in dogs with various underlying causes of cough. *J Am Vet Med Assoc* 2019; 255(5):574–583. DOI: 10.2460/javma.255.5.574

CHAPTER 5 – CASE 6

ENGLISH SETTER SUBJECTED TO PROLONGED ANESTHESIA AND ABNORMAL LUNG FUNCTION

Chiara Di Franco and Angela Briganti

HISTORY, TRIAGE, AND STABILIZATION

A 14-year-old female spayed English Setter was anesthetized for dental procedures. The dental procedure was prolonged and prior to extubating the patient the anesthetist assessed the dog's lung function.

Triage exam findings (vitals):
- Respiratory rate: the dog was spontaneously ventilating with a respiratory rate of between 12 and 18 breaths per minute with supplemental oxygen (FiO_2 80%)
- Heart rate: 117 beats per minute
- Mucous membranes: pale pink, <2 sec capillary refill time
- Temperature: 36.4°C (97°F)
- Invasive blood pressure monitoring: systolic, diastolic, and mean arterial pressures of 112, 47, and 63 mmHg, respectively. Blood pressure was low at times during anesthesia and repeat IV crystalloid boluses were administered as needed (2 mL/kg). Noradrenaline was also administered during the last 2 h of anesthesia.
- SpO_2: 94%.

POCUS exam(s) to perform:
- PLUS

Abnormal POCUS exam results:
See **Figure 5.6.1** and **Video 5.6.1**.

Additional point-of-care diagnostics and initial management:
None.

QUESTIONS AND ANSWERS

1. What are the differentials to rule in or out with POCUS based on history and physical exam?
2. What are the sonographic findings?

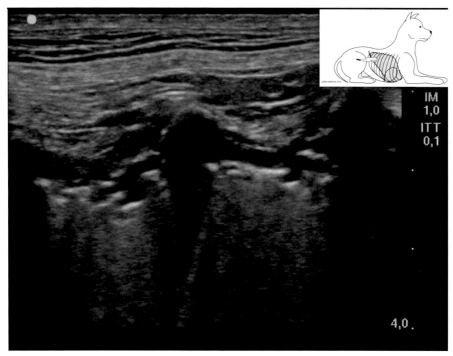

FIGURE 5.6.1 PLUS of the right hemithorax at the caudal pleural border near the curtain sign with the patient in sternal recumbency, using a linear probe oriented perpendicular to the ribs; depth set at 4 cm. Findings were similar on the left hemithorax. Video 5.6.1: Video obtained near the starting site of Figure 5.6.1, with sliding of the probe cranially from the caudal PLUS border to the cranial PLUS border.

3. What is the sonographic diagnosis?
4. If necessary, what additional sonographic examination or findings would help rule in or out the differential diagnoses?

1. Differential diagnosis to rule in or rule out with POCUS:
   - Atelectasis, aspiration pneumonia, hemorrhage.

2. Describe your sonographic findings:
   - Hypoechoic areas are present below the pleural line, which move synchronously with the respiratory cycle.
   - The pleural line is irregular and thickened at various locations during the PLUS scan (**Figure 5.6.2, Video 5.6.2**).
   - Hyperechoic ring down artifact is visible arising from the distal borders of the hypoechoic regions, which extend to the far field of the image. These findings are present at multiple locations (**Video 5.6.2**).

3. What is your sonographic diagnosis?
   - Bilateral partial lung lobe consolidation with pleural thickening and irregularities, consistent with small shred signs.

4. What additional sonographic examination or finding would help you rule in or rule out your differential diagnoses (if necessary)?
   - None.

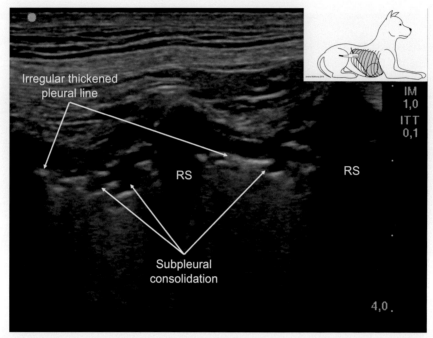

FIGURE 5.6.2 Note the pleural line has become thickened and irregular with subpleural consolidations visible. RS: rib shadow. Video 5.6.2 (Video 5.6.1 labelled): Note that the presence of air bronchograms often seen with atelectasis are static in nature; there is no visible movement of air within the small airways, which appear as white punctate to linear structures within the region of onsolidation. RS: rib shadow.

CLINICAL INTEGRATION

Suspected diagnosis based on ultrasound findings:

Atelectasis was suspected given the length of the dental procedure, prolonged recumbency, and the serial development and resolution of lung pathology that was absent on pre-anesthetic PLUS evaluation.

Sonographic interventions, monitoring, and outcome:

- Both hemithoraces were investigated and found to have similar PLUS findings.
- The SpO$_2$ value was slightly decreased following the dental procedure.
- In this case, PLUS allowed the lungs to be rapidly assessed for potential atelectasis.
- Although aspiration and other forms of pneumonia may cause similar PLUS findings, the pre-anesthetic PLUS evaluation was normal and the distribution of lesions was not classic for aspiration pneumonia.
- Given the suspicion of atelectasis, and the decreased SpO$_2$ value, recruitment maneuvers using incremental positive end-expiratory pressure (PEEP) were performed while assessing the lung with ultrasonography.
- The improved sonographic appearance of the lungs seen in real time during lung recruitment maneuvers ruled out other causes of lung lobe consolidation (**Figure 5.6.3**).
- Lung ultrasound was also useful in guiding PEEP levels, and helped guide the decision of when to stop applying recruitment maneuvers.

Important concepts regarding the sonographic diagnosis of atelectasis and recruitment maneuvers:

- Lung lobe consolidation can have numerous etiologies. Assessing patients for atelectasis following prolonged anesthetic procedures, prior to extubation, allows the clinician to re-expand the alveoli and recover animals with improved lung function.
- Atelectasis is a common side effect of anesthesia and may result from a combination of muscle relaxation, intra-abdominal pressure increase (e.g., from surgical procedures), cranial shift of the diaphragm due to pressure from the abdominal organs, increased FiO_2, and gravitational effects on the thoracic viscera. The amount of atelectasis is case dependent.
- Atelectasis generally resolves spontaneously after placing the animal in sternal recumbency, stressing the importance of good patient care, which includes rotating the patient and placing them in sternal recumbency as needed.
- The systematic evaluation of all anesthetic patients with POCUS allows the clinician to individually assess and tailor postoperative patient management.
- Including pre-anesthetic POCUS assessment may identify subclinical findings that can alter anesthetic protocols and allows any pre-existing pathology to be carefully monitored for changes.
- Knowing what baseline POCUS changes are present prior to induction also helps narrow differential diagnoses should patients decompensate during or following anesthesia.
- Real-time sonographic monitoring of recruiting maneuvers is described in human medicine and helps select positive end-expiratory pressure levels that optimally keep the alveoli open while simultaneously avoiding atelectasis.[1,2]
- In this case, ultrasound helped select a maximum PEEP of 10 cm H_2O, which was the PEEP level where atelectasis was seen to sonographically resolve (**Figure 5.6.3**).
- Ultrasound can also be used during sustained inflation techniques, where continuous pressures of 40 cmH_2O are applied to the airways for up to 60 sec. Using ultrasound during this technique to monitor the resolution of B-lines and regions of lung consolidation allows the duration of the maneuver to be adjusted and discontinued as soon as the lung is recruited. It is particularly important to balance the degree of lung recruitment against the drop in preload which occurs during this technique because of the high airway pressure (40 cm H_2O) applied during the inspiratory phase of positive pressure ventilation.[3] Ultrasound helps determine when recruitment has occurred, avoiding unnecessarily long positive pressure times that can negatively impact cardiac output.[4]
- In the present case, monitoring blood pressure during recruitment was considered inadequate due to the variation in blood pressure that occurred during anesthesia. Stepwise recruitment of lung based on sonographic assessment was considered safer. Lung changes noted with ultrasound dictated when PEEP increments should be discontinued (**Video 5.6.3**).

Techniques and tips to identify the appropriate ultrasound images/views:

- Place the probe, oriented perpendicular to the ribs, in the caudo-dorsal region of the thorax near the abdominal curtain sign and slide the probe between the cranial and caudal borders while moving ventrally in an S-shaped pattern.[5]
- Assess both hemithoraces, depending on the recumbency of the animal, as atelectasis

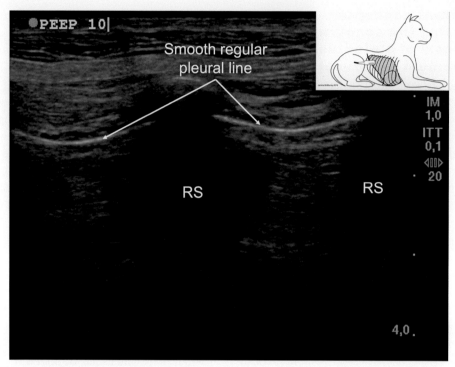

FIGURE 5.6.3 PLUS findings at a similar location to Figure 5.6.1 showing resolution of the abnormalities when 10 cm H$_2$O positive end-expiratory pressure (PEEP) was applied. Video 5.6.3: Split screen video of the PLUS findings following the dental procedure when atelectasis is visible and following recruitment maneuvers, with normal lung surface findings.

should be more evident in the gravity-dependent regions.
- Atelectasis appears as hypoechoic regions of lung, originating at the lung surface, with B-lines and static air bronchograms.
- Atelectasis resolves following recruitment maneuvers.

Take-home message
- If PLUS is applied to all patients undergoing anesthesia, the immediate recognition of lung and pleural abnormalities during and after anesthesia is rapidly facilitated; peri-anesthetic use of PLUS allows the rapid identification of causes of lung impairment and is useful to help guide pulmonary recruitment.

REFERENCES

1. Hartland BL, Newell TJ, Damico N. Alveolar recruitment maneuvers under general anesthesia: A systematic review of the literature. *Respir Care*. 2015;60(4):609–620.
2. Cylwik J, Buda N. Lung ultrasonography in the monitoring of intraoperative recruitment maneuvers. *Diagnostics (Basel)*. 2021 Feb 10;11(2):276. doi: 10.3390/diagnostics11020276. PMID: 33578960. PMCID: PMC7916700.
3. García-Sanz V, Aguado D, Gómez de Segura IA, Canfrán S. Individualized positive end-expiratory pressure following alveolar recruitment

manoeuvres in lung-healthy anaesthetized dogs: A randomized clinical trial on early postoperative arterial oxygenation. *Vet Anaesth Analg.* 2021 Nov;48(6):841–853. doi: 10.1016/j.vaa.2021.03.019. Epub 2021 Jul 20. PMID: 34391669.

4. Godet T, Constantin JM, Jaber S, Futier E. How to monitor a recruitment maneuver at the bedside. *Curr Opin Crit Care.* 2015 Jun;21(3):253–258. doi: 10.1097/MCC.0000000000000195. PMID: 25827586.

5. Armenise A, Boysen RS, Rudloff E, Neri L, Spattini G, Storti E. Veterinary-focused assessment with sonography for trauma-airway, breathing, circulation, disability and exposure: A prospective observational study in 64 canine trauma patients. *J Small Anim Pract.* 2019 Mar;60(3):173–182. doi: 10.1111/jsap.12968. Epub 2018 Dec 13. PMID: 30549049.

CHAPTER 5 – CASE 7

BEAGLE WITH ACUTE ONSET OF RESPIRATORY DISTRESS AND ANURIA FOLLOWING A LAPAROTOMY

Andrea Armenise and Anna De Nitto

HISTORY, TRIAGE, AND STABILIZATION

A 4-year-old male neutered Beagle was referred for acute onset of respiratory distress and anuric renal failure that developed 24 h following an orchiectomy and cystotomy for bladder stones.

Triage exam findings (vitals):
- Mentation: stuporous, lateral recumbency
- Respiratory rate: 60 breaths per minute
- Heart rate: 160 beats per minute
- Mucous membranes: cyanotic, >2 sec capillary refill time
- Femoral and dorsal pedal arterial pulses: weak
- Temperature: 37.7°C (99.9°F)
- Systolic blood pressure (SBP): 90 mmHg
- Electrocardiography: sustained ventricular tachycardia

POCUS exam(s) to perform:
- Triage-performed PLUS

Abnormal POCUS exam results:
See **Figures 5.7.1** and **5.7.2** as well as **Videos 5.7.1** and **5.7.2**.

Additional point-of-care diagnostics and initial management:
Helmet continuous positive airway pressure (H-CPAP) with a positive end-expiratory pressure (PEEP) at 7.5 cm H_2O and an oxygen flow rate of 10 L/min mixed with high-pressure medical air was applied to improve respiratory function immediately following point-of-care diagnostics (**Figure 5.7.1**).

QUESTIONS AND ANSWERS

1. What are the differentials to rule in or out with POCUS based on history and physical exam?
2. What are the sonographic findings?
3. What is the sonographic diagnosis?
4. If necessary, what additional sonographic examination or findings would help rule in or out the differential diagnoses?

DOI: 10.1201/9781003436690-14

Respiratory Distress and Anuria Following a Laparotomy

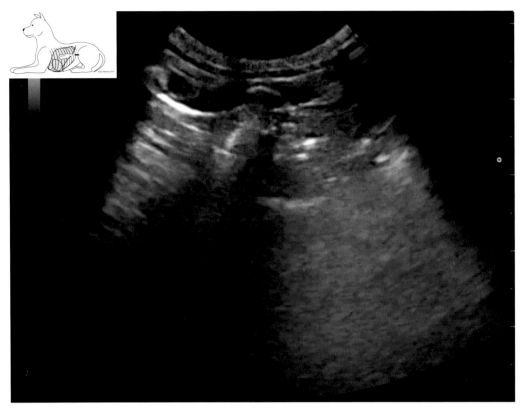

FIGURE 5.7.1 PLUS of the left perihilar region with the patient in sternal recumbency, using a microconvex probe oriented perpendicular to the ribs with the depth set at 5 cm and the frequency set at 11 MHz. Similar lesions were also seen over the right middle hemithoraces. Fur was not clipped, and alcohol was used as the coupling agent. Video 5.7.1: Video obtained from the same window as Figure 5.7.1.

1. **Differential diagnosis to rule in or rule out with POCUS:**
   - Fluid overload, cardiogenic pulmonary edema, aspiration pneumonia, pneumonia, non-cardiogenic pulmonary edema, and acute lung injury (ALI)/acute respiratory distress syndrome (ARDS) are all possible given the presenting history and clinical findings (**Figure 5.7.2**).
2. **Describe your sonographic findings:**
   - A mixed lung pattern is noted with individual to coalescent B-lines, spared normal-appearing lung regions, and areas with varying degrees of lung consolidation with and without air and fluid bronchograms (**Figures 5.7.3** and **5.7.4**, **Videos 5.7.3** and **5.7.4**).
3. **What is your sonographic diagnosis?**
   - Increased individual to coalescing B-lines, an irregular pleural line, lung consolidation with air bronchograms, and spared lung regions.
4. **What additional sonographic examination or finding would help you rule in or rule out your differential diagnoses?**
   - Cardiac POCUS was normal, ruling out congestive heart failure.

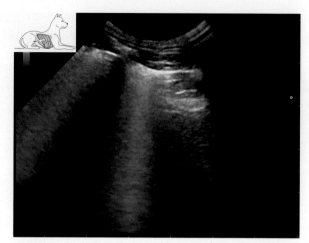

FIGURE 5.7.2 PLUS of the right perihilar region with the patient in sternal recumbency using a microconvex probe oriented perpendicular to the ribs with the depth set at 5 cm and the frequency set at 11 MHz. Similar lesions were also seen over the right middle hemithoraces. Fur was not clipped, and alcohol was used as the coupling agent. Video 5.7.2: Video obtained from the same window as Figure 5.7.2.

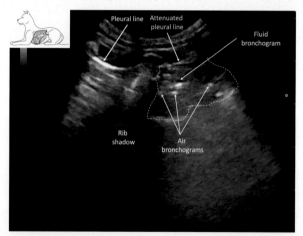

FIGURE 5.7.3 To the right side of the rib/rib shadow, lung consolidation is visible (region outlined by curved dotted white line) with occasional air bronchograms; on the left of the rib, a bright hyperechoic pleural line is visible at the soft tissue to air interface created by the thorax and air-filled lung (lung sliding was present, although not appreciated in the still image – see Video 5.7.1). To the right of the rib, the pleural line between the chest wall and consolidation becomes attenuated and less bright as the soft tissue to air interface is lost, being replaced by a soft tissue to soft tissue interface. Video 5.7.3 (Video 5.7.1 labelled): As the probe is slid across the intercostal spaces from caudal to cranial, normal-looking lung, focal subpleural consolidations (region outlined by curved dotted white line), coalescent multiple B-lines, and larger areas of lung consolidation without sliding are visible. Note that the pleural line is hyperechoic and bright where aerated lung contacts the chest wall between the ribs on the left side of the image, and becomes attenuated where the region of lung consolidation is in contact with the chest wall. In real time, there will also be a loss of lung sliding between the consolidated region and the chest wall where the attenuation of the pleural line occurs (not visible in a still image).

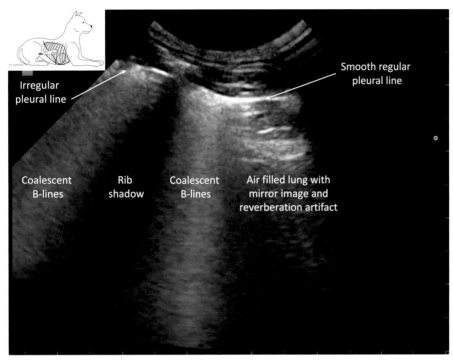

FIGURE 5.7.4 From the left to the center of the image, multiple coalescent B-lines with an irregular pleural line are seen. On the right of the image a smooth hyperechoic pleural line is present. Video 5.7.4 (Video 5.7.2 labelled): As the probe is slid across the intercostal spaces from caudal to cranial, coalescent B-lines, focal subpleural consolidations, lung consolidation, and an irregular pleural line with coalescent multiple B-lines are seen. Note also that the pleura is thickened and irregular at several regions where coalescing B-lines occur.

CLINICAL INTEGRATION

Suspected diagnosis based on ultrasound findings:

Based on the history, clinical findings, and sonographic findings, ARDS is the most likely final diagnosis. Fine needle aspiration of the lung consolidations revealed the presence of neutrophilic infiltration, supporting VetARDS criteria.[1]

Sonographic interventions, monitoring, and outcome:

- Butorphanol (0.2 mg/kg) and propofol (constant rate infusion [CRI] at 0.1 mg/kg/min) were administered for sedation.
- H-CPAP with PEEP between 2.5 and 5 cm H_2O was discontinued after 12 h when arterial oxygenation, lung ultrasound appearance, and respiratory rate/effort improved.
- The patient was switched to supplemental oxygen and monitored daily, which included serial PLUS evaluation.
- The dog eventually recovered and was discharged.

Important concepts regarding the sonographic diagnosis of ARDS:

- Multiple sonographic patterns are seen in patients with ARDS, including individual

B-lines, coalescent B-lines, spared lung regions, and lung consolidation.[2]
- Even though current VetARDS criteria do not include PLUS,[1] because ARDS PLUS criteria has been established in people, it is reasonable to include in small animals.

Techniques and tips to identify the appropriate ultrasound images/views:
- One of the published small animal PLUS protocols that scans an extended lung surface area and multiple lung regions over both hemithoraces should be used.[3,4]
- Cardiac POCUS should always be performed to evaluate left ventricle contractility, volume, and an La:Ao ratio to rule in/out left atrial enlargement, and to rule out a cardiac cause for increased B-lines.
- A serial PLUS exam together with serial arterial blood gas helps monitor the clinical evolution of patients with ALI/ARDS.

Take-home messages
- Discrete B-lines, coalescent B-lines, an irregular pleural line, spared lung regions, and lung consolidations with an acute onset of signs should raise suspicion of VetARDS if other supportive findings are also present.
- Serial LUS examinations help to monitor the clinical evolution of the disease.

REFERENCES

1. Wilkins PA, Otto CM, Baumgardner JE, et al. Acute lung injury and acute respiratory distress syndromes in veterinary medicine: Consensus definitions: The Dorothy Russell Havemeyer working group on ALI and ARDS in veterinary medicine. *JVECC* 2007; 17(4):333–339.
2. Chiumello D, Froio S, Bouhemad B, et al. Clinical review: Lung imaging in acute respiratory distress syndrome patients – An update. *Critical Care* 2013; 17(6):243.
3. Armenise A, Boysen RS, Rudloff E, et al. Veterinary-focused assessment with sonography for trauma-airway, breathing, circulation, disability and exposure: A prospective observational study in 64 canine trauma patients. *JSAP* 2019; 60(3):173–182.
4. Vezzosi T, Mannucci A, Pistoresi F, et al. Assessment of lung ultrasound B-Lines in dogs with different stages of chronic valvular heart disease. *JVIM* 2017; 31(3):700–704.

CHAPTER 5 – CASE 8

MIXED BREED K9 WITH PROGRESSIVE LETHARGY AND HYPOREXIA FOR 3 WEEKS

Serge Chalhoub

HISTORY, TRIAGE, AND STABILIZATION

A 9-year-old male neutered medium-sized mixed breed dog presented for progressive lethargy and hyporexia over the previous 3 weeks. There was travel history from Alberta to Northern Ontario, Canada, every summer.

Triage exam findings (vitals):
- Mentation: quiet, alert, and responsive (QAR)
- Respiratory rate: 40 breaths per minute, mild expiratory effort
- Heart rate: 120 beats per minute
- Mucous membranes: pink, <2 sec capillary refill time
- Femoral and dorsal pedal arterial pulses: strong and synchronous
- Temperature: 38.9°C (103.8°F)

POCUS exam(s) to perform:
- Because of the increased respiratory rate and expiratory effort, PLUS was the first POCUS exam chosen

Abnormal POCUS exam results:
See **Figure 5.8.1** and **Video 5.8.1**.

Additional point-of-care diagnostics and initial management:
- Abdominal POCUS did not indicate any visible pathology.
- Flow-by oxygen was provided.
- Packed cell volume/total solids (PCV/TS) 53%, 7.8 g/dL, lactate 1.2 mmol/L.

QUESTIONS AND ANSWERS

1. What are the differentials to rule in or out with POCUS based on history and physical exam?
2. What are the sonographic findings?
3. What is the sonographic diagnosis?
4. If necessary, what additional sonographic examination or findings would help rule in or out the differential diagnoses?

DOI: 10.1201/9781003436690-15

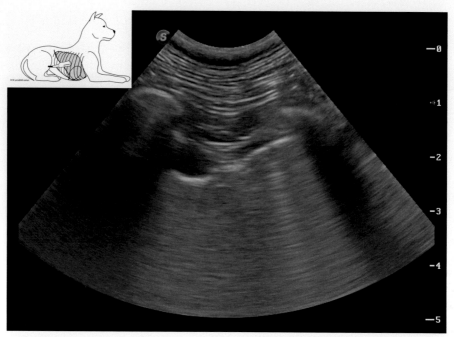

FIGURE 5.8.1 **PLUS** scan at the mid-thoracic region of the right hemithorax with the dog in sternal recumbency, using a microconvex probe oriented perpendicular to the ribs, with the depth set at 5 cm. Fur was not clipped and alcohol was used as the coupling agent. Similar lesions were found throughout both hemithoraces. Video 5.8.1: Video obtained from the same location as Figure 5.8.1.

1. **Differential diagnosis to rule in or rule out with POCUS:**
   - Based on the history and clinical findings, metastatic neoplasia, primary neoplasia, fungal disease, and lung abscess are all possible differential diagnoses.
2. **Describe your sonographic findings:**
   - Lung sliding (aka a glide sign) is present (**Video 5.8.1**).
   - Several partial lung consolidations with smooth, regular, and well-defined deep borders, and an absence of air bronchograms were seen bilaterally at multiple PLUS locations (**Figure 5.8.2, Video 5.8.2**).
3. **What is your sonographic diagnosis?**
   - Multiple bilateral diffuse nodule signs and occasional B-lines.
4. **What additional sonographic examination or finding would help you rule in or rule out your differential diagnoses (if necessary)?**
   - Multiple nodules were identified with PLUS.
   - Considering the age and presentation of the patient, neoplasia should be considered the most likely cause. However, other causes of nodule signs, which are not discernible with ultrasound alone, cannot be ruled out.
   - With the travel history to a region known to have fungal disease (Northern Ontario, Canada), blastomycosis was also a strong possibility.
   - To help differentiate the cause, fine needle aspiration (FNA) of the nodules along with a search for pathology in other organs should be undertaken.

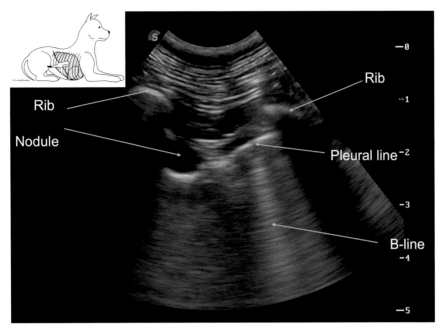

FIGURE 5.8.2 Several well-delineated, round, hypoechoic structures visible arising from the lung surface, with an absence of air and fluid bronchograms were visible throughout both hemithoraces bilaterally, consistent with a nodule sign. Occasional individual B-lines could be seen between nodules. Video 5.8.2 (Video 5.8.2 labelled).

CLINICAL INTEGRATION

Suspected diagnosis based on ultrasound findings:
Neoplasia and blastomycosis were the two most likely differential diagnoses in this case.

Sonographic interventions, monitoring, and outcome:
- Ultrasound-guided FNA and the cytology of the nodules was consistent with blastomycosis.
- The dog went home with antifungal therapy as further diagnostics and care were not feasible due to financial constraints.
- The dog was rechecked at 1- and 2-months post diagnosis and was doing well, with visible improvement in nodule number on PLUS and thoracic radiographs.

Important concepts regarding the sonographic diagnosis of pulmonary fungal disease:
- The key PLUS finding was multiple nodule signs, which are one type of partial lung lobe consolidation characterized as hypoechoic, rounded lesions with smooth, well-defined distal borders/boundaries.[1]
- In contrast to shred signs, they tend to lack air bronchograms and have a smooth homogeneous appearance on ultrasound.
- By completing a comprehensive "S"-shaped regional PLUS scan of both hemithoraces, multiple nodules were found.
- In veterinary medicine, nodules should prompt consideration of neoplasia although other causes of nodule signs cannot be

ruled out, emphasizing the importance of geographical or travel history, patient examination, and other clinical findings.
- B-lines may or may not be present between nodules depending on if there is associated decreased air at the lung periphery.
- If B-lines are present they may represent early nodule formation prior to the consolidation being visible sonographically, or secondary to another cause of decreased lung aeration (associated inflammation/cell infiltrate/edema).

Techniques and tips to identify the appropriate ultrasound images/views:
- Identifying the pleural line is key.
- Starting over the lung is primordial to not confuse abdominal findings with PLUS findings.
- Beginning at the most caudo-dorsal region of each hemithorax, the hemithorax is scanned using an "S"-shaped pattern to scan as much pleural space and lung as possible.
- Consolidations are tissue-like patterns originating from the pleural line.

- Partial consolidations do not traverse the entire lung lobe because they encounter aerated lung below the region of consolidation. This prevents the ultrasound beam from traversing the deeper-lying aerated lung tissue.
- Nodules have smooth distal/deep boundaries compared to other partial consolidations such as the shred sign.
- Partial consolidations are distinguished from pleural effusion because pleural effusion separates the parietal and visceral pleura.

Take-home messages
- Nodule signs are a type of partial lung consolidation which often represent neoplastic or fungal disease.[1]
- Comprehensive regional PLUS scanning ensures that if nodules are at the lung surface, then they will likely be found. However, not all nodules are visible on PLUS because they may not reach the lung surface.
- Hence, complimentary exams such as thoracic radiographs or CT scans may be necessary.

REFERENCE

1. Boysen S, Gommeren K, Chalhoub S. Clinical applications of PLUS: Is there lung consolidation, yes/no? In: Boysen S, Gommeren K, Chalhoub S, eds. *The Essentials of Veterinary Point-of-Care Ultrasound: Pleural Space and Lung*. Edra publishing, Laval, Quebec; 2022, pp. 139–182.

CHAPTER 5 – CASE 9

DOGUE DE BORDEAUX WITH EXERCISE INTOLERANCE AND A 5-DAY HISTORY OF TACHYPNEA

Emilie Van Renterghem and Anne-Christine Merveille

HISTORY, TRIAGE, AND STABILIZATION

A 6-year-old female Dogue de Bordeaux presented for exercise intolerance (1 month duration) and a 5-day history of tachypnea. Last check-up 6 months previously showed no abnormalities.

Triage exam findings (vitals):
- Respiratory rate: 80 breaths per minute, restrictive respiratory distress, muffled heart and lung sounds ventrally, increased lung sounds dorsally
- Heart rate: 160 beats per minute, 1/6 left apical systolic murmur, cardiac rhythm regular
- Mucous membranes: pale pink, 2.5 sec capillary refill time
- Femoral and dorsal pedal arterial pulses: weak and regular
- Temperature: 37.0°C (98.6°F)
- Mild abdominal distension
- Doppler blood pressure: 90 mmHg systolic

POCUS exam(s) to perform:
- PLUS
- Cardiac POCUS
- Abdominal POCUS

Abnormal POCUS exam results:
See **Figures 5.9.1** through **5.9.5** and **Videos 5.9.1** through **5.9.4**.

Additional point-of-care diagnostics and initial management:
Supplemental oxygen and administration of butorphanol (0.2 mg/kg IV). Thoracocentesis bilaterally (modified transudate).

QUESTIONS AND ANSWERS

1. What are the differentials to rule in or out with POCUS based on history and physical exam?
2. What are the sonographic findings?

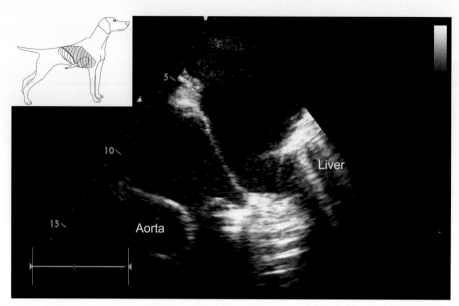

FIGURE 5.9.1 **PLUS** with the dog standing and the probe directed transthoracically at the pericardio-diaphragmatic (PD) window. The probe was oriented perpendicular to the ribs with the heart (left of image) and the diaphragm and liver (right of image) visible. A phased-array probe was used at a frequency of 3 Hz and a depth of 15 cm. The PD window is part of the PLUS protocol. The fur was not clipped, only parted, and alcohol was used as the coupling agent. Video 5.9.1: Video obtained from the same window as Figure 5.9.1.

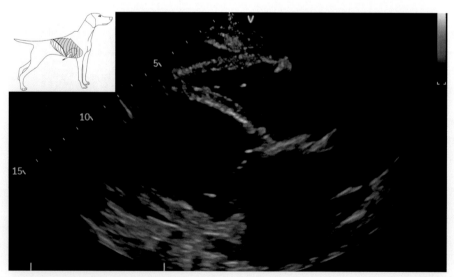

FIGURE 5.9.2 Cardiac POCUS; right parasternal long-axis view of the heart. The phased-array probe is oriented with the marker directed toward the right shoulder. The frequency and depth of the probe are 3 Hz and 15 cm, respectively. The fur was not clipped, only parted, and a combination of alcohol and gel was used as the coupling agent. Video 5.9.2: Video obtained from the same window as Figure 5.9.2.

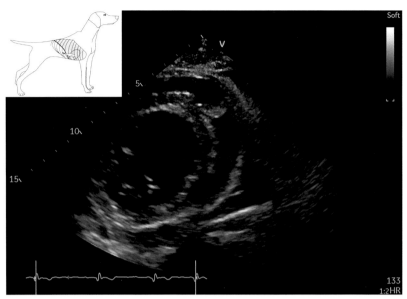

FIGURE 5.9.3 Cardiac POCUS; right parasternal short-axis view of the heart. The probe is rotated 90° clockwise from the right parasternal long-axis view to obtain this view. The frequency and depth of the probe are 3 Hz and 15 cm, respectively. The fur was not clipped, only parted, and a combination of alcohol and gel was used as the coupling agent. Video 5.9.3: Video obtained from the same window as Figure 5.9.3.

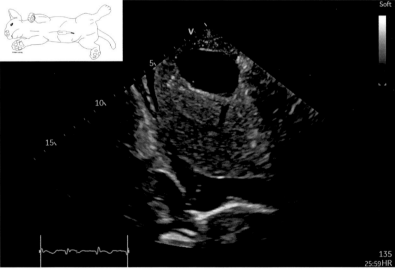

FIGURE 5.9.4 Abdominal POCUS at the subxiphoid view with the dog in right lateral recumbency assessing the caudal vena cava (CVC). The probe is placed longitudinally under the subxiphoid process to obtain this view. If the CVC is not initially seen, the probe can be inclined (fanned) left and right from the midline until the CVC, crossing the diaphragm, which is clearly visualized. Pressure on the probe should be minimal while performing this view to avoid collapse of the CVC. The frequency and depth of the phased-array probe are 3 Hz and 15 cm, respectively. The fur was not clipped, only parted, and a combination of alcohol and gel was used as the coupling agent. Video 5.9.4: Video obtained from the same window as Figure 5.9.4.

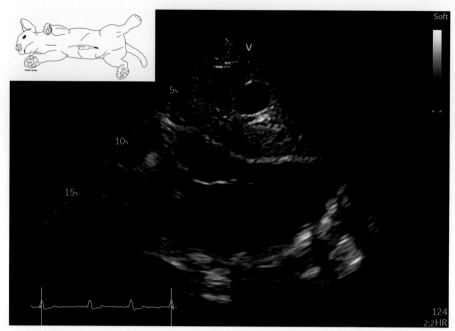

FIGURE 5.9.5 Abdominal POCUS at the subxiphoid view, with the depth extended beyond the diaphragm to assess the pleural space. To obtain this view, the probe is inclined cranially (rocked) from the previous view. Greater pressure on the probe might be required to see beyond the diaphragm. The frequency and depth of the phased-array probe are 3 Hz and 15 cm, respectively. The fur was not clipped, only parted, and a combination of alcohol and gel was used as the coupling agent.

3. What is the sonographic diagnosis?
4. If necessary, what additional sonographic examination or findings would help rule in or out the differential diagnoses?

1. **Differential diagnosis to rule in or rule out with POCUS**
   - The differential diagnoses based on history and physical exam include pleural effusion (PE) due to right-sided congestive heart failure (R-CHF), pericardial effusion (PCE) causing tamponade, chylothorax, neoplasia, lung lobe torsion, hemorrhage, pyothorax, and less likely diaphragmatic hernia or hypoalbuminemia. PLUS is the first sonographic exam to perform to rule in the presence of PE. If PE is present, cardiac POCUS must be performed to assess global systolic function, right heart chambers, and the presence of PCE. Abdominal POCUS including the subxiphoid view must be performed to diagnose the presence of ascites and distended non-compliant caudal vena cava (CVC), and help differentiate PE from PCE.

2. **Describe your sonographic findings:**
   - Presence of a large amount of PE noted transthoracically at the pericardio-diaphragmatic window of PLUS, and at the subxiphoid view (**Figures 5.9.6, 5.9.7,** and **5.9.10, Videos 5.9.1, 5.9.2,** and **5.9.4**).
   - Enlarged left and right heart chambers with thinned ventricular walls and cardiac hypo-contractility seen on the parasternal long- and short-axis views (**Figures 5.9.7** and **5.9.8, Videos 5.9.2** and **5.9.3**).

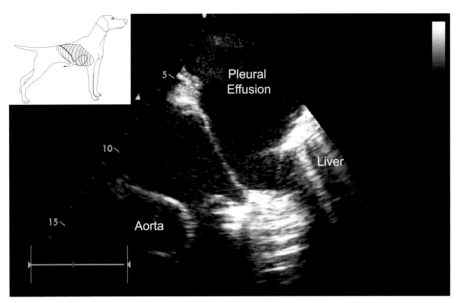

FIGURE 5.9.6 Pleural effusion is easily differentiated from pericardial effusion at the pericardio-diaphragmatic window as pleural effusion tracks along the diaphragm and fills the costo-phrenic recess in this window, while pericardial effusion will track away from the diaphragm and outlines the contours of the heart.

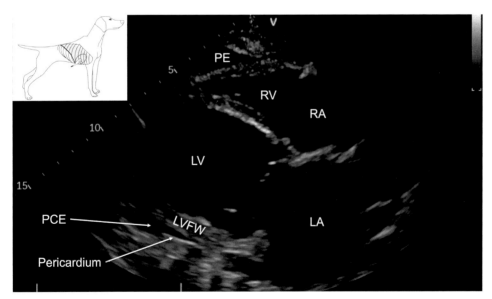

FIGURE 5.9.7 Severely enlarged right and left heart chambers. Note the presence of pleural effusion around the heart in the near field, which can be mistaken for pericardial effusion (PCE). The echoic pericardium should be used as a landmark to differentiate PE from PCE. To not confound cardiac chambers as effusion, the probe can be slid caudally and cranially to the heart while using the echoic pericardium as a landmark. LV, left ventricle; LA, left atrium; LVFW, left ventricular free wall; RV, right ventricle; RA, right atrium.

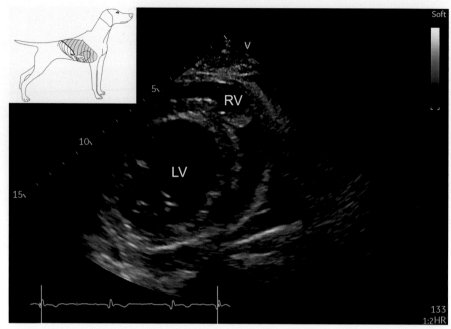

FIGURE 5.9.8 The left ventricle (LV) with a part of the mitral valve and right ventricle (RV) are seen. Pleural fluid can be seen surrounding the heart. Pericardial effusion is not visible in this window. It can be difficult to differentiate pleural and pericardial effusion from a single window, and the four-chamber view, subxiphoid window, and pericardio-diaphragmatic window should be assessed if doubt exists regarding the type of fluid present.

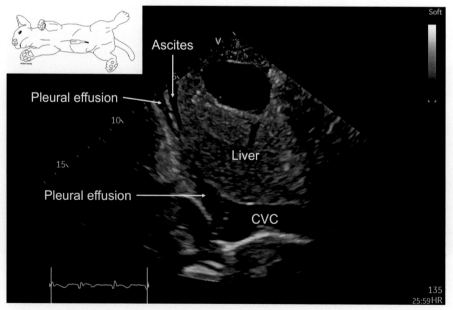

FIGURE 5.9.9 The distended caudal vena cava (CVC) is visualized. There is a mild amount of ascites (A) present. Pleural effusion (PE) can be seen cranial to the diaphragm. Note that the near wall of the CVC is difficult to visualize in this image, making the CVC appear to blend with the pleural effusion.

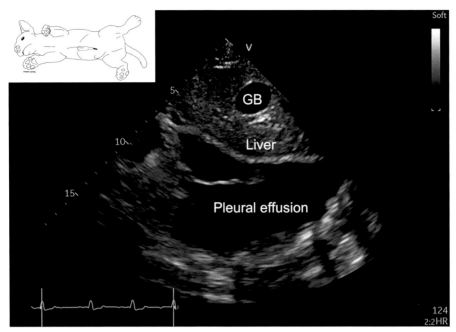

FIGURE 5.9.10 Pleural effusion can be seen cranial to the diaphragm. GB, gallbladder.

- Distended non-compliant CVC at the subxiphoid view with mild ascites (**Figure 5.9.9**, **Video 5.9.4**).

3. **What is your sonographic diagnosis?**
   - Generalized heart chamber enlargement with severe systolic dysfunction associated with scant PCE, significant PE, and a distended non-compliant CVC secondary to R-CHF.

4. **What additional sonographic examination or finding would help you rule in or rule out your differential diagnoses?**
   - Once the dog is stable, and following thoracocentesis, the lungs should be assessed to document the extent and distribution of B-lines compatible with signs of concomitant left-sided congestive heart failure (L-CHF).

CLINICAL INTEGRATION

Suspected diagnosis based on ultrasound findings:

Based on the history, presenting signs, and clinical findings, dilated cardiomyopathy (DCM; primary or secondary) with secondary R-CHF is suspected.

Sonographic interventions, monitoring, and outcome:

- Due to severe respiratory distress, supplemental oxygen was provided and butorphanol was administered before obtaining PLUS and cardiac POCUS.
- Cardiac tamponade was ruled out and a large quantity of PE and severe systolic dysfunction were noted.
- Abdominal POCUS including the subxiphoid view were compatible with R-CHF.
- Thoracocentesis was performed and the dog was treated with furosemide, pimobendane, and spironolactone.
- Concomitant L-CHF is possible but difficult to evaluate when PE is present.

- Serial PLUS exams are recommended following thoracocentesis to look for B-lines which can be secondary to cardiogenic pulmonary edema or atelectasis of the lungs.
- An echocardiogram confirmed the diagnosis of primary DCM.

Important concepts regarding the sonographic diagnosis of DCM and right heart failure (presenting with pleural effusion):

- PE is visible in the lower/ventral regions of the chest cavity (gravity and thus recumbency dependent).
- Dogs with DCM have systolic dysfunction with decreased left ventricular fractional shortening (<20%) compared to normal dogs (25%–45%). Both left and right heart chambers can be enlarged without septal flattening.[1]
- In most cases, the normal ratio between the right and left ventricle is preserved with a ratio of 1:3. A non-compliant CVC and cavitary effusion without evidence of tamponade was necessary to diagnose R-CHF in this case.[1,2]
- R-CHF manifests mostly with the presence of ascites, but PE and PCE without tamponade are possible.[1]
- CVC:Ao >0.63 and CVC collapsibility index <30% were more than 90% sensitive and specific for a diagnosis of R-CHF compared with non-cardiac causes of cavitary effusion in dogs.[3]

Techniques and tips to identify the appropriate ultrasound images/views:

- The heart is best imaged entirely so that the echoic pericardium can be used as a landmark to differentiate PE from PCE.[4]
- Due to the crescent shape of the right ventricle (RV), the latter can be mistaken for PE/PCE. Additional views beyond the right short-axis view including a long-axis four-chamber view, subxiphoid view, and the pericardio-diaphragmatic window can be used. At the pericardio-diaphragmatic window PCE makes a rounded shape surrounding the heart while PE follows the diaphragm, filling the costophrenic recess.[5]

Take-home message

- The sonographic findings that support a diagnosis of DCM and R-CHF are decreased systolic function, distended non-compliant CVC, ascites, and less often PE and/or PCE.

REFERENCES

1. DeFrancesco, TC, Ward, JL. Focused canine cardiac ultrasound. *Vet Clin Small Anim* 51(6) (2021) 1203–1216.
2. Bonagura, JD, Visser, MSC. Echocardiographic assessment of dilated cardiomyopathy in dogs. *J Vet Cardiol* 40 (2022) 15–50.
3. Chou, YY, Ward, JL, Barron, LZ, et al. Diagnostic utility of caudal vena cava measurements in dogs with cavitary effusions or heart failure. *J Vet Intern Med* 34 (2020) 2846 (abstract).
4. Lisciandro, GR. TFAST accurate diagnosis of pleural and pericardial effusion, caudal Vena cava in dogs and cats. *Vet Clin Small Anim* 51(6) (2021) 1169–1182.
5. Boysen S, Gommeren K, Chalhoub S. Clinical applications of PLUS: Is there pleural effusion, yes/no? In: Boysen S, Gommeren K, Chalhoub S, editors. *The Essentials of Veterinary Point-of-Care Ultrasound: Pleural Space and Lung*. Edra publishing, Laval, Quebec; 2022, pp. 91–109.

CHAPTER 5 – CASE 10

ABYSSINIAN WITH PYREXIA, COUGHING, AND PRIOR HISTORY OF PYOTHORAX

Tove M Hultman and Ivayla Yozova

HISTORY, TRIAGE, AND STABILIZATION

A 1-year-old male neutered Abyssinian cat presented with pyrexia and a 1-week history of coughing. The patient had been diagnosed with pyothorax 6 months earlier, which responded to treatment with chest drains and antimicrobials.

Triage exam findings:
- Mentation: able to stand unassisted, responsive but dull
- Respiratory rate and effort: 40 breaths per minute, mildly increased respiratory effort, and increased bronchovesicular sounds over the right hemithorax
- Heart rate: 200 beats per minute
- Mucous membranes: pink, 1–2 sec capillary refill time
- Femoral pulses: normal
- Temperature: 40.0°C (104°F)

POCUS exam(s) to perform:
- PLUS
- Cardiac POCUS

Abnormal POCUS exam results:
See **Figure 5.10.1** and **Video 5.10.1**.

Additional point-of-care diagnostics and initial management:
Blood work showed a mild neutrophilia and a severely elevated serum amyloid A. Initial treatment included oxygen therapy and Robenacoxib at 2 mg/kg subcutaneously.

QUESTIONS AND ANSWERS

1. What are the differentials to rule in or out with POCUS based on history and physical exam?
2. What are the sonographic findings?
3. What is the sonographic diagnosis?
4. If necessary, what additional sonographic examination or findings would help rule in or out the differential diagnoses?

DOI: 10.1201/9781003436690-17

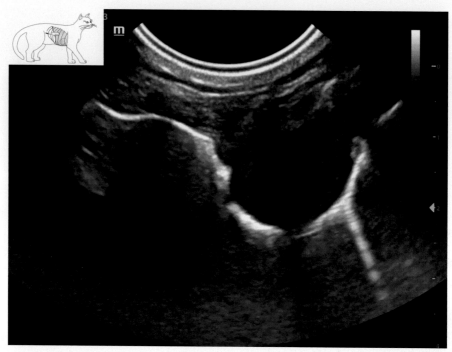

FIGURE 5.10.1 **PLUS** image two-thirds of the way up the right hemithorax with the cat standing, using a microconvex probe oriented perpendicular to the ribs, probe indicator marker directed cranially, and the depth set at 4 cm. Fur was not clipped, only parted, and alcohol was used as the coupling agent. Video 5.10.1 obtained from the same window as Figure 5.10.1.

1. **Differential diagnosis to rule in or rule out with POCUS:**
   - The most likely differential diagnosis based on history and physical exam is recurrent pyothorax complicated by an abscess. Differentials for the rounded lung lesion include neoplasia, granuloma, and hematoma. The first step would be PLUS, and if pleural effusion is detected, a diagnostic tap should be performed for analysis and microbiological culture.
2. **Describe your sonographic findings:**
   - A 3 × 3 × 1.5 cm round structure with a hypoechoic core is visible in the right caudal lung lobe (**Figure 5.10.2, Video 5.10.2**).
   - Moderate number of B-lines surrounding the soft tissue structure (**Figure 5.10.2, Video 5.10.2**).
   - Pleural irregularity (**Figure 5.10.2, Video 5.10.2**).
   - There was a small quantity of pleural effusion in the most ventral thoracic regions which were only visible when the probe was oriented parallel to the ribs (not shown).
   - Cardiac POCUS was unremarkable (not shown).
3. **What is your sonographic diagnosis?**
   - Soft tissue structure with a hypoechoic center in the right caudal lung lobe, resembling a nodule sign, and irregular pleura.
4. **What additional sonographic examination or finding would help you rule in or rule out your differential diagnoses (if necessary)?**
   - Use of color Doppler ultrasound to investigate vascularization of the structure could be performed.

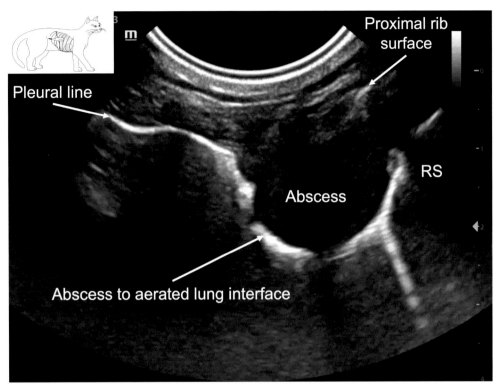

FIGURE 5.10.2 A roughly 3 cm hypoechoic rounded structure is visible arising from the lung surface (pleural line as the lung is in contact with the chest wall) with a clear well-circumscribed hyperechoic (white) interface between the abscess and the underlying aerated lung. Because there is a loss of the soft tissue–air interface between the chest wall and abscess, the pleural line is not visible and lung sliding is absent where the abscess is in contact with the chest wall. As the abscess contains fluid and is not aerated lung there are no visible air bronchograms within the region of the abscess, which helps differentiate it from shred signs and other pulmonary consolidations. Video 5.10.2 (Video 5.10.2 labelled).

CLINICAL INTEGRATION

Suspected diagnosis based on ultrasound findings:
Pulmonary abscess.

Sonographic interventions, monitoring, and outcome:
- Computed tomography was performed, showing a 3 × 3 × 1.5 cm round structure, localized to the right caudal lung lobe.
- The structure was contrast enhancing and contained gas bubbles.
- Ultrasound-guided thoracocentesis with cytology and culture of the pleural effusion confirmed a septic exudate secondary to bacteroides heparinolyticus.
- Antimicrobial therapy was initiated.
- Surgical lung lobectomy of the right caudal lung lobe was performed.
- Histopathology of the abscess revealed foreign plant material.
- The cat was hospitalized and received analgesia and supportive care for 2 days prior to discharge.

Important concepts regarding the sonographic diagnosis of pulmonary abscess:

- Pulmonary abscesses appear as rounded hypoechoic lesions with outer margins and look like nodule signs (**Figure 5.10.1**). If a cavity is present, additional hyperechoic lines are generated by the gas–tissue interface surrounding the abscess.[1]
- PLUS provides a detailed assessment of the peripheral lung only. If the abscess is not in contact with the pleura, it will be missed using ultrasound alone.[2]

Technique and tips to identify a pulmonary abscess:

- To screen for an abscess, a classical PLUS protocol can be used.[2]
- To identify vascularization of the structure, color Doppler placed over the structure can be used. If there is vascularization, an abscess is less likely and neoplasia should be considered.[3]

Take-home messages

- PLUS provides a detailed assessment of the peripheral lungs.
- Pulmonary abscesses without connection to the pleura cannot be detected by ultrasound. Therefore, if clinical suspicion of a lung abscess is high in cases with negative PLUS findings, three-view chest radiographs and/or computed tomography should be considered.

REFERENCES

1. Bouhemad B, Zhang M, Lu Q, Rouby JJ. Clinical review: Bedside lung ultrasound in critical care practice. *Crit Care*. 2007;11(1):205. doi: 10.1186/cc5668.
2. Boysen S, Gommeren K, Chalhoub S. Clinical applications of PLUS: Is there lung consolidation, yes/no? In: Boysen S, Gommeren K, Chalhoub S, editors. *The Essentials of Veterinary Point-of-Care Ultrasound: Pleural Space and Lung*. Edra publishing, Laval, Quebec; 2022, 139–182.
3. Kraft C, Lasure B, Sharon M, Patel P, Minardi J. Pediatric lung abscess: Immediate diagnosis by point-of-care ultrasound. *Pediatr Emerg Care*. 2018 Jun;34(6):447–449. doi: 10.1097/PEC.0000000000001547.

CHAPTER 5 – CASE 11

ENGLISH SPRINGER SPANIEL WITH ACUTE ONSET OF RESPIRATORY DISTRESS AND FEVER

Julie Menard

HISTORY, TRIAGE, AND STABILIZATION

A 2-year-old female intact English Springer Spaniel presented to the emergency service for evaluation of an acute onset of respiratory distress and fever. The dog is a field hunting dog and travels extensively for hunting trials.

Triage exam findings (vitals):
- Mentation: responsive but dull
- Respiratory rate: 80 breaths per minute, restrictive breathing pattern with tachypnea, absent lung sounds ventrally, harsh dorsally
- Heart rate: 150 beats per minute, no murmur or arrhythmia
- Mucous membranes: hyperemic, 1 sec capillary refill time
- Femoral and dorsal pedal arterial pulses: bounding synchronous pulses
- Temperature: 39.6°C (103.4°F)
- Pulse ox: 91% on room air, 95% with flow by O_2
- Blood pressure: 110/60 (78) mmHg.
- Rest of examination was within normal limits

POCUS exam(s) to perform:
- Triage POCUS while receiving supplemental O_2
- PLUS
- Cardiac POCUS

Abnormal POCUS exam results:
See **Figures 5.11.1** and **5.11.2** as well as **Videos 5.11.1** and **5.11.2**.

Additional point-of-care diagnostics and initial management:
Packed cell volume (PCV): 47%, total solids (TS): 7 g/dL, Azo: 5-15, blood glucose: 82 mg/dL (4.6 mmol/L).

O_2 prongs placed at 100 mL/min, methadone 0.2 mg/kg IV, isotonic crystalloid bolus 20 mL/kg over 20 min.

QUESTIONS AND ANSWERS

1. What are the differentials to rule in or out with POCUS based on history and physical exam?
2. What are the sonographic findings?

DOI: 10.1201/9781003436690-18

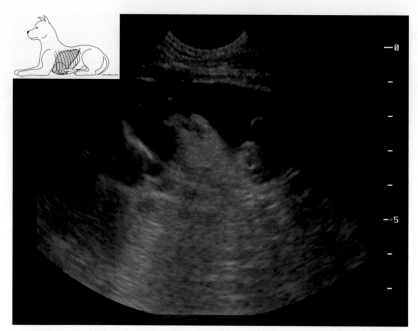

FIGURE 5.11.1 PLUS exam at the ventral caudal region of the left hemithorax with the patient in sternal recumbency, using a microconvex probe perpendicular to the ribs with the depth set at 7 cm. Fur was not clipped and alcohol was used as the coupling agent. Video 5.11.1: Comprehensive examination of the ventral pleural regions of the left hemithorax similar to Figure 5.11.1, which involves turning the probe parallel to the ribs at times. The depth setting of the probe was adjusted throughout the scan to highlight findings.

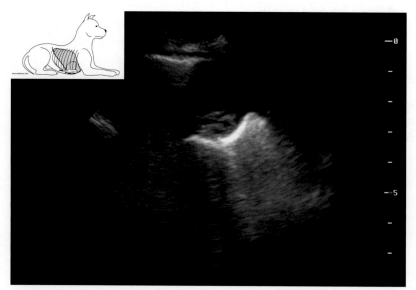

FIGURE 5.11.2 PLUS exam at the ventral cranial region of the right hemithorax with the patient in sternal recumbency, using a microconvex probe perpendicular to the ribs with the depth set at 7 cm. Fur was not clipped and alcohol was used as the coupling agent. Video 5.11.2: Right parasternal short-axis window of the cardiac POCUS showing the mushroom view, fish mouth view, and left atrial to aortic root ratio, with the dog in sternal recumbency, using a microconvex probe with the depth set at 8.4 cm.

3. What is the sonographic diagnosis?
4. If necessary, what additional sonographic examination or findings would help rule in or out the differential diagnoses?

1. **Differential diagnosis to rule in or rule out with POCUS:**
   - Based on the presenting clinical findings, pyothorax, lung lobe torsion, hemothorax (secondary to coagulopathy), chylothorax, less likely neoplasia, right-sided congestive heart failure.
2. **Describe your sonographic findings:**
   - Large volume of anechoic accumulation between the parietal and visceral pleural in the ventral thorax with a jellyfish sign showing the presence of an atelectatic lung (**Figures 5.11.3** and **5.11.4**, **Video 5.11.3**).
   - Pleural effusion at the pericardio-diaphragmatic window which confirms that the fluid is pleural effusion and not pericardial effusion (**Video 5.11.3**).
   - Several fibrinous strands can also be seen within the pleural effusion, particularly at the pericardio-diaphragmatic window (**Video 5.11.3**).
   - Cardiac evidence of hypovolemia (**Video 5.11.4**).
3. **What is your sonographic diagnosis?**
   Pleural effusion and suspected hypovolemia.
4. **What additional sonographic examination or finding would help you rule in or rule out your differential diagnoses?**
   - No additional sonographic examination needed at this time.
   - Cytology of the pleural effusion is essential for diagnosis.

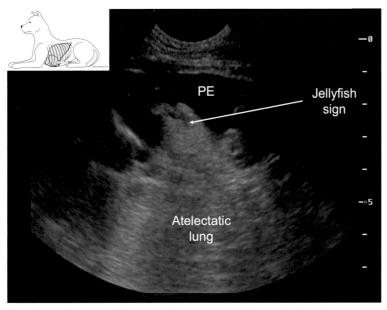

FIGURE 5.11.3 Pleural effusion (PE) is evident causing compressive atelectasis of the lung which may appear as lung consolidation, as noted in this image. When the lung becomes consolidated and appears to "float" within pleural effusion it is termed the "jellyfish sign." The lung parenchyma should not be assessed when compressive atelectasis secondary to pleural effusion or to pleural space–occupying lesions are present as consolidation may resolve as pleural space pathology is corrected and the lung reinflates. Video 5.11.3 (Video 5.11.1 labelled): Pleural effusion is visible at the pericardio-diaphragmatic window, filling the costophrenic recess and tracking along the diaphragm.

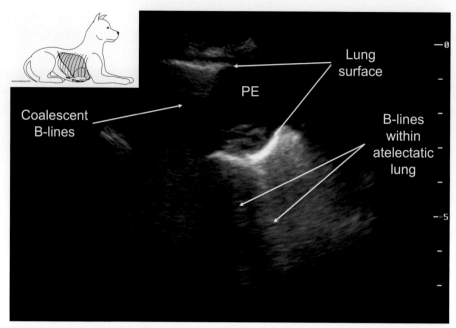

FIGURE 5.11.4 Pleural effusion (PE) is evident causing compressive atelectasis of the lung which may resemble B-lines, as noted in this image. The lung parenchyma should not be assessed when compressive atelectasis secondary to pleural effusion or to the pleural space–occupying lesions is present as B-lines may resolve as pleural space pathology is corrected and the lung reinflates. Video 5.11.4 (Video 5.11.2 labelled): Cardiac POCUS right parasternal short-axis view. Note the left ventricular lumen and left atrial size are small while the ventricular walls appear thickened, which is consistent with hypovolemia in this case.

CLINICAL INTEGRATION

Suspected diagnosis based on ultrasound findings:

Suspected pyothorax based on the history of the hunting dog and the large volume of pleural effusion with the presence of fibrin strands within the effusion. In addition, clinically, there are signs of hypovolemic shock with left ventricular pseudohypertrophy, fever, and hyperemic mucous membranes.

Sonographic interventions, monitoring, and outcome:

- Given the unstable nature of the patient, supplemental oxygen was provided, and a fluid bolus was administered as the PLUS and cardiac POCUS were performed simultaneously.
- The presence of a large volume of pleural effusion is the likely source of respiratory distress and hypoxemia.
- A diagnostic thoracocentesis showed TS: 4.2 g/dL, with septic suppurative inflammation and intracellular and extracellular filamentous rods.
- A 14 G small bore chest tube was placed and 1200 mL of fluid was retrieved. The dog was started on broad spectrum antimicrobials (ampicillin sulbactam 30 mg/kg IV TID and enrofloxacin 10 mg/kg IV SID).
- A computed tomography scan showed pleural thickening, cranial mediastinal lymphadenopathy, and consolidation of the right cranial ventral lung lobe with a moderate amount of pleural effusion.

- No obvious foreign body or masses were detected. A mediastinal thoracotomy was performed for tissue debridement and lavage.
- Numerous sulfur granules were found on the pleural surfaces. Anaerobic culture yielded *Actinomyces* spp.
- After 5 days of hospitalization, the dog was discharged and made a full recovery.

Important concepts regarding the sonographic diagnosis of pyothorax:

- Diagnosis of pyothorax relies on the identification of hypo to hyperechoic fluid between the parietal and visceral pleura. Atelectatic lung with a "jellyfish" sign can sometimes be seen.
- In addition, if concurrent lung pathology is present (pneumonia and parapneumonic pyothorax), B-lines and shred or wedge signs can be seen adjacent to the pleural effusion, although it is not easy to differentiate lung parenchymal pathology from simple compression atelectasis when significant pleural effusion is present as both can cause increased B-lines and lung consolidation.
- In some cases, when the effusion is highly suppurative and inflammatory in nature, it can appear isoechoic or have fibrin strands visible as floating white lines, like "inflatable long balloons floating in the wind."
- The sonographic appearance of pleural effusion is not diagnostic of pyothorax and ultrasound-guided thoracocentesis with fluid analysis and identification of an exudate and suppurative inflammation with intracellular bacteria is confirmatory.[1]
- Placement of thoracostomy tubes under ultrasound guidance is recommended for therapeutic management, in addition to correcting fluid deficits and initial broad spectrum antimicrobial therapy.
- In humans, the hypoechogenicity index (HI) has been described, which is a quantitative marker of pleural fluid echogenicity.[2] Although a small study, HI was correlated with common biomarkers of pleural inflammation in humans which include pH, glucose, and lactate dehydrogenase (LDH).

Techniques and tips to identify pleural effusion/pyothorax:

- To rapidly rule out clinically significant pleural effusion, the probe can be placed in the ventral one-third of the thorax with a depth of 2–6 cm depending on the animal's body condition score and size, with the probe oriented perpendicular or parallel to the ribs.
- To determine if visible fluid is pleural, pericardial, or both, the pericardio-diaphragmatic window and/or the sub-xiphoid window can be assessed: at both windows pleural effusion will appear uncontained, track along the diaphragm, and will form sharp angles or triangles, filling the costophrenic recess, compared to pericardial effusion which will appear contained as a spherical shape outlining the heart and curving away from the diaphragm.
- To avoid missing smaller quantities of pleural effusion which are unlikely to cause clinical signs of dyspnea, but may have diagnostic significance, the probe should be turned parallel to the ribs and slid ventrally until the sternal muscles fill one-third to half of the image.
- When positive for pleural effusion with the probe parallel to the ribs at the most ventral pleural space border, a sail sign will be seen forming between the visceral pleural (outer lining of the lung) and the sternal/pectoral muscles.

Take-home messages
- Pyothorax is identified with the presence of pleural effusion and the possible identification of concurrent lung pathology.
- Definitive diagnosis is obtained via cytology of the fluid. POCUS-guided thoracocentesis may help in obtaining a sample in small volume pleural effusions.

REFERENCES

1. Stillion JR, Letendre JA. A clinical review of the pathophysiology, diagnosis, and treatment of pyothorax in dogs and cats. *J Vet Emerg Crit Care*. 2015;25(1):113–129. doi:10.1111/vec.12274

2. Varsamas C, Kalkanis A, Gourgoulianis KI, Malli F. The use of a novel quantitative marker of echogenicity of pleural fluid in parapneumonic pleural effusions. *Can Respir J*. 2020;2020. doi:10.1155/2020/1283590

CHAPTER 5 – CASE 12

NON-AMBULATORY LABRADOR WITH TACHYPNOEA AND MODERATE RESPIRATORY EFFORT FOLLOWING MOTOR VEHICULAR TRAUMA

Marta Garcia Arce and Daria Starybrat

HISTORY, TRIAGE, AND STABILIZATION

A 3-year-old female entire Labrador presented non-ambulatory with tachypnoea and moderate respiratory effort following motor vehicular trauma.

Triage exam findings (vitals):
- Mentation: mildly obtunded
- Respiratory rate: 60 breaths per minute, moderate increase in expiratory effort, generalized harsh lung sounds bilaterally
- Heart rate: 120 beats per minute
- Mucous membranes: pale pink, 1.5 sec capillary refill time
- Femoral and dorsal pedal arterial pulses: normodynamic and synchronous
- Temperature: 36.8°C (98.24°F)

POCUS exam(s) to perform:
- PLUS
- Cardiac POCUS
- Abdominal POCUS

Abnormal POCUS exam results:
See **Figure 5.12.1** and **Video 5.12.1**.

Additional point-of-care diagnostics and initial management:
- Supplemental oxygen.
- Analgesia: IV methadone 0.2 mg/kg.
- ECG: sinus rhythm.
- SpO_2 92% (on room air).
- Packed cell volume (PCV) 50%, total solids 5.0 g/dL (50 g/L).
- Blood glucose 113.4 mg/dL (6.4 mmol/L).
- Lactate 1.8 mmol/L.

QUESTIONS AND ANSWERS

1. What are the differentials to rule in or out with POCUS based on history and physical exam?
2. What are the sonographic findings?
3. What is the sonographic diagnosis?
4. If necessary, what additional sonographic examination or findings would help rule in or out the differential diagnoses?

DOI: 10.1201/9781003436690-19

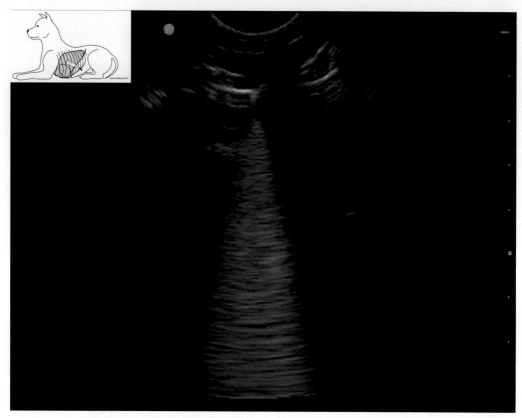

FIGURE 5.12.1 PLUS at the fifth and sixth intercostal space of the left mid-thorax with the dog in sternal recumbency using a microconvex probe, perpendicular to and between the ribs, with the depth set at 8.4 cm. Multiple lung fields were assessed bilaterally. Fur was not clipped, only parted, and ultrasound gel was used as the coupling agent. Video 5.12.1: Video obtained from the same starting point as shown in Figure 5.12.1. From the starting point the probe is moved caudally from one intercostal space to another to assess several lung regions.

1. **Differential diagnosis to rule in or rule out with POCUS:**
   - In trauma patients presenting with respiratory distress, it is important to rule out lung contusions, hemothorax, pneumothorax, and diaphragmatic hernia.
2. **Describe your sonographic findings:**
   - Multiple to coalescing B-lines (**Figure 5.12.2**, **Video 5.12.2**).
3. **What is your sonographic diagnosis?**
   - B-lines and confluent B-lines.
4. **What additional sonographic examination or finding would help you rule in or rule out your differential diagnoses (if necessary)?**
   - Presence of a normal abdominal curtain sign, B-lines, and a glide sign rule out pneumothorax in the assessed lung field.
   - Normal left atrial to aorta ratio helps to rule out cardiogenic pulmonary edema.

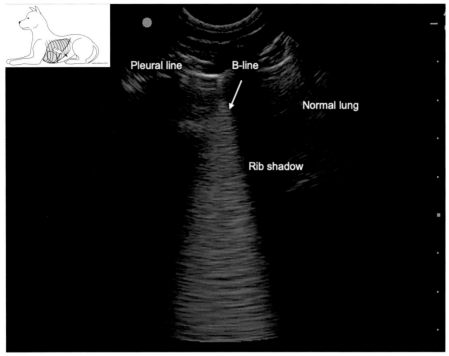

FIGURE 5.12.2 B-line appears as a bright hyperechoic vertical line originating from the pleural line (see arrow). Video 5.12.2 (Video 5.12.1 labelled): Confluent B-lines can be seen as wide white vertical lines occupying most of the intercostal space. This occurs where too many B-lines are present and fuse creating a wider hyperechoic shadow also called white lung. Glide sign (lung sliding) is present; this is the movement of the parietal pleura against the visceral pleura and indicates the absence of air within the thoracic cavity at that probe location. Although mild, at times the pleura looks slightly irregular and thickened.

CLINICAL INTEGRATION

Suspected diagnosis based on ultrasound findings:

The most likely diagnosis for the presence of B-lines in a previously healthy, young animal following trauma is pulmonary contusions.

Sonographic interventions, monitoring, and outcome:
- PLUS showed bilateral B-lines and confluent-B lines.
- Serial PLUS exams were performed to assess the progression and resolution of the B-lines and pulmonary contusions.
- The dog was treated supportively with oxygen therapy and analgesia and eventually discharged on Day 5 following resolution of clinical signs.

Important concepts regarding the sonographic diagnosis of lung contusions:
- B-lines are defined as hyperechoic vertical lines arising from the pleural line. They are most often formed at the interface of fluid and air contained within alveoli in the outermost 1–3 mm of the lung.[1]

- The presence of B-lines can be evaluated by placing the ultrasound probe perpendicular to and between ribs over different lung fields.
- In patients with pulmonary contusions, the presence of an irregular pleural line suggests evidence of lung consolidation.[2,3]
- PLUS has been proven to be a reliable diagnostic imaging tool for the identification of lung contusions in both dogs and humans.[2-4] Although B-lines are the most common ultrasonographic findings in patients with lung contusions, the step sign (abnormal glide sign) has also been described.[5]
- In the unstable trauma patient, PLUS can be used to assess for the absence or presence of lung contusions, pleural effusion, and pneumothorax without the need to perform thoracic radiographs or a CT scan which, in most cases, would require sedation.
- Other differential diagnoses for the presence of B-lines, such as pulmonary edema or pneumonia, are unlikely given this dog's acute history and normal cardiac POCUS.
- Lung contusions can progress over time (24–48 h); therefore, serial PLUS examinations are recommended.[6]

Techniques and tips to identify the appropriate ultrasound images/views:

- Ensure a minimum of four sites per hemithorax are evaluated for the presence of B-lines or pneumothorax in trauma patients.
- Consistent probe location and orientation on the thorax (longitudinal or transverse) is recommended for serial monitoring.

Take-home messages

- Presence of B-lines on PLUS confirms the diagnosis of lung contusions in previously healthy patients presenting following trauma.
- Consistent probe location and orientation on the thorax can help assess the progression or resolution of B-lines.

REFERENCES

1. Lisciandro GR, Fosgate GT, Fulton RM. Frequency and number of ultrasound lung rockets (B-lines) using a regionally based lung ultrasound examination named vet blue (veterinary bedside lung ultrasound exam) in dogs with radiographically normal lung findings. *Vet Radiol Ultrasound*. 2014;55(3):315–322.
2. Armenise A, Boysen RS, Rudloff E, et al. Veterinary-focused assessment with sonography for trauma-airway, breathing, circulation, disability and exposure: A prospective observational study in 64 canine trauma patients. *J Small Anim Pract*. 2019;60(3):173–182.
3. Dicker SA, Lisciandro GR, Newell SM, Johnson JA. Diagnosis of pulmonary contusions with point-of-care lung ultrasonography and thoracic radiography compared to thoracic computed tomography in dogs with motor vehicle trauma: 29 cases (2017–2018). *J Vet Emerg Crit Care*. 2020;30(6):638–646.
4. Hyacinthe A, Broux C, Francony G, et al. Diagnostic accuracy of ultrasonography in the acute assessment of common thoracic lesions after trauma. *Chest*. 2012;141(5):1177–1183.
5. Lisciandro GR, Lagutchick MS, Mann KA, et al. Evaluation of a thoracic focused assessment with sonography for trauma (TFAST) protocol to detect pneumothorax and concurrent thoracic injury in 145 traumatized dogs. *J Vet Emerg Crit Care*. 2008;18(3):258–269.
6. Rendeki S, Molnar TF. Pulmonary contusion. *J Thorac Dis*. 2019;11(2):141–151.

CHAPTER 5 – CASE 13

BORDER COLLIE WITH ACUTE ONSET OF RESPIRATORY DISTRESS AFTER HAVING A LEASH WRAPPED AROUND HER NECK

Julia Delle Cave and Jo-Annie Letendre

HISTORY, TRIAGE, AND STABILIZATION

A 4-month-old intact female Border Collie presented with acute onset of respiratory distress after being found outside with a leash wrapped around her neck.

Triage exam findings (vitals):
- Mentation: bright
- Respiratory rate: 64 breaths per minute, significant respiratory effort, increased broncho-vesicular sounds bilaterally
- Heart rate: 150 beats per minute, no heart murmur
- Mucous membranes: pale pink, <2 sec capillary refill time
- Femoral and dorsal pedal arterial pulses: strong and regular
- Temperature: 38.3°C (100.9°F)

POCUS exam(s) to perform:
- Triage-performed PLUS

Abnormal POCUS exam results:
See **Figures 5.13.1** and **5.13.2** as well as **Video 5.13.1**.

Additional point-of-care diagnostics and initial management:
- SpO_2 of 87%, which increased to 97% with flow-by oxygen supplementation.
- Butorphanol 0.2 mg/kg IV.
- Arterial blood gas while the patient was receiving oxygen through nasal prongs (100 mL/kg/min) showed a $PaCO_2$ of 42.3 mmHg and a PaO_2 of 95.6 mmHg.

QUESTIONS AND ANSWERS

1. What are the differentials to rule in or out with POCUS based on history and physical exam?
2. What are the sonographic findings?
3. What is the sonographic diagnosis?
4. If necessary, what additional sonographic examination or findings would help rule in or out the differential diagnoses?

DOI: 10.1201/9781003436690-20

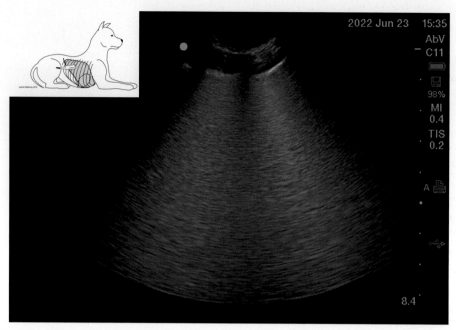

FIGURE 5.13.1 PLUS exam at the caudo-dorsal region of the right hemithorax with the patient in sternal recumbency, using a microconvex probe placed perpendicular to the ribs with the depth set at 8.4 cm. The marker is directed cranially (blue dot). The rest of the exam was performed at a depth of 4–6.6 cm (not shown). Fur was not clipped, and alcohol was used as the coupling agent. Video 5.13.1: Video obtained from the same window as Figure 5.13.1.

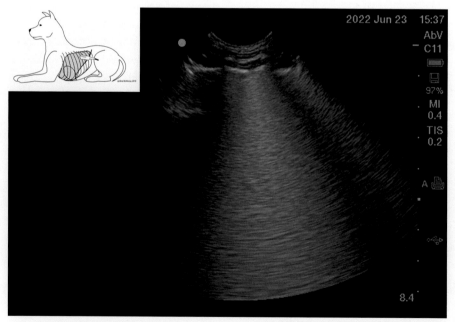

FIGURE 5.13.2 PLUS exam at the caudo-dorsal region of the left hemithorax with the patient in sternal recumbency, using a microconvex probe placed perpendicular to the ribs with the depth set at 8.4 cm. The marker is directed cranially (blue dot). The rest of the exam was performed at a depth of 4–6.6 cm (not shown). Fur was not clipped, and alcohol was used as the coupling agent.

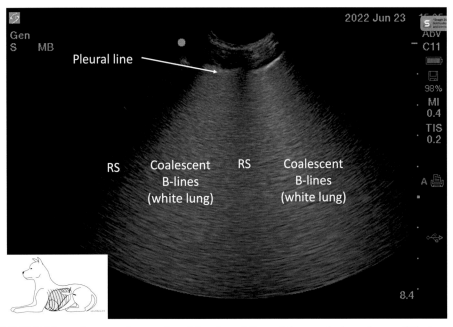

FIGURE 5.13.3 Note the smooth, regular thin pleural line without subpleural consolidations. RS: rib shadow. Video 5.13.2 (Video 5.13.1 labelled): Note the transition from the pleural line to the stomach wall at the caudal border of the lung at the abdominal curtain sign (labelled freeze frame). The vertical edge artifact that defines the abdominal curtain sign is hidden under rib shadow, making the transition from the aerated lung to the gas-filled stomach more difficult to identify. The patient's respiratory effort is too shallow to shift the edge artifact of the abdominal curtain sign from under the rib shadow. Gas shadowing in the stomach can closely mimic the coalescent B-lines in this case. RS: rib shadow.

1. **Differential diagnosis to rule in or rule out with POCUS:**
   - The most likely differential diagnosis based on the history is negative pressure pulmonary edema (NPPE) secondary to strangulation. Other differentials to consider in a young dog with acute respiratory distress include negative pressure pulmonary edema secondary to upper airway obstruction with a foreign body or food, pneumonia, cardiogenic pulmonary edema especially if a heart murmur is auscultated or if there is a history of heart disease, pulmonary contusions, pneumothorax, hemothorax, or diaphragmatic hernia if there is a history of trauma, and less likely inflammatory respiratory disease, pulmonary hemorrhages, other causes of non-cardiogenic pulmonary edema including neurogenic pulmonary edema, pulmonary embolism, and neoplasia.
2. **Describe your sonographic findings:**
   - Bilateral coalescent B-lines, worse on the right hemithorax, and in the caudo-dorsal lung fields (**Figures 5.13.3** and **5.13.4**, **Video 5.13.2**).
3. **What is your sonographic diagnosis?**
   - Regional (caudo-dorsal) bilateral pulmonary edema.
4. **What additional sonographic examination or finding would help you rule in or rule out your differential diagnoses?**
   - Although the history and physical examination do not suggest cardiogenic pulmonary edema, cardiac POCUS to

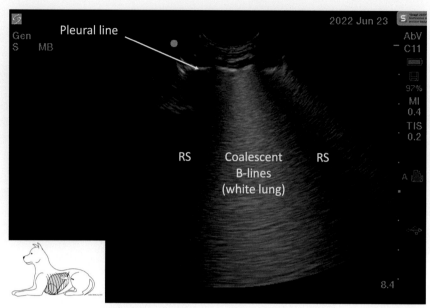

FIGURE 5.13.4 Note the smooth, regular thin pleural line without subpleural consolidations. RS: rib shadow.

evaluate the size of the left atrium and left atrial to aortic ratio (LA:Ao ratio) would help rule out congestive heart failure.
- LA:Ao was normal (less than 1.5:1).

CLINICAL INTEGRATION

Suspected diagnosis based on ultrasound findings:
Based on the history and caudo-dorsal, bilateral pulmonary edema, negative pressure pulmonary edema secondary to strangulation was diagnosed.

Sonographic interventions, monitoring, and outcome:
- Given the unstable nature of the patient, supplemental oxygen was provided, and triage PLUS and cardiac POCUS were immediately performed on admission.
- The dog was treated conservatively with oxygen supplementation through nasal prongs until hypoxemia resolved.
- Butorphanol was given as needed for sedation. Serial PLUS was performed to monitor the progression/resolution of B-lines over time.
- The patient was weaned from oxygen and discharged after complete resolution of clinical signs 3 days after presentation.

Important concepts regarding the sonographic diagnosis of negative pressure pulmonary edema (e.g., choke):
- NPPE is classified as a mixed cause of pulmonary edema (high hydrostatic pressure and increased permeability) and occurs due to vigorous inspiratory efforts against an obstructed airway.[1]
- Pulmonary edema can be identified on lung ultrasound (LUS) through detection of an increased number of B-lines. When B-lines become numerous they can coalesce forming a solid "white sheet" of B-lines, sometimes referred to as "white lung."

- Given that the identification of increased B-lines does not discern the underlying cause, their distribution, the presence of consolidation, and the appearance of the pleural line should be assessed to narrow the differential diagnoses.
- In a retrospective study of 35 dogs with NPPE, the radiographic pattern of lung infiltration (alveolar and/or interstitial) was caudo-dorsal in 65.7% of cases.
- In another retrospective study evaluating the radiographic appearance of non-cardiogenic edema in 60 patients, changes were bilateral in 76.7% of the cases and increased pulmonary opacity was more often asymmetric, unilateral, and dorsal for NPPE compared to other causes of non-cardiogenic pulmonary edema.[2]
- Therefore, the identification of an increased number of B-lines regionally (caudo-dorsal) should lead to a suspicion of NPPE, especially when the history suggests possible choking. A thickened irregular pleura seems more likely with inflammatory pulmonary pathology and may therefore be less likely in cases of NPPE.[3]
- Lung consolidation also seems less likely in cases of NPPE, though research is lacking.
- Cardiac POCUS is important to rule out cardiogenic causes of pulmonary edema.
- There are no veterinary studies using LUS to diagnose and monitor NPPE. Its use has been described in human medicine, although research is lacking.[4]

Techniques and tips to identify the appropriate ultrasound images/views:

- The distribution of B-lines is important for the sonographic diagnosis.
- Evaluation of the caudo-dorsal lung fields is important in suspected cases of NPPE.
- When assessing the caudo-dorsal lung fields, ensure that the probe is positioned over the lungs to avoid misinterpreting gastric gas/abdominal content as pulmonary pathology.
- Patients in respiratory distress can decompensate during restraint/handling. Use low stress handling.
- Dorsal recumbency should be avoided.
- Sternal recumbency or standing are the preferred positions for dyspneic patients.
- Supplemental oxygen +/− anxiolytics can be provided as needed during PLUS.

Take-home messages

- With appropriate history and clinical findings, the detection of diffuse bilateral B-lines, especially in the caudo-dorsal lung fields, can support the suspicion of NPPE.
- The use of LUS allows early diagnosis, preventing the need for pulmonary radiographs in a patient with respiratory distress.
- Serial PLUS can be performed during the patient's hospitalization to monitor the progression of pulmonary infiltrates.

REFERENCES

1. Adamantos S, Hughes D. Chapter 21 - Pulmonary edema. In: Silverstein DC, Hopper K, editors. *Small Animal Critical Care Medicine* (2nd Edition). St. Louis: W.B. Saunders; 2015, pp. 116–20. https://www.sciencedirect.com/science/article/pii/B9781455703067000210.

2. Bouyssou S, Specchi S, Desquilbet L, Pey P. Radiographic appearance of presumed noncardiogenic pulmonary edema and correlation with the underlying cause in dogs and cats. *Vet Radiol Ultrasound*. 2017;58(3):259–65.

3. Boysen S, Gommeren K, Chalhoub S. Clinical applications of PLUS: Are there increased B-Lines, yes/no? In: Boysen, Gommeren, Chalhoub eds. *The Essentials of Veterinary Point-of-Care Ultrasound: Pleural Space and Lung.* Edra publishing, Laval, Quebec;; 2022, pp. 111–37.

4. Zhang G, Huang X, Wan Q, Zhang L. Ultrasound guiding the rapid diagnosis and treatment of negative pressure pulmonary edema: A case report. *Asian J Surg.* 2020 Oct 1;43(10):1047–8.

CHAPTER 5 – CASE 14

TOY POODLE WITH HEMOLYTIC ANEMIA AND SEVERE RESPIRATORY DISTRESS FOLLOWING PREDNISONE AND A WHOLE BLOOD TRANSFUSION

Andrea Armenise

HISTORY, TRIAGE, AND STABILIZATION

A 4-year-old female spayed Toy Poodle was referred for severe anemia and shock secondary to immune-mediated hemolytic anemia (IMHA). Triage POCUS at presentation revealed signs of hypovolemia (pseudohypertrophy) and unremarkable abdominal POCUS and pleural and lung ultrasound (LUS). The dog was treated with prednisone and given a whole blood transfusion. Three days after being hospitalized the dog developed severe respiratory distress.

Triage exam findings (vitals):
- Mentation: alert and responsive
- Respiratory rate: 48 breaths per minute, severe respiratory distress
- Heart rate: 152 beats per minute
- Mucous membranes: pale pink/icteric, <2 sec capillary refill time
- Femoral and dorsal pedal arterial pulses: strong and regular
- Temperature: 38.1°C (100.6°F)

POCUS exam(s) to perform:
- PLUS
- Cardiac POCUS
- Abdominal POCUS

Abnormal POCUS exam results:
See **Figures 5.14.1 – 5.14.5** as well as **Videos 5.14.1–5.14.3.**

Additional point-of-care diagnostics and initial management:
Arterial blood gas (ABG) on room air showed pH 7.13, PCO_2 79 mmHg, HCO_3 24.2 mmol/L, PaO_2 80 mmHg, SO_2 88%, PaO_2/FiO_2 200, PA-a 177 mmHg. Systemic POCUS was repeated.

QUESTIONS AND ANSWERS

1. What are the differentials to rule in or out with POCUS based on history and physical exam?
2. What are the sonographic findings?

DOI: 10.1201/9781003436690-21

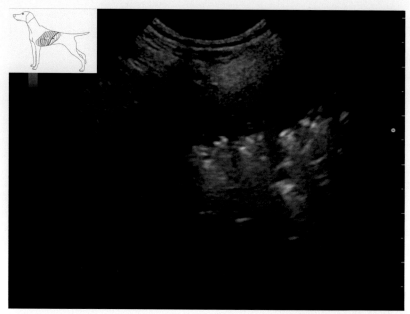

FIGURE 5.14.1 **PLUS** at the left sixth to seventh dorsal intercostal spaces with the dog standing, midway up the thorax, using a microconvex curvilinear probe oriented perpendicular to the ribs, with the depth set at 6 cm and the frequency set at 11 MHz. The fur was not clipped, only parted, and alcohol was used as the coupling agent. Video 5.14.1: Video obtained from the same window as Figure 5.14.1.

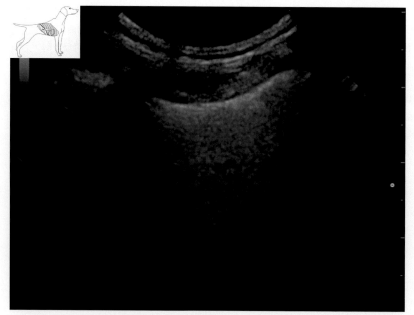

FIGURE 5.14.2 **PLUS** at the right fifth to sixth middle intercostal spaces with the dog standing, using a microconvex curvilinear probe oriented perpendicular to the ribs, with the depth set at 4 cm and the frequency set at 11 MHz. The fur was not clipped, only parted, and alcohol was used as the coupling agent. Video 5.14.2: Video obtained from the same window as Figure 5.14.2.

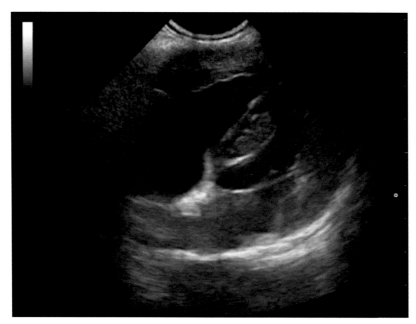

FIGURE 5.14.3 Cardiac POCUS at the left apical four-chamber view obtained with a microconvex curvilinear probe, frequency set at 9 MHz, and the probe situated slightly toward the right side of the heart. The patient was in left lateral recumbency, the probe was oriented close to the sternum toward the heart base from the cardiac apex, the marker was oriented cranially. The dog was in a standing position and fur was not clipped, only parted, and alcohol was used as the coupling agent. Video 5.14.3: Video obtained from the same window as Figure 5.14.3.

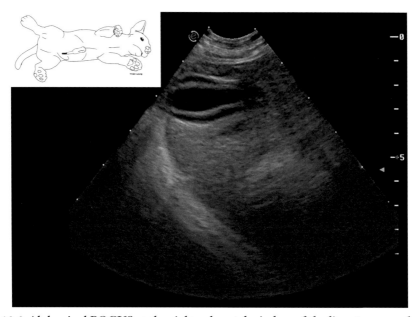

FIGURE 5.14.4 Abdominal POCUS at the right subcostal window of the liver to assess the gallbladder with the dog standing. The probe is a microconvex curvilinear probe, the marker is directed cranially, the depth is set at 6 cm and the frequency is set at 6.6 MHz. If the gallbladder cannot be seen, slide the probe toward the xiphoid process. The fur was not clipped, only parted, and alcohol was used as the coupling agent.

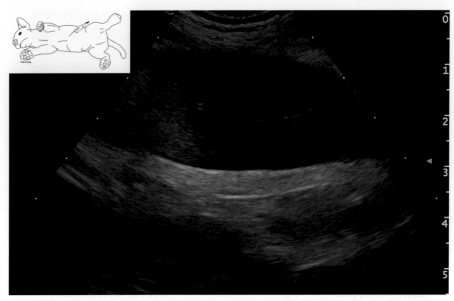

FIGURE 5.14.5 Abdominal POCUS at the left paralumbar site for evaluation of the spleen with the dog standing. The probe is a microconvex curvilinear probe, the marker is directed cranially, the depth is set at 6 cm and the frequency is set at 6.6 MHz. The fur was not clipped, only parted, and alcohol was used as the coupling agent.

3. What is the sonographic diagnosis?
4. If necessary, what additional sonographic examination or findings would help rule in or out the differential diagnoses?

1. **Differential diagnosis to rule in or rule out with POCUS:**
   - The most likely differential diagnosis based on clinical evolution post transfusion was systemic thromboembolism and delayed transfusion-related acute lung injury with splenic thromboembolism.
2. **Describe your sonographic findings:**
   - Hypoechoic partial lung lobe consolidation with central hyperechoic air bronchograms (**Figures 5.14.1** and **5.14.6**, **Videos 5.14.1** and **5.14.4**), which were previously not present.
   - Coalescent B-lines with pleural line irregularities (**Figures 5.14.2** and **5.14.7**, **Video 5.14.2**), which were previously not present.
   - Right heart chambers are subjectively larger than the left heart chambers, which were normal 3 days previously.
   - In healthy dogs the left cardiac chambers should contain two-thirds of the total cardiac volume (**Figures 5.14.3** and **5.14.8**, **Video 5.14.3**).
   - In the right parasternal long-axis view, the lumen of the left ventricle (LV) should be three to four times larger than the right ventricle (RV) and the two atria should be roughly the same size with a neutral interatrial septum.
     – In this case the right ventricular lumen is at least equal to and appears slightly larger than the left ventricle.

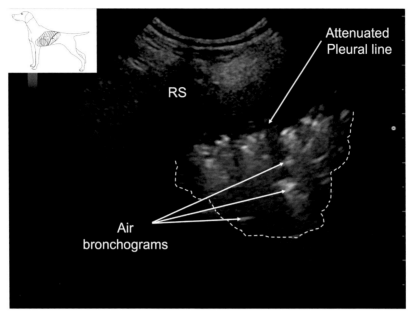

FIGURE 5.14.6 A partial lung consolidation (shred sign) is visible (white dotted line). Air bronchograms (punctate to linear hypoechoic structures) are visible within the region of lung consolidation. The pleural line is attenuated (less visible) where the lung consolidation contacts the chest wall due to loss of the normal soft tissue–air interface seen when the air-filled lung is in contact with the chest wall. Video 5.14.4 (Video 5.14.1 labeled, which corresponds to Figure 5.14.6).

- The right atrium (RA) is significantly larger than the left atrium.
- The interatrial septum appears to deviate and bow toward the LA, likely due to increased RA pressure.[1]
- Gallbladder wall thickening with a medial hyperechoic band surrounded by a hypoechoic band (aka a gallbladder halo sign) (**Figures 5.14.4** and **5.14.9**).
- The central region of the spleen parenchyma appears hypoechoic and dishomogeneous compared to the cranial and caudal portions of the spleen. On color Doppler (not shown) there was no evidence of blood flow within the vasculature of the hypoechoic splenic region, which supports a diagnosis of infarction (**Figures 5.14.5** and **5.14.10**).

3. **What is your sonographic diagnosis?**
   - Dishomogeneous alveolar interstitial syndrome with diffuse lung consolidations, subjective dilation of the right cardiac chambers with interatrial bowing, gallbladder wall edema, and splenic thromboembolism.

4. **What additional sonographic examination or finding would help you rule in or rule out your differential diagnoses (if necessary)?**
   None.

CLINICAL INTEGRATION

Suspected diagnosis based on ultrasound findings:

Based on history, clinical findings, clinical evolution, and serial comparison of systemic POCUS findings, systemic thromboembolic disease was suspected.

Sonographic interventions, monitoring, and outcome:

- Given the significant POCUS findings, supplemental oxygen was provided, low molecular weight heparin was administered, and a CT scan was performed, which confirmed the POCUS finding of systemic thromboembolic disease.
- Serial systemic POCUS examinations together with serial ABGs were performed to monitor the clinical evolution.
- Lung consolidations improved within 12 h and a repeat ABG (on room air) showed the following: pH 7.48, PCO_2 47 mmHg, HCO_3 32.3 mmol/L, PaO_2 96 mmHg, SO_2 95%, PaO_2/FiO_2 395, P_{A-a} 8.3 mmHg.

Important concepts regarding the sonographic diagnosis of thromboembolic disease:

- The sudden appearance of acute hypoechoic lung consolidation with central hyperechoic air bronchograms suggests pulmonary thromboembolism (PTE),[2] particularly in a dog predisposed to hypercoagulable states with previously normal lung finding on POCUS.
- Although pleural effusion can be present in many patients with PTE,[2] it was not seen in this case.
- The concurrent acute development of right cardiac chamber dilation, interatrial bowing, and thickening of the gallbladder wall support increased right-sided cardiac pressures secondary to massive PTE.
- Obstruction of the pulmonary vasculature increases pulmonary vascular resistance leading to right ventricular pressure overload and dysfunction.
- When severe this can subsequently restrict left cardiac chamber filling and ultimately reduce left ventricular output.[3]
- Assessment of the left apical four-chamber view, although not considered routine during cardiac POCUS, can help confirm right-sided heart changes.[1]

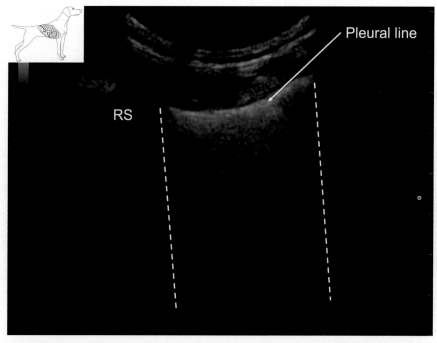

FIGURE 5.14.7 Coalescent B-lines are visible between the two dotted vertical white lines. The pleural line is slightly thickened. RS: rib shadow.

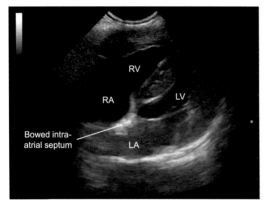

FIGURE 5.14.8 The right-sided cardiac chambers are visible in the near wall and appear larger than the left-sided cardiac chambers. The atrial septum (which should be neutral) is bowed toward the left atrium. RV: right ventricle; RA: right atrium; LV: left ventricle; LA: left atrium.

- Gallbladder wall edema occurs secondary to increased hydrostatic pressure as a result of pulmonary obstruction and subsequent increased right-sided cardiac pressures and venous congestion.[4]
- Splenic infarcts confirm the presence of systemic thromboembolic disease and support the PLUS findings of PTE.
- After lung function improved, a splenectomy was performed, and a histology of the spleen confirmed parenchymal infarction and necrosis consistent with embolic disease.

Techniques and tips to identify the appropriate ultrasound images/views:

- Left apical four-chamber views help evaluate cardiac function if the patient is in left lateral recumbency.
- To obtain the correct window, position the probe perpendicular to the ribs, with the marker cranially, above to the sternum on the left side of the patient and slide

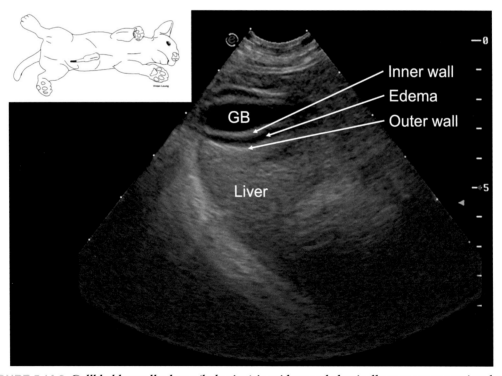

FIGURE 5.14.9 Gallbladder wall edema (halo sign) is evident and classically appears as a striated pattern of white black white. GB: gallbladder.

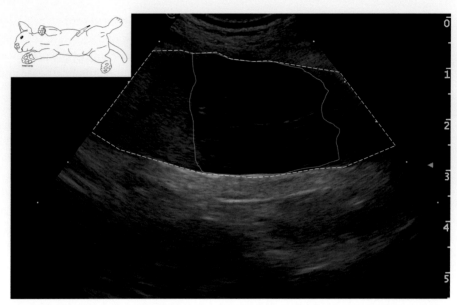

FIGURE 5.14.10 The spleen is visible (larger outlined white dotted region). Within the spleen, a more hypoechoic region (smaller white dotted region) is visible which represents an infarct in this case.

cranially until the liver disappears and the heart becomes visible.
- Fan the transducer dorsally along the chest wall until the heart base is visible at the bottom of the image and the apex is visible at the top of the image.
- The left ventricle should be vertical when correctly imaged from the left apical four-chamber view.[5]
- Although this is a more challenging skill and not part of routine cardiac POCUS, it is an important window to obtain as it can assist in making a correct cardiac diagnosis in some patients.

Take-home messages
- Pulmonary thromboembolism and systemic thromboembolism are a reported complications of IMHA.
- Systemic POCUS findings, including the acute appearance of partial lung lobe consolidations, alveolar-interstitial syndrome, right cardiac chamber dilation, gallbladder wall edema, and splenic infarcts in dogs with predisposed hypercoagulable states should prompt strong consideration of thromboembolic disease. These POCUS findings can be used to guide management, particularly if other specific diagnostic tests such as ROTEM® are not available.

REFERENCES

1. DeFrancesco TC, Ward JL. Focused canine cardiac ultrasound. *Vet Clin Small Anim* 2021; 51(6): 1203–1216.

3. Mathis G, Blank W, Reißig A, et al. Thoracic ultrasound for diagnosis of pulmonary embolism: A prospective multicenter study of 352 patients. *Chest* 2005; 128(3): 1531–1538.

2. Zhu, R, Ma, X-C. Clinical value of ultrasonography in diagnosis of pulmonary embolism in critically ill patients. *J Transl Int Med* 2017 Dec; 5(4): 200–204.
4. Larson MM, Mattoon JS, Lawrence Y, et al. Liver. In: Mattoon JS, Sellon RK, Berry CR (Eds.), *Small Animal Diagnostic Ultrasound*, 4th Edition. Elsevier, 2020; pp. 355–421.
5. Boon JA. Imaging planes: Techniques in the dog and cat. In: Boon JA (Ed.), *Two-Dimensional and M-Mode Echocardiography for the Small Animal Practitioner*, 2nd Edition. Wiley; 2017, pp. 51–74.

CHAPTER 5 – CASE 15

MIXED BREED K9 WITH ACUTE LETHARGY AND ANOREXIA OF 3 DAYS' DURATION

Serge Chalhoub

HISTORY, TRIAGE, AND STABILIZATION

A 12-year-old male neutered medium-sized mixed breed dog presented for acute lethargy and anorexia of 3 days' duration. No travel history. No prior medical history.

Triage exam findings (vitals):
- Mentation: bright, alert, and responsive
- Respiratory rate: 38 breaths per minute, mild expiratory effort
- Heart rate: 90 beats per minute
- Mucous membranes: pink, <2 sec capillary refill time
- Femoral and dorsal pedal arterial pulses: strong and synchronous
- Temperature: 40.2°C (104.3°F)
- Enlarged popliteal and mandibular lymph nodes

POCUS exam(s) to perform:
- Because of the increased respiratory rate and expiratory effort, PLUS was the first POCUS exam chosen

Abnormal POCUS exam results:
See **Figures 5.15.1** and **5.15.2** as well as **Videos 5.15.1–5.15.3**.

Additional point-of-care diagnostics and initial management:
- Abdominal POCUS did not indicate any visible pathology.
- Flow-by oxygen was provided.
- Packed cell volume/total solids 48%, 6.8 g/dL, lactate 1.4 mmol/L.

QUESTIONS AND ANSWERS

1. What are the differentials to rule in or out with POCUS based on history and physical exam?
2. What are the sonographic findings?

DOI: 10.1201/9781003436690-22

Mixed Breed K9 with Acute Lethargy and Anorexia of 3 Days' Duration

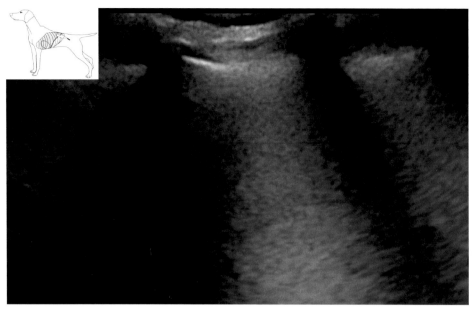

FIGURE 5.15.1 **PLUS** scan at the caudo-dorsal region of the left hemithorax with the dog standing, using a microconvex probe oriented perpendicular to the ribs, with the depth set at 2 cm. Similar findings were noted throughout both hemithoraces using a regional "S"-shaped PLUS scan. Video 5.15.1: Video obtained at the same site as Figure 5.15.1.

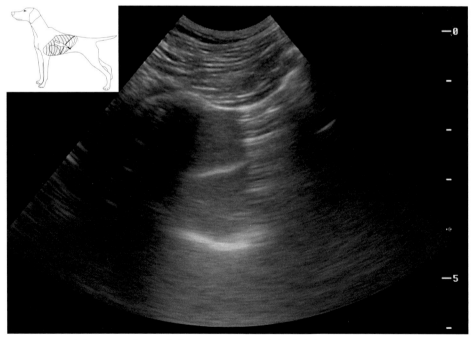

FIGURE 5.15.2 **PLUS** scan at the mid-thoracic region of the left hemithorax with the dog standing, using a microconvex probe oriented perpendicular to the ribs, with the depth set at 5 cm. Similar lesions of varying size from 1 to 3 cm were found throughout both hemithoraces. Video 5.15.2: Ultrasound clip of typical findings in multiple locations of both hemithoraces.

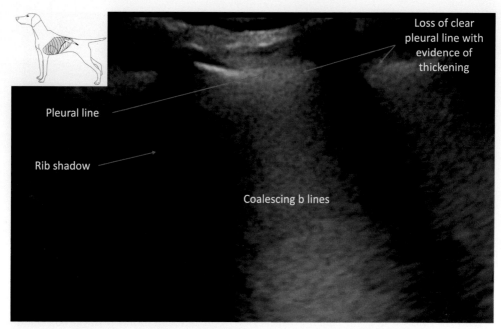

FIGURE 5.15.3 Note the formation of "white lung" due to multiple coalescing B-lines, which in this case are "dry" in origin, as a result of decreased aeration at the lung periphery due to infiltrative pulmonary pathology. Video 5.15.3 (Video 5.15.1 labelled).

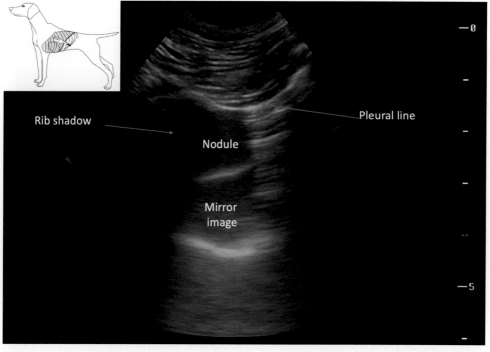

FIGURE 5.15.4 Note the distinct absence of obvious air bronchograms within the region of lung consolidation. Video 5.15.4 (Video 5.15.2 labelled).

3. What is the sonographic diagnosis?
4. If necessary, what additional sonographic examination or findings would help rule in or out the differential diagnoses?

1. **Differential diagnosis to rule in or rule out with POCUS:**
   - Differentials to consider given the history and ultrasound findings include metastatic neoplasia, primary neoplasia, fungal disease, and lung abscesses.
2. **Describe your sonographic findings:**
   - Lung sliding (aka glide sign) was present in all lung fields (**Videos 5.15.1** and **5.15.2**).
   - There were sections of lung with multiple to coalescing B-lines (**Figure 5.15.3, Video 5.15.4**).
   - Several partial lung consolidations with smooth, regular, and well-defined deep borders and an absence of air bronchograms were seen bilaterally at multiple PLUS locations (**Figure 5.15.4, Videos 5.15.5** and **5.15.6**.)
3. **What is your sonographic diagnosis?**
   - Multiple to coalescing B-lines and nodule signs.

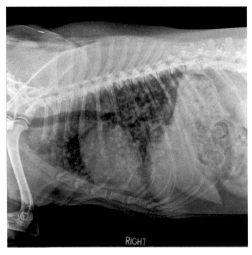

FIGURE 5.15.5 Left lateral thoracic radiograph of the dog in question. Multiple diffuse lung nodules are visible through all lung fields.

4. **What additional sonographic examination or finding would help you rule in or rule out your differential diagnoses?**
   - Ultrasound-guided fine-needle aspiration with cytology can help with the diagnosis, as well as a comprehensive abdominal ultrasound for further staging.

CLINICAL INTEGRATION

Suspected diagnosis based on ultrasound findings:

Considering the age and presentation of the patient, neoplasia was the most likely cause. The most likely diagnosis was metastatic neoplasia. The presence of multiple coalescing B-lines is not always seen with metastatic neoplasia, and a dry "infiltrative" cause of decreased lung aeration at the lung periphery was the suspected cause of the increased B-lines in this case.

Sonographic interventions, monitoring, and outcome:

- Thoracic radiographs were taken (**Figure 5.15.5**) which supported the diagnosis of metastatic neoplasia. No further sonographic evaluation was performed.
- Following discussion with the owners, the patient was euthanized due to financial constraints and a guarded prognosis.

Important concepts regarding the sonographic diagnosis of pulmonary neoplasia with dry B-lines:

- By completing a comprehensive "S"-shaped regional PLUS scan of both hemithoraces, multiple nodules were found.
- Because nodules can be identified anywhere over the lung surface and can be variable

- in number, more comprehensive protocols that scan larger lung surface areas may be required to identify nodules that only occur in isolated locations.[1]
- In veterinary medicine, nodules most likely represent neoplasia secondary to metastasis.
- Because ultrasound cannot differentiate neoplasia from other causes of nodules (e.g., fungal disease), the history, geographical location, and other clinical findings are important to help narrow the differential diagnoses.
- The increased B-lines in this case were likely "dry" in origin; in other words, regions of decreased air at the lung periphery occurred due to an increased number of neoplastic cells (instead of fluid in the case of wet lung).[2]
- Progressing from normal lung surface areas to regions with increased infiltrative cells will decrease the ratio of aerated lung, leading to the development of B-lines.
- As the number of infiltrative cells continues to increase, the ratio of aerated lung will fall further (lung density increases) until consolidation is seen.

Techniques and tips to identify the appropriate ultrasound images/views:
- Identifying the pleural line is key.
- Starting over lung is primordial to not confuse abdominal findings with PLUS findings.
- Beginning at the most caudo-dorsal region of each hemithorax, the hemithorax is scanned using an "S"-shaped pattern to scan as much pleural space and lung as possible.
- Consolidations are tissue-like patterns originating from the pleural line.
- Partial consolidations do not traverse the entire lung lobe because they encounter aerated lung below the region of consolidation. This prevents the ultrasound beam from traversing the deeper-lying aerated lung tissue.
- Nodules have smooth distal/deep boundaries compared to other partial consolidations such as the shred sign.
- Partial consolidations are distinguished from pleural effusion because pleural effusion separates the parietal and visceral pleura.
- B-lines are a type of artifact often caused by a decrease in the ratio of aerated to non-aerated lung at the lung surface, which is often secondary to wet lung pathology such as hemorrhage, aspiration pneumonia, non-cardiogenic pulmonary edema, and cardiogenic pulmonary edema. However, because B-lines can occur due to any decrease in aerated lung at the lung periphery, they can also be created by an increase in pathologic cells at the lung surface (neoplasia, fibrosis, etc.), or even atelectasis. Therefore, B-lines are not always secondary to wet lung.

Take-home messages
- B-lines may represent increased lung surface fluid, cells, or even atelectasis.[2] Nodule signs are a type of partial lung consolidation which often represent neoplastic and fungal disease, or less commonly abscesses.[1]
- Comprehensive regional PLUS scanning ensures that if nodules are at the lung surface, they will likely be found.
- However, not all nodules are visible on PLUS because they may not reach the lung surface. Hence, complimentary exams such as thoracic radiographs or CT scan may be necessary.

REFERENCES

1. Boysen S, Gommeren K, Chalhoub S. Clinical applications of PLUS: Is there lung consolidation, yes/no? In: Boysen S, Gommeren K, Chalhoub S, editors. *The Essentials of Veterinary Point-of-Care Ultrasound: Pleural Space and Lung.* Edra publishing, Laval, Quebec; 2022, pp. 139–182.

2. Boysen S, Gommeren K, Chalhoub S. Clinical applications of PLUS: Are there increased B lines, yes/no? In: Boysen S, Gommeren K, Chalhoub S, editors. *The Essentials of Veterinary Point-of-Care Ultrasound: Pleural Space and Lung.* Edra publishing, Laval, Quebec; 2022, pp. 111–137.

CHAPTER 5 – CASE 16

LABRADOR RETRIEVER WITH RESPIRATORY DISTRESS AND SUPERFICIAL ABRASIONS FOLLOWING MOTOR VEHICULAR TRAUMA

Søren Boysen

HISTORY, TRIAGE, AND STABILIZATION

A 4-year-old neutered male Labrador Retriever presented with respiratory distress and superficial abrasions following motor vehicular trauma.

Triage exam findings (vitals):
- Respiratory rate: 76 breaths per minute, significant respiratory distress, difficult to auscult
- Temperature: not assessed
- Heart rate: 164 beats per minute
- Mucous membranes: pink, <2 sec capillary refill time
- Femoral and dorsal pedal arterial pulses: strong and regular
- Doppler blood pressure: 126 mmHg systolic

POCUS exam(s) to perform:
- PLUS

Abnormal POCUS exam results:
See **Figure 5.16.1** and **Video 5.16.1**.

Additional point-of-care diagnostics and initial management:
- Supplemental oxygen administered via flow by.
- An intravenous catheter was placed and methadone (0.1 mg/kg IV) administered.

QUESTIONS AND ANSWERS

1. What are the differentials to rule in or out with POCUS based on history and physical exam?
2. What are the sonographic findings?
3. What is the sonographic diagnosis?
4. If necessary, what additional sonographic examination or findings would help rule in or out the differential diagnoses?

Respiratory Distress and Superficial Abrasions

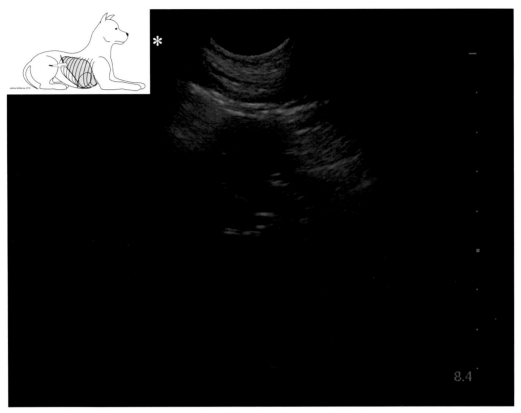

FIGURE 5.16.1 PLUS exam at the caudal border of the right hemithorax two-thirds of the way up the thorax with the dog standing, using a microconvex robe orientation perpendicular to the ribs, with the depth set at 8 cm. Fur was not clipped and alcohol was used as the coupling agent. Video 5.16.1: Video recorded at the same location as Figure 5.16.1.

1. **Differential diagnosis to rule in or rule out with POCUS:**
   - Pulmonary contusions, hemothorax, diaphragmatic hernia, rib fractures, and pneumothorax are all likely causes of dyspnea in trauma patients.
2. **Describe your sonographic findings:**
   - Double curtain sign.
   - Two vertical air–soft tissue edge artifacts are visible in the still image (**Figure 5.16.2**), which converge toward each other on inspiration and away from each other on expiration (**Video 5.16.2**).
3. **What is your sonographic diagnosis?**
   - Double curtain sign.
4. **What additional sonographic examination or finding would help you rule in or rule out your differential diagnoses (if necessary)?**
   - Absence of the following five sonographic findings should raise suspicion of pneumothorax at a specific probe location: (1) lung sliding (aka the glide sign), (2) a lung pulse, (3) B-lines, (4) lung consolidations, and (5) pleural effusion.[1]
   - Identification of a lung point would also confirm pneumothorax.[1–4]
   - A barcode sign using time motion mode may also be used to confirm pneumothorax.

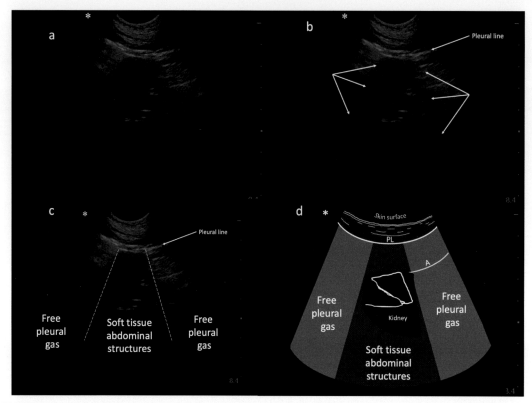

FIGURE 5.16.2 Still ultrasound images and schematic showing a double curtain sign where the underlying soft tissue includes the kidney. (a) Unlabelled image shown in Figure 5.16.1. (b) The pleural line is identified by the single white arrow. The two borders of the double curtain sign are indicated by the three diverging white arrows. (c) The cranial and caudal vertical edges of the double curtain sign are highlighted by the white dotted lines. Free pleural air is seen cranially and caudally to the visible soft tissue structures, giving rise to the double curtain sign. (d) Schematic labelled image of the structures seen in the ultrasound still images. PL: pleural line; A: A-line. Video 5.16.2 (Video 5.16.1 labelled). Image courtesy Søren Boysen, with permission

CLINICAL INTEGRATION

Suspected diagnosis based on ultrasound findings:

Pneumothorax. Pulmonary contusions cannot be ruled out as the lung surface cannot be seen in any of the images provided.[1,4]

Sonographic interventions, monitoring, and outcome:

- The left hemithorax was assessed and revealed similar findings.
- Bilateral thoracocentesis was performed.
- Cardiac and abdominal POCUS were performed following thoracocentesis; both were unremarkable.
- Subsequent PLUS (results not shown) following bilateral thoracocentesis revealed lung sliding and increased B-lines (likely the result of pulmonary contusions – see other cases).
- The dog was managed with oxygen via nasal prongs, analgesia, and fluid therapy.

He was monitored closely, which included serial PLUS. Pneumothorax did not recur and the B-lines resolved. The dog was discharged.

Important concepts regarding the sonographic diagnosis of pneumothorax using abnormal curtain signs:

- Two abnormal curtain signs are reported in dogs with pneumothorax: (1) double curtain signs (**Videos 5.16.2** and **5.16.3**) and (2) asynchronous curtain signs (**Video 5.16.4**). The latter is also seen with pneumothorax and is created when the vertical edge artifact of the abdominal curtain sign moves in the opposite direction to visible abdominal structures during the phases of respiration).[2,4]
- Abnormal curtain signs have been identified in dogs positioned in sternal, standing, and both lateral recumbencies.[2]
- In dogs with confirmed pneumothorax, normal, asynchronous, and double signs have been identified in the same patient on the same side of the thorax.[2]
  - In sternal or standing patients with partial pneumothorax, normal curtain signs are found ventrally, below the region of pneumothorax, and abnormal curtain signs are identified dorsally within the region of pneumothorax.[2]
  - If pneumothorax is suspected, sweep the probe dorsally along the curtain sign to look for asynchronous and double curtain signs.[1,2,4]
- Abnormal curtain signs can vary in appearance depending on the underlying soft issue structures visible and how the lung, free pleural gas, and costophrenic recess interact with each other throughout the respiratory cycle (**Video 5.16.5**, variations

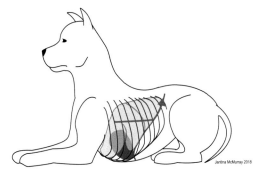

FIGURE 5.16.3 **Probe movements to assess patients for pneumothorax in a standing/sternal position. Start over lung and move the probe caudally until the curtain sign is identified (red arrow). Assess if the curtain sign is normal or abnormal. The probe is then moved along the curtain sign to the most caudo-dorsal site searching for abnormal curtain signs. If no abnormal curtain signs are seen, the most caudo-dorsal site is assessed for lung sliding. Video 5.16.3: Double curtain sign is created where soft tissues of the abdomen break into a region of pneumothorax and free air without interposed lung along the costophrenic recess.**

on the appearance of the double and asynchronous curtain signs).[2,4]

- Abnormal curtain signs have only been reported in a small canine case series and in patients with a mild to moderate pneumothorax[2]: the appearance or ability to detect abnormal curtain signs with massive or tension pneumothorax is unknown, and further research is required to determine the accuracy, sensitivity, and specificity of abnormal curtain signs.
- It is unknown if the detection of only normal curtain signs along the entire length of the sonographic thoraco-abdominal interface can be used to rule out pneumothorax.
- Anecdotal evidence suggests abnormal curtain signs can be identified in cats with pneumothorax, although publications are lacking.

Techniques and tips to identify the appropriate ultrasound images/views:

- Start with the probe oriented perpendicular to the ribs in the mid-thoracic region, just caudal to the forelimb.[4]
- Slide the probe caudally until the abdominal curtain sign is identified (**Figure 5.16.3**). If a double or asynchronous curtain sign is identified, thoracocentesis should be considered.[4]
- If the curtain sign is normal, assess for abnormal curtain signs by sweeping the probe caudo-dorsally along the curtain sign, maintaining a perpendicular probe orientation until the pleural line disappears in the hypaxial muscles (caudo-dorsal site).[2,4]
- Interpreting the curtain sign is often easier with the depth increased to 6 cm or deeper.
- The freeze and track functions to "slow" cineloop speed often make interpretation of the curtain sign easier. In contrast, freeze and track options are less helpful in assessing lung sliding as they require dynamic "real-time" assessment of the pleural line to see the shimmer of lung sliding.[4]
- The double curtain sign can be diagnosed on still images.

Take-home message

- With appropriate history and clinical findings, the detection of abnormal curtain signs quickly supports the diagnosis of pneumothorax, allowing early initiation via thoracocentesis.

REFERENCES

1. Boysen SR. Lung ultrasonography for pneumothorax in Dogs and cats. *Vet Clin North Am Small Anim Pract* 2021 Nov;51(6):1153–1167.
2. Boysen S, McMurray J, Gommeren K. Abnormal curtain signs identified with a novel lung ultrasound protocol in six dogs with pneumothorax. *Front Vet Sci* 2019;6:291.
3. Hwang TS, Yoon YM, Jung DI, Yeon SC, Lee HC. Usefulness of transthoracic lung ultrasound for the diagnosis of mild pneumothorax. *J Vet Sci* 2018;19(5):660–666.
4. Boysen S, Chalhoub S, Gommeren K. Clinical applications of PLUS: Is there pneumothorax, yes/no? In: Boysen S, Chalhoub C, Gommeren K, eds. *The Essentials of Veterinary Point-of-Care Ultrasound: Pleural Space and Lung*. Edra publishing, Laval, Quebec; 2022, pp. 63–90.

CHAPTER 5 – CASE 17

CHIHUAHUA WITH SEVERE DYSPNEA AND TREMORS FOLLOWING MILBEMYCIN TOXICITY AND INTRA-LIPID THERAPY

Elizabeth Gribbin and Rachael Birkbeck

HISTORY, TRIAGE, AND STABILIZATION

A 6-year-old female neutered Chihuahua presented with tremors following a ten-fold (5 mg/kg) overdose of milbemycin. The referring vet administered intra-lipid (1.5 mL/kg bolus, followed by 0.25 mL/kg/min over 30 min) and within 5 h severe dyspnea developed. No history of vomiting, regurgitation, seizures, electrocution, choking, or near drowning.

Triage exam findings (vitals):
- Mentation: quiet, alert, and responsive
- Respiratory rate: 140 breaths per minute, increased respiratory effort, harsh lung sounds bilaterally
- Heart rate: 170 beats per minute, regular rhythm, no cardiac murmur detected
- Mucous membranes: pink, moist, <2 sec capillary refill time
- Femoral and dorsal pedal arterial pulses: strong and synchronous
- Doppler blood pressure: 100 mmHg systolic
- Temperature: 35.7°C (96.3°F)

POCUS exam(s) to perform:
- PLUS
- Cardiac POCUS

Abnormal POCUS exam results:
See **Figure 5.17.1** and **Video 5.17.1**.

Additional point-of-care diagnostics and initial management:
- Arterial blood gas analysis revealed a PaO_2 of 58 mmHg on room air.
- Supplemental oxygen was provided.

QUESTIONS AND ANSWERS

1. What are the differentials to rule in or out with POCUS based on history and physical exam?
2. What are the sonographic findings?

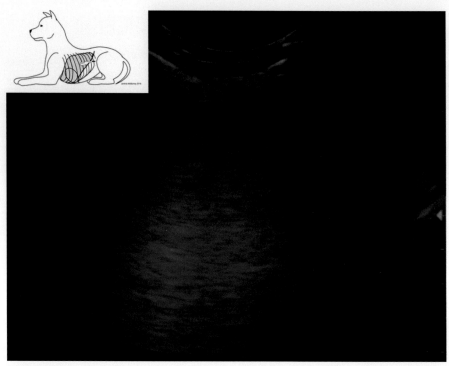

FIGURE 5.17.1 PLUS exam at the mid-caudal thoracic region of the left hemithorax with the patient standing, using a microconvex probe in a transverse orientation, with the depth set at 8 cm. Similar views were obtained in all lung quadrants. Fur was not clipped and alcohol was used as the coupling agent. Video 5.17.1: Video obtained from the same window as Figure 5.17.1.

3. What is the sonographic diagnosis?
4. If necessary, what additional sonographic examination or findings would help rule in or out the differential diagnoses?

1. **Differential diagnosis to rule in or rule out with POCUS:**
   - Differential diagnoses to consider in this patient include non-cardiogenic pulmonary edema (NCPE) (acute lung injury [ALI] vs acute respiratory distress syndrome [ARDS]), iatrogenic volume overload, aspiration pneumonia, left-sided congestive heart disease, spontaneous pneumothorax, pleural space disease, pulmonary hypertension, pulmonary thromboembolism, *Angiostronylus vasorum* infestation, and neoplasia.
2. **Describe your sonographic findings:**
   - Multiple to coalescing B-lines with an irregular pleural surface were noted throughout multiple lung regions of both hemithoraces (See **Figure 5.17.2** and **Video 5.17.2**).
3. **What is your sonographic diagnosis?**
   - Non-cardiogenic pulmonary edema.
4. **What additional sonographic examination or finding would help you rule in or rule out your differential diagnoses (if necessary)?**
   - Normal left atrial to aorta ratio (LA:Ao) of <1.6.[1]
   - Absence of pericardial effusion.
   - No evidence of pleural space pathology.
   - Presence of a glide sign bilaterally in the dorsal lung fields.

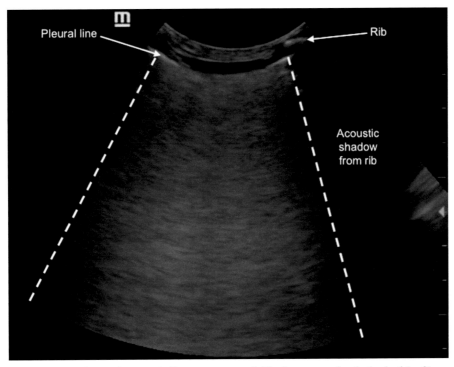

FIGURE 5.17.2 Multiple, coalescing B-lines are appreciable, between the dashed white lines, giving the lung a "white-out" appearance. Video 5.17.2 (Video 5.17.1 labelled).

CLINICAL INTEGRATION

Suspected diagnosis based on ultrasound findings:
Acute lung injury vs acute respiratory distress syndrome associated with intralipid administration.[2]

Sonographic interventions, monitoring, and outcome:
- Mechanical ventilation was initiated within 1 h following admission due to severe persistent hypoxemia and patient welfare concerns.
- Arterial blood gas analysis revealed a PaO_2/FiO_2 ratio of 164, indicating inefficient gas exchange consistent with ARDS.[3]
- Echocardiography was within normal limits.
- Transtracheal wash was performed following intubation. Fluid cytology revealed a neutrophilic cellular population with no evidence of bacterial sepsis.
- Thoracic radiography confirmed bilateral pulmonary infiltrates. The history and diagnostic findings were deemed consistent with ARDS.[3]
- Serial PLUS revealed the progression of pulmonary changes with lung consolidation noted on Day 3 of hospitalization (**Figure 5.17.3A,B**).
- Due to worsening hypoxemia and cardiovascular instability, euthanasia was performed on the fifth day of hospitalization.

Important concepts regarding the sonographic diagnosis of ARDS with consolidation:
- Lung ultrasound and thoracic radiography for the diagnosis of ARDS are considered complimentary imaging modalities rather than interchangeable.[5]

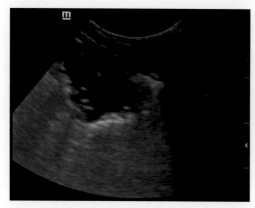

FIGURE 5.17.3 (A) Image taken with the probe placed in a transverse position on the left hemithorax. Lung consolidation can be appreciated ultrasonographically as hypoechoic tissue at the lung periphery. The hyperechoic "white" foci within the consolidated tissue indicate air bronchograms and partially aerated lung, consistent with a shred sign. Lung consolidation can also have a complete lack of aeration causing the lung to appear like tissue such as the liver, this is known as the tissue sign.[4]

- Imaging criterion for diagnosing ARDS on lung ultrasound includes multiple B-lines and/or lung consolidation affecting at least one region in each hemithorax.
- Diffuse or bilaterally distributed B-lines have been found on PLUS in cats and dogs diagnosed with ARDS.[6]
- Echocardiography can reliably distinguish NCPE from cardiogenic edema.[8]
- In humans, consolidation and an irregular pleural line are used to help distinguish NCPE from cardiogenic pulmonary edema on PLUS when B-lines are present.[9]
- B-lines and lung consolidation may be appreciated in cases of aspiration pneumonia; however, this distribution is typically in the right middle lung region.[10]
- Serial LUS is useful in monitoring for ARDS development and progression in hospitalized patients.
- Consolidation may be seen as lung pathology worsens.[11]

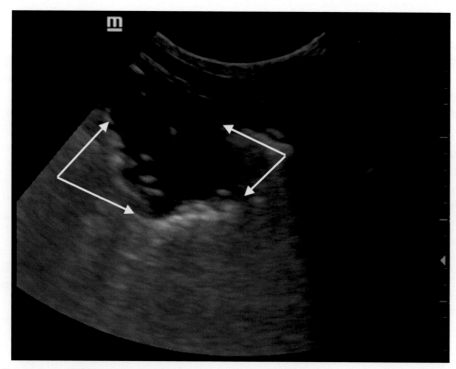

FIGURE 5.17.3 (B) White arrows indicating the area of consolidation.

Techniques and tips to identify the appropriate ultrasound images/views:

- PLUS and cardiac POCUS can be performed with the patient in either a standing or sternal position.[7,11]
- During cardiac POCUS, extending the right shoulder cranially can optimize probe positioning.

Take-home messages

- ARDS is a clinical diagnosis.[3] PLUS and cardiac POCUS can raise the index of suspicion for ARDS and rule out other causes of dyspnea.
- Serial POCUS can track the progression or resolution of pulmonary pathology in cats and dogs with ARDS.

REFERENCES

1. Rishniw M, Caivano D, Dickson D, et al. Two-dimensional echocardiographic left-atrial-to-aortic ratio in healthy adult dogs: A re-examination of reference intervals. *J Vet Cardiol* 2019;26:29–38.
2. Friedman T, Feld Y, Adler Z, Bolotin G and Bentur Y. Acute respiratory distress syndrome associated with intravenous lipid emulsion therapy for verapamil toxicity, successfully treated with veno-venous ECMO. *J Dev Drugs* 2017;6(3):2.
3. Wilkins PA, Otto CM, Baumgardner JE, et al. Acute lung injury and acute respiratory distress syndromes in veterinary medicine: Consensus definitions: The Dorothy Russell Havemeyer working group on ALI and ARDS in veterinary medicine. *J Vet Emerg Crit Care* Aug 2007;17(4):333–339.
4. Lisciandro GR. *Focused Ultrasound Techniques for the Small Animal Practitioner*. 2nd Ed. John Wiley & Sons; 2014.
5. See KC, Ong V, Tan YL, Tan YL, Sahagun J and Taculod J. Chest radiography versus lung ultrasound for identification of acute respiratory distress syndrome: A retrospective observational study. *Crit Care* Aug 2018;22(1):203.
6. Ward JL, Lisciandro GR and DeFrancesco TC. Distribution of alveolar-interstitial syndrome in dogs and cats with respiratory distress as assessed by lung ultrasound versus thoracic radiographs. *J Vet Emerg Crit Care* Sept 2018;28(5):415–428.
7. Murphy SD, Ward JL, Viall AK, et al. Utility of point-of-care lung ultrasound for monitoring cardiogenic pulmonary edema in dogs. *J Vet Intern Med* Jan 2021;35(1):68–77.
8. Hezzell MJ, Ostroski C, Oyama M, et al. Investigation of focused cardiac ultrasound in the emergency room for differentiation of respiratory and cardiac causes of respiratory distress in dogs. *J Vet Emerg Crit Care* Mar 2020;30(2):159–164.
9. Soldati G and Demi M. The use of lung ultrasound images for the differential diagnosis of pulmonary and cardiac interstitial syndrome. *J Ultrasound* 2017;20(2):91–96.
10. Fernandes Rodrigues N, Giraud L, Bolen G, et al. Comparison of lung ultrasound, chest radiographs, C-reactive protein, and clinical findings in dogs treated for aspiration pneumonia. *J Vet Intern Med* 2022;36(2):743–752.
11. Cole L, Pivetta M and Humm K. Diagnostic accuracy of a lung ultrasound protocol (Vet BLUE) for detection of pleural fluid, pneumothorax and lung pathology in dogs and cats. *J Small Anim Pract* Mar 2021;62(3):178–186.

CHAPTER 5 – CASE 18

WEST HIGHLAND WHITE TERRIER WITH PROGRESSIVE RESPIRATORY DISTRESS, DRY COUGH, AND EXERCISE INTOLERANCE OF 2 MONTHS' DURATION

Elodie Rizzoli and Géraldine Bolen

HISTORY, TRIAGE, AND STABILIZATION

A 12-year-old male neutered West Highland White Terrier presented with progressive respiratory distress, dry cough, and exercise intolerance noticed 2 months previously.

Triage exam findings (vitals):
- Mentation: alert
- Respiratory rate: 48 breaths per minute, slight expiratory dyspnea, and pulmonary crackles auscultated in the whole lung field and audible from the mouth without stethoscope
- Heart rate: 104 beats per minute, no audible heart murmur
- Mucous membranes: pink, <2 sec capillary refill time
- Femoral and dorsal pedal arterial pulses: strong and regular
- Temperature: 38.7°C (101.7°F)

POCUS exam(s) to perform:
- PLUS
- Cardiac POCUS: normal

Abnormal POCUS exam results:
See **Figures 5.18.1** and **5.18.2** as well as **Video 5.18.1**.

Additional point-of-care diagnostics and initial management:
- Supplemental oxygen.

QUESTIONS AND ANSWERS

1. What are the differentials to rule in or out with POCUS based on history and physical exam?
2. What are the sonographic findings?
3. What is the sonographic diagnosis?
4. If necessary, what additional sonographic examination or findings would help rule in or out the differential diagnoses?

DOI: 10.1201/9781003436690-25

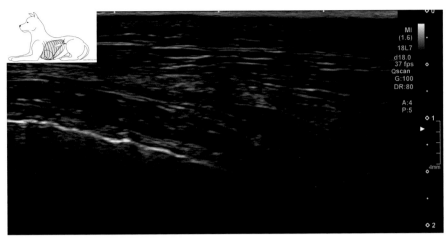

FIGURE 5.18.1 PLUS exam at the caudo-dorsal region of the right hemithorax with the dog in sternal recumbency using a linear probe located two-thirds of the way up the thorax at the seventh intercostal space, oriented parallel to the ribs with the depth set at 2 cm and the frequency set at 18 Hz. The patient was not clipped, and the coupling agent was alcohol. Video 5.18.1: Ultrasound clip of typical findings from a similar case shown in Figures 5.18.1 and 5.18.2.

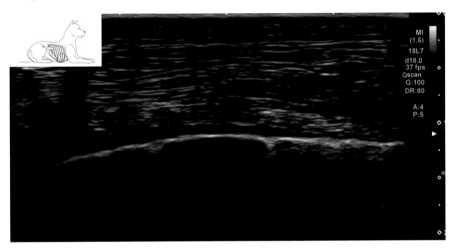

FIGURE 5.18.2 PLUS exam at the caudo-dorsal region of the left hemithorax with the dog in sternal recumbency using a linear probe located two-thirds of the way up the thorax at the ninth intercostal space, oriented parallel to the ribs with the depth set at 2 cm and the frequency set at 18 Hz. The patient was not clipped, and the coupling agent was alcohol.

1. Differential diagnosis to rule in or rule out with POCUS:
   - Pulmonary fibrosis, bronchopneumopathy (eosinophilic bronchopneumopathy or aspecific chronic bronchitis)/bronchomalacia, left-sided congestive heart failure, less likely non-cardiogenic pulmonary edema, mucopurulent bronchopneumopathy, neoplasia, other.

2. Describe your sonographic findings:
   - Irregular visceral pleural surface (PS) (Figures 5.18.1–5.18.4).
   - Generalized individual B-lines (BL) over all lung surfaces assessed (up to 5–6 cm in depth) (Figures 5.18.1–5.18.4).
   - Peripheral subpleural millimetric hypoechoic nodules (Figures 5.18.2 and 5.18.4).

3. **What is your sonographic diagnosis?**
   Generalized peripheral pulmonary lesions consisting of B-lines, an irregular pleural line, and subpleural millimetric hypoechoic nodules.
4. **What additional sonographic examination or finding would help you rule in or rule out your differential diagnoses (if necessary)?**
   Normal cardiac POCUS rules out congestive heart failure as a cause of B-lines.

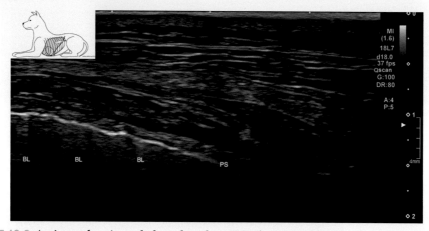

FIGURE 5.18.3 An irregular visceral pleural surface (PS) (horizontal hyperechoic line) is visible. The irregularities move with lung sliding (not appreciated on a still image), similar to B-lines, indicating that the lesions are arising from the lung surface and not the parietal pleura. Generalized individual B-lines (BL) over all lung surfaces are visible. They extend up to 5–6 cm in depth, although only 2 cm is shown in the still image. B-lines appear as vertical hyperechoic ring down artifacts originating from the visceral surface of the pleural line and extending to the bottom of the field, sliding synchronously with the lung (not appreciated on a still image).

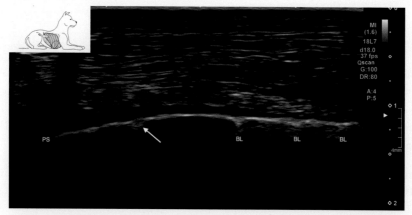

FIGURE 5.18.4 Generalized individual B-lines (BL) over all lung surfaces are visible. They extend up to 5–6 cm in depth, although only 2 cm is shown in the still image. B-lines appear as vertical hyperechoic ring down artifacts originating from the visceral surface of the pleural line and extending to the bottom of the field, sliding synchronously with the lung (not appreciated on a still image). Peripheral subpleural millimetric hypoechoic nodules are also visible throughout both lung fields bilaterally (arrow).

CLINICAL INTEGRATION

Suspected diagnosis based on ultrasound findings:

Based on the breed, history, and clinical findings, canine idiopathic pulmonary fibrosis (CIPF) was suspected. The absence of a heart murmur and normal cardiac POCUS ruled out myxomatous mitral valve disease (MMVD) with secondary left-sided congestive heart failure. However, at this stage, severe bronchopneumopathy with bronchomalacia could not be ruled out.

Sonographic interventions, monitoring, and outcome:

Thoracic CT scan confirmed the suspicion of CIPF. The dog was discharged with a long-term follow-up plan including cardiologist echocardiograms to investigate secondary pulmonary arterial hypertension.

Important concepts regarding the sonographic diagnosis of pulmonary fibrosis:

- CIPF predominantly affects older West Highland White Terrier dogs.[1]
  - Symptoms include progressive respiratory distress, exercise intolerance, cough, cyanosis, and crackles on lung auscultation.
  - The disease is fatal, characterized by progressive respiratory insufficiency.
  - Diagnosis is based on clinical findings, diagnostic imaging, and the exclusion of other cardiac or respiratory diseases.
  - If the patient is stable for sedation or anesthesia and if available, further investigation with CT scan allows the identification of parenchymal lesions of CIPF.
  - However, a definitive diagnosis can only be obtained with post-mortem histopathology.[1,2]
- PLUS is not specific for the diagnosis of CIPF but is useful to detect peripheral pulmonary lesions and subtle changes in the subpleural space, as in human IPF.[3] Sonographic findings can be identified with no preferential location[4]:
  - Bilateral, diffuse, multiple (one to five per field), separated B-lines indicate parenchymal disease.[4,5]
  - Thickened, irregular pleural surfaces are seen due to subpleural fibrotic tissue.[3,4] In humans, the number of B-lines and the thickness of the pleural line are correlated with the severity of fibrosis.[3,5]
  - Subpleural nodules may also be seen.[4,6]
- Depending on the patient's state of respiratory distress, oxygen supplementation and anxiolytics (such as butorphanol) can be used prior to any further examination.

Techniques and tips to identify the appropriate ultrasound images/views:

- Orienting the probe parallel to the ribs is helpful to avoid acoustic shadow produced by the ribs.
- Linear probes provide better resolution, but decent images can be obtained with a microconvex curvilinear probe if a linear probe is not available or when penetration is not optimal.
- In dogs with high body condition scores, the frequency may need to be decreased and the depth setting increased to allow the ultrasound beam to traverse fat and to optimize visualization of the pleural line.

Take-home messages

- Clinical context is primordial with CIPF as PLUS is not specific for lung fibrosis. CIPF is common in older West Highland White Terriers with progressive lower respiratory symptoms.
- Sonographic findings featuring pulmonary fibrosis are diffuse multiple B-lines, thick and irregular pleural line, and occasional subpleural nodules.

REFERENCES

1. Clercx, C., Fastrès, A. & Roels, E. Idiopathic pulmonary fibrosis in West Highland white terriers: An update. *Veterinary Journal* 242, 53–58 (2018).
2. Laurila, H. P. & Rajamäki, M. M. Update on canine idiopathic pulmonary fibrosis in West Highland white terriers. *Veterinary Clinics of North America: Small Animal Practice* 50(2), 431–446 (2020).
3. Manolescu, D., Davidescu, L., Traila, D., Oancea, C. & Tudorache, V. The reliability of lung ultrasound in assessment of idiopathic pulmonary fibrosis. *Clinical Interventions in Aging* 13, 437–449 (2018).
4. Soliveres, E. *et al*. Thoracic ultrasonography for assessment of pulmonary fibrosis in west highland white terriers with or without canine idiopathic pulmonary fibrosis. *Proceedings of the European Veterinary Diagnostic Imaging Online Congress* 2021, 42 (2021).
5. Manolescu, D. *et al*. Ultrasound mapping of lung changes in idiopathic pulmonary fibrosis. *The Clinical Respiratory Journal* 14(1), 54–63 (2020).
6. Soldati, G. & Demi, M. The use of lung ultrasound images for the differential diagnosis of pulmonary and cardiac interstitial pathology. *Journal of Ultrasound* 20(2), 91–96 (2017).

CHAPTER 5 – CASE 19

CHOCOLATE LAB WITH RESPIRATORY DISTRESS AND HIND LIMB LAMENESS FOLLOWING MOTOR VEHICULAR TRAUMA

Søren Boysen

HISTORY, TRIAGE, AND STABILIZATION

A 2-year-old female spayed Chocolate Labrador presented with respiratory distress and hind limb lameness following motor vehicular trauma.

Triage exam findings (vitals):
- Respiratory rate: 62 breaths per minute, increased effort, absent breath sound dorsally, possible crackles ventrally
- Heart rate: 148 beats per minute
- Mucous membranes: pink, <2 sec capillary refill time
- Femoral and dorsal pedal arterial pulses: strong and regular
- Temperature: 39.6°C (103.3°F)

POCUS exam(s) to perform:
- PLUS

Abnormal POCUS exam results:
See **Figure 5.19.1** and **Video 5.19.2**.

Additional point-of-care diagnostics and initial management:
- Supplemental oxygen and hydromorphone (0.1 mg/kg IV).

QUESTIONS AND ANSWERS

1. What are the differentials to rule in or out with POCUS based on history and physical exam?
2. What are the sonographic findings?
3. What is the sonographic diagnosis?
4. If necessary, what additional sonographic examination or findings would help rule in or out the differential diagnoses?

DOI: 10.1201/9781003436690-26

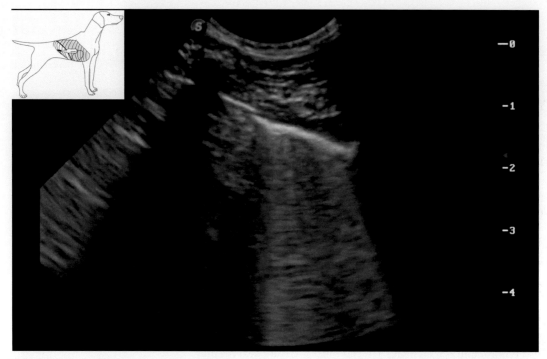

FIGURE 5.19.1 PLUS at the right middle hemithorax with the dog standing, probe orientation perpendicular to the ribs, and depth set at 4 cm. Fur was not clipped and alcohol was used as the coupling agent. Video 5.19.1: Video obtained at the same location as shown in Figure 5.19.1.

1. **Differential diagnosis to rule in or rule out with POCUS:**
   - Pulmonary contusions, hemothorax, diaphragmatic hernia, rib fractures, and pneumothorax are all likely causes of dyspnea in trauma patients.
2. **Describe your sonographic findings:**
   - Lung point (**Figure 5.19.2**, **Video 5.19.2**) with underlying B-lines where the lung recontacts the chest wall.
   - In the video, the lung point results in lung sliding only being visible in a portion of the ultrasound window, in this case the right half of the cineloop (**Video 5.19.2**).
3. **What is your sonographic diagnosis?**
   - Lung point with underlying B-lines.
4. **What additional sonographic examination or finding would help you rule in or rule out your differential diagnoses (if necessary)?**
   - Pneumothorax should be suspected when all the following sonographic findings are *absent* at a specific probe location: (1) lung sliding (aka the glide sign), (2) a lung pulse, (3) B-lines, (4) lung consolidations, and (5) pleural effusion.[1–3]
   - Pneumothorax can be sonographically confirmed when asynchronous and double curtain signs are identified.[1,3]
   - Once the dog is stable, and following thoracocentesis the lung should be assessed to document the extent and distribution of B-lines and to look for any lung consolidation (see other cases).[1,3]

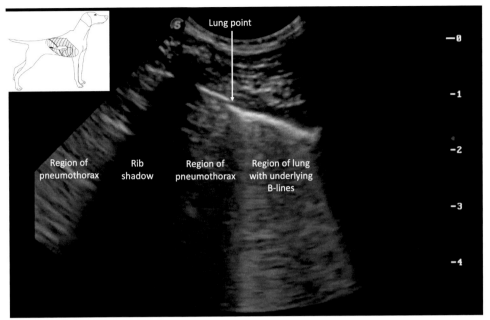

FIGURE 5.19.2 The left half of the image shows reverberation artifact and A-lines due to soft tissue–pleural free gas interface, and a rib shadow. The right half of the image shows reverberation artifact and vertical B-lines originating from the pleural line. In this example, the underlying lung has B-lines present due to concurrent pulmonary contusions. Lung sliding (which cannot be detected in a still image) would be visible to the right where lung contacts the chest wall and absent to the left where pneumothorax is present (Video 4.2.2). Note that where the lung separates away from the chest wall (left side of the image), B-lines are no longer visible, as air impedes ultrasound beams, preventing the underlying lung from being interpreted. Although this image depicts the lung point, without assessing the dynamic cineloop it is not possible to determine if the section to the left of the image is due to air-filled lung or pneumothorax. The presence or absence of lung sliding must be assessed with interpretation of the cineloop to conclude that this is the lung point.

CLINICAL INTEGRATION

Suspected diagnosis based on ultrasound findings:

Pneumothorax with underlying pulmonary contusions.

Sonographic interventions, monitoring, and outcome:

- The left hemithorax was assessed similarly, while emergency right-sided thoracocentesis was performed. Increased B-lines were noted on the left hemithorax.
- Cardiac and abdominal POCUS were performed following thoracocentesis, when the patient was more stable, and were found to be unremarkable (not shown).
- Serial PLUS was performed to monitor the degree/recurrence of pneumothorax and to monitor the progression/resolution of increased B-lines over time.
- The dog was treated conservatively, monitored closely, and eventually discharged.

Important concepts regarding the sonographic diagnosis of pneumothorax and the lung point:

- The lung point is defined as the site within the thorax where the visceral pleura of the lung recontacts the parietal pleura of the thoracic wall in patients with pneumothorax.[1-3]
- The lung point becomes visible when the probe is situated such that it is partially over a region of pneumothorax and partially over a region of lung in contact with the chest wall.[1,3]
- If absence of lung sliding is identified, a search for the lung point can be undertaken, but only provided the patient is sufficiently stable to tolerate such evaluation.
- Interpretation of the still image in this case may represent one of two scenarios (**Figure 5.19.2**): (1) a mixed pattern of aerated lung (left side of the image) and pulmonary contusions (right side of the image) within the same lung lobe or at the transition between two lung lobes or (2) a pneumothorax (left side of the image) and pulmonary contusions (right side of the image), which is in fact the situation encountered in this case. It is not possible to differentiate these two situations on a still image alone. The cineloop must be assessed to determine if lung sliding is present across the entire ultrasound window (no pneumothorax at that probe location, Scenario 1 above) or lung sliding is only visible in a portion of the ultrasound window (pneumothorax at that probe location, Scenario 2 above). In other words, the lung point is dynamic and occurs when the probe is situated directly over the regions where the lung recontacts the chest wall in cases with pneumothorax.
- If underlying pathology is present (e.g., pulmonary contusions), B-lines or lung consolidation will be identified with the lung point (**Video 5.19.2**).[1,3]

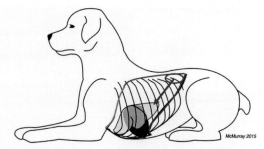

FIGURE 5.19.3 If an absence of lung sliding is noted, the probe is moved toward the hilus of the lung looking for the return of lung sliding. The area where a return of lung sliding is identified is the lung point. In this image, the probe is moved from the most caudo-dorsal site (red circle; most sensitive area for free pleural air to accumulate in the sternal patient) toward the elbow (lung hilus).

Techniques and tips to identify the appropriate ultrasound images/views:

- The lung point is identified when the probe is moved from an area of absent lung sliding to an area where lung sliding returns (**Figure 5.19.3**).[1-3]
- With pleural space pathology, the lung tends to compress around the hilus of the lung, which is the direction the probe should be moved to detect the lung point.
- Given that the caudo-dorsal location is the most sensitive site for free pleural gas to accumulate in sternal or standing patients, and sternal or standing is the preferred position to assess patients with respiratory distress, moving the probe from the caudo-dorsal site of the PLUS protocol toward the elbow will usually detect the lung point (**Figure 5.19.3**).

Take-home message

- The lung point is considered definitive for the diagnosis of pneumothorax, but thoracocentesis takes precedence in unstable patients with supporting evidence of pneumothorax on history and clinical assessment.

REFERENCES

1. Boysen SR. Lung Ultrasonography for Pneumothorax in Dogs and Cats. *Vet Clin North Am Small Anim Pract* 2021;51(6):1153–1167.
2. Hwang TS, Yoon YM, Jung DI, Yeon SC, Lee HC. Usefulness of Transthoracic Lung Ultrasound for the Diagnosis of Mild Pneumothorax. *J Vet Sci* 2018;19(5):660–666.
3. Boysen S, Chalhoub S, Gommeren K. Clinical Applications of PLUS: Is There Pneumothorax, Yes/No? In: Boysen S, Chalhoub C, Gommeren K, eds. *The Essentials of Veterinary Point-of-Care Ultrasound: Pleural Space and Lung*. Edra publishing, Laval, Quebec; 2022, pp. 63–90.

CHAPTER 5 – CASE 20

BLOODHOUND WITH SEVERE MYOSITIS SECONDARY TO A SNAKE BITE THAT OCCURRED 4 DAYS PREVIOUSLY

Chiara Di Franco and Angela Briganti

HISTORY, TRIAGE, AND STABILIZATION

An 8-year-old female spayed Bloodhound presented with severe myositis secondary to a snake bite that occurred 4 days previously.

Triage exam findings (vitals):
- Mentation: depressed, unable to stand, and laying in left lateral recumbency
- Respiratory rate: 16 breaths per minute
- Heart rate: 95 beats per minute
- Mucous membranes and: pale, 2 sec capillary refill time
- Femoral and dorsal pedal arterial pulses: normal
- Temperature: 38.2°C (100.8°F)
- Sedated with dexmedetomidine infusion 1 mcg/kg/h

POCUS exam(s) to perform:
- PLUS

Abnormal POCUS exam results:
See **Figure 5.20.1** and **Video 5.20.1**.

Additional point-of-care diagnostics and initial management:
Arterial blood gas analysis on room air: PaO_2 78 mmHg, $PaCO_2$ 38 mmHg, lactate 4.5 mmol/L, pH 7.29, Hct 28%.

QUESTIONS AND ANSWERS

1. What are the differentials to rule in or out with POCUS based on history and physical exam?
2. What are the sonographic findings?
3. What is the sonographic diagnosis?
4. If necessary, what additional sonographic examination or findings would help rule in or out the differential diagnoses?

DOI: 10.1201/9781003436690-27

Severe Myositis Secondary to a Snake Bite that Occurred 4 Days Previously

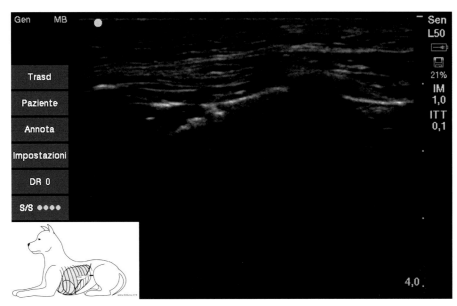

FIGURE 5.20.1 PLUS exam over the left mid-hemithorax near the curtain sign after the patient was gently rolled into sternal recumbency, using a linear probe placed perpendicular to the ribs with the depth set at 4 cm. The marker is directed cranially (blue dot). Fur was not clipped, and alcohol was used as the coupling agent. Video 5.20.1: Video obtained at the starting point of Figure 5.20.1, near the curtains sign, sliding from the caudal PLUS border to the cranial PLUS border.

1. **Differential diagnosis to rule in or rule out with POCUS:**
   - Atelectasis, aspiration pneumonia, hemorrhage, and acute respiratory distress syndrome (ARDS) are all possible in this case.
2. **Describe your sonographic findings:**
   - Increased to coalescing B-lines and lung consolidation, specifically shred signs, with a thickened, irregular, and non-linear pleural line were visible over multiple intercostal spaces of the left hemithorax (**Figures 5.20.1** and **5.20.2**, **Videos 5.20.1** and **5.20.2**).
   - The right hemithorax was assessed and had an occasional sparse B-line but was otherwise unremarkable.
3. **What is your sonographic diagnosis?**
   - Lung consolidation, B-lines, and irregular pleural thickening (**Figure 5.20.2** and **Video 5.20.2**).
4. **What additional sonographic examination or finding would help you rule in or rule out your differential diagnoses?**
   - Classifying air bronchograms within consolidated lung regions as static or dynamic helps differentiate atelectasis from other causes of lung consolidation in human patients.[1]
     - When bronchioles remain patent and continuous with the larger airways, airway movement associated with the respiratory cycle can create visible movement of air within the smaller airways of consolidated lung regions.
     - This movement of air in the smaller airways may create a visible "shimmer" referred to as a dynamic air bronchogram.[1]
     - Dynamic air bronchograms are often seen with pneumonia and

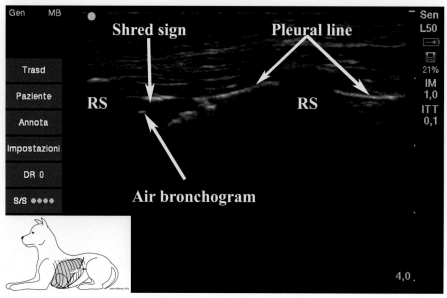

FIGURE 5.20.2 On the left side of the image a shred sign is visible to the right side of the rib shadow (RS). An air bronchogram is visible within the shred sign, appearing as a bright white hyperechoic slightly linear-shaped structure. Video 5.20.2 (Video 5.20.1 labelled).

- inflammatory causes of lung consolidation.
- In contrast, static air bronchograms are often seen with atelectasis but can also be seen with inflammatory causes of lung consolidation, including ARDS and aspiration pneumonia.[1]
- Classification of air bronchograms in veterinary medicine is largely lacking.
- Echocardiography may help rule out cardiogenic causes of increased B-lines, although a unilateral distribution, thickened pleura, and lung consolidation are unlikely with congestive heart failure.

CLINICAL INTEGRATION

Suspected diagnosis based on ultrasound findings:

Atelectasis. The ultrasonographic findings were only present on the left hemithorax and in both cranial and caudal lung lobes, and both the dorsal and ventral lung regions. A history of prolonged left lateral recumbency also supports a diagnosis of atelectasis.

Sonographic interventions, monitoring, and outcome:

- The dog was placed in sternal recumbency but tended to shift into left lateral recumbency.
- The dog was supported in sternal recumbency using towels, and Helmet-continuous positive airway pressure (H-CPAP) was used to recruit the lung with a positive end-expiratory pressure (PEEP) of 13 cm H_2O (**Figures 5.20.2** and **5.20.3**).[2,3]
- The fact that the lung of the right hemithorax was normal makes ARDS and hemorrhage very unlikely.
- Given that the abnormal lung findings were diffuse over the entire left hemithorax made atelectasis more likely than aspiration pneumonia.

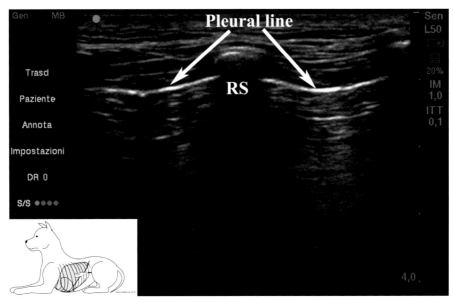

FIGURE 5.20.3 Still image obtained from the same thoracic location as Figure 5.20.1 following a recruitment maneuver. Video 5.20.3: Similar thoracic location and site to Video 5.20.1, following a recruitment maneuver.

- A recruitment maneuver with H-CPAP was performed and the lung reassessed 15 min later.
- PLUS revealed normal lung with minimal pleural irregularities (**Figures 5.20.1 and 5.20.3, Videos 5.20.1 and 5.20.3**), further confirming the diagnosis of atelectasis.

Important concepts regarding the sonographic diagnosis of atelectasis and the recruitment maneuver:

- Ultrasound is rapid and accurate for the assessment of atelectasis.[3,4]
- Sonographic lung consolidation has many etiologies. Rapidly differentiating atelectasis from ARDS or hemorrhage is important to direct therapy and re-establish lung functionality.
- Atelectasis is a common side effect of sedation, anesthesia, and lateral recumbency in hospitalized animals. Contributing factors include muscle relaxation, intra-abdominal pressure increase (e.g., due to abdominal effusion), cranial shift of the diaphragm due to pressure from the abdominal organs, increased FiO_2, and gravitational effects on the thoracic viscera. The amount of atelectasis is case dependent.
- Atelectasis generally resolves spontaneously after placing animals in sternal recumbency, stressing the importance of good patient care, which includes rotating the patient and placing them in sternal recumbency as needed.
- In significantly compromised patients atelectasis can result in severe impairment of lung function.

Techniques and tips to identify the appropriate ultrasound images/views:

- Place the probe, oriented perpendicular to the ribs, in the caudo-dorsal region of the thorax at the abdominal curtain sign and slide the probe cranially using an S-shaped pattern.[1,5]

- Assess both hemithoraces and, depending on the recumbency of the animal, atelectasis should be greater in the gravity-dependent regions.
- Atelectasis appears as hypoechoic regions of lung, originating at the lung surface, with B-lines and static air bronchograms.[1,4]
- Atelectasis resolves following recruitment maneuvers.

Take-home message
- Immediate recognition of lung and pleural abnormalities in hospitalized animals that experience compromised lung function allows rapid confirmation of lung impairment and is helpful in guiding pulmonary recruitment in cases of atelectasis.

REFERENCES

1. Boysen S, Gommeren K, Chalhoub S. Clinical applications of PLUS: Are there increased B-Lines, yes/no? In: Boysen, Gommeren, Chalhoub eds. *The Essentials of Veterinary Point-of-Care Ultrasound: Pleural Space and Lung.* Eds Edra publishing, Laval, Quebec; 2022, 112–137.
2. Lista G, Castoldi F, Cavigioli F, Bianchi S, Fontana P. Alveolar recruitment in the delivery room. *J Matern Fetal Neonatal Med* 2012 Apr;25;Suppl 1:39–40. doi: 10.3109/14767058.2012.663164. Epub 2012 Mar 5. PMID: 22313342.
3. MacIntyre NR. Physiologic effects of noninvasive ventilation. *Respir Care* 2019 Jun;64(6):617–628. doi: 10.4187/respcare.06635. PMID: 31110031.
4. Acosta CM, Maidana GA, Jacovitti D, et al. Accuracy of transthoracic lung ultrasound for diagnosing anesthesia-induced atelectasis in children. *Anesthesiology* 2014;120(6):1370–1379. doi: 10.1097/ALN.0000000000000231.
5. Armenise A, Boysen RS, Rudloff E, Neri L, Spattini G, Storti E. Veterinary-focused assessment with sonography for trauma-airway, breathing, circulation, disability and exposure: A prospective observational study in 64 canine trauma patients. *J Small Anim Pract* 2019 Mar;60(3):173–182. doi: 10.1111/jsap.12968. Epub 2018 Dec 13. PMID: 30549049.

CHAPTER 5 – CASE 21

STABLE FEMALE SPAYED DOMESTIC SHORTHAIR FOLLOWING MOTOR VEHICLE TRAUMA

Søren Boysen

HISTORY, TRIAGE, AND STABILIZATION

A 3-year-old female spayed domestic shorthair presented following motor vehicle trauma.

Triage exam findings (vitals):
- Mentation: quiet, alert, and responsive
- Gait: ambulatory
- Respiratory rate: 28 breaths per minute
- Heart rate: 180 beats per minute, regular rhythm, no murmur
- Mucous membranes: pale pink, moist, 1 sec capillary refill time
- Femoral and dorsal pedal arterial pulses: synchronous and strong
- Temperature: 38°C (100.4°F)

POCUS exam(s) to perform:
- PLUS
- Cardiovascular POCUS
- Abdominal POCUS

Abnormal POCUS exam results:
See **Figure 5.21.1** and **Video 5.21.1**.

Additional point-of-care diagnostics and initial management:
- Packed cell volume (PCV): 45%, total solids 6.2 g/dL, blood lactate 1.2 mmol/L, blood glucose 6.3 mmol/L.
- Methadone 0.2 mg/kg IV.
- Cardiac POCUS: unremarkable.
- Abdominal POCUS: unremarkable.

QUESTIONS AND ANSWERS

1. What are the differentials to rule in or out with POCUS based on history and physical exam?
2. What are the sonographic findings?
3. What is the sonographic diagnosis?
4. If necessary, what additional sonographic examination or findings would help rule in or out the differential diagnoses?

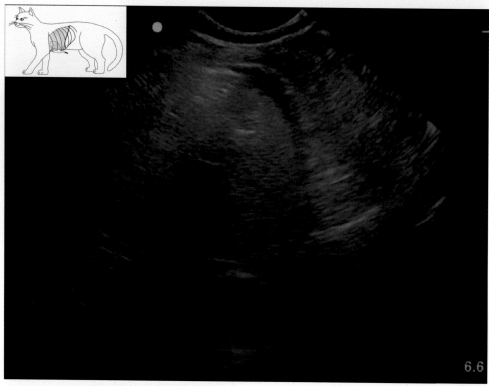

FIGURE 5.21.1 PLUS at the ventral lung to sternal muscle interface with the cat standing, using a microconvex probe oriented parallel to the ribs, depth set at 6.6 cm. The fur was not clipped, and alcohol was used as the coupling agent. Video 5.21.1: Video obtained at the same window as Figure 5.21.1.

1. **Differential diagnosis to rule in or rule out with POCUS:**
   The cat was stable on physical examination and a systemic POCUS examination, as an extension of the physical exam, including PLUS, abdominal, and cardiovascular POCUS was performed. The primary concern was occult hemorrhage or abdominal trauma such as uroabdomen or gallbladder rupture that may not be immediately evident on physical examination findings alone.
2. **Describe your sonographic findings:**
   - Scant pleural effusion which was only noted when the probe was placed over the most ventral regions of the thorax and oriented parallel to the ribs (**Figure 5.21.2**, **Video 5.21.2**).
3. **What is your sonographic diagnosis?**
   Scant pleural effusion.
4. **What additional sonographic examination or finding would help you rule in or rule out your differential diagnoses?**
   If the origin of the effusion is uncertain, fluid aspiration and analysis is recommended. In this case, with the history of trauma, hemorrhage was suspected and serial POCUS was used to monitor the progression/resolution of the effusion.

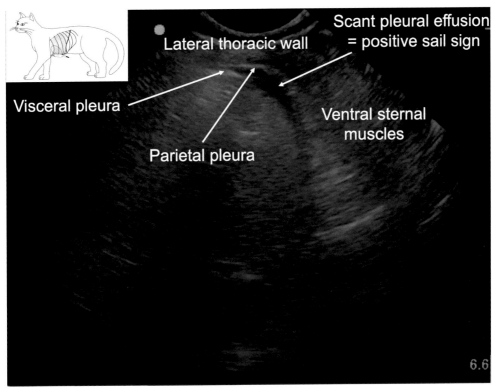

FIGURE 5.21.2 Pleural effusion can be seen at the most gravity-dependent site between the ventral lung surface and the sternal muscles, which is best appreciated when the probe is oriented parallel to the ribs and the probe slid ventrally from the lung until the sternal muscles fill one-third to one-half of the ultrasound image. Pleural effusion is noted separating the visceral and parietal surfaces. Video 5.21.2 (Video 5.21.1 labelled): The heart is visible in the video.

CLINICAL INTEGRATION

Suspected diagnosis based on ultrasound findings:

Pleural hemorrhage, likely trauma induced.

Sonographic interventions, monitoring, and outcome:

- The right hemithorax was assessed similarly and an equivalent scant volume of pleural effusion was identified.
- The remainder of the PLUS exam was unremarkable.
- The caudal vena cava and heart were closely assessed for volume status and found to be unremarkable.

Important concepts regarding pleural effusion/hemothorax:

- Pleural effusion accumulates at the most ventral sites between the lungs and the sternal muscles in standing/sternal patients, and at the widest gravity-dependent areas in patients in lateral recumbency.
- The accuracy of ultrasound to detect pleural effusion is poor to moderate compared to computed tomography, likely because the most gravity-dependent sites are not assessed with many thoracic-focused assessment with sonography for trauma (FAST) and lung ultrasound protocols, and the same sites

are often assessed regardless of patient positioning.[1-4]
- Patients in respiratory distress should be scanned in the position they are most comfortable, which will often be in a sternal or standing position.[5]
- When patients are standing or sternal, large volumes of pleural effusion, resulting in clinical signs of respiratory distress, are easily identified with the probe situated perpendicular or parallel to the ribs in the ventral third of the thorax.[4]
- Assessing the subxiphoid window, pericardio-diaphragmatic window, and costo-phrenic recess in these patients will help differentiate pericardial from pleural effusion.[5]
- When scant volumes of pleural effusion are present and the patient is sternal/standing, turning the probe parallel to the ribs and scanning the most gravity-dependent regions of the pleural space between the lungs and sternal muscles ventrally increase the sensitivity of detecting pleural effusion.[4]
- When scant fluid is present in a standing/sternal patient with the probe parallel to the ribs at the most ventral regions of the pleural space, it tends to have a crescent-shaped appearance, termed the "sail" sign, which can be seen between the lung and sternal muscles and/or between the apex of the heart and sternal muscles depending on where the probe is situated (**Video 5.21.3**)
- Small volume pleural effusions are not likely to cause clinical signs of respiratory distress but may be of diagnostic importance.
- The nature of the pleural fluid can affect its sonographic appearance; however, fluid aspiration and analysis is required to identify the type of effusion that is present.
- Lung sliding will be absent secondary to pleural effusion due to the separation of the visceral and parietal pleural surfaces.[5]
- As pleural space–occupying pathology can result in atelectasis, the lung parenchyma should be interpreted cautiously when significant pleural effusion is present as compressive atelectasis can result in the formation of B-lines and lung consolidations.[5]

Techniques and tips to identify pleural effusion/hemothorax:
- Positioning the probe parallel to the ribs in the ventral regions, such that the sternal muscles fill the ventral one-third to one-half of the ultrasound image, and scanning its entire ventral pleural border, from cranial to caudal, will maximize the chances of detecting small volume pleural effusion.
- Consistent patient positioning and probe location/orientation on the thorax (perpendicular or parallel to the ribs) are recommended for serial monitoring.
- Ensure depth settings are adequate so that the full extent of the pleural effusion is evaluated.

Take-home messages
- Consider patient position and the effect of gravity on where fluid will accumulate when searching for pleural effusion and adapt the ultrasound protocol accordingly.
- To rule out small volume pleural effusion it may be necessary to turn the probe parallel to the ribs and scan the pleural space between the ventral lungs and sternal muscles when the patient is sternal or standing and assess the widest gravity-dependent region when the patient is in lateral recumbency.

REFERENCES

1. Boysen S, McMurray J, Gommeren K. Abnormal curtain signs identified with a novel lung ultrasound protocol in six dogs with pneumothorax. *Front Vet Sci*, 2019, 6:291.
2. Cole L, Pivetta M, Humm K. Diagnostic accuracy of a lung ultrasound protocol (Vet BLUE) for detection of pleural fluid, pneumothorax and lung pathology in dogs and cats. *J Small Anim Pract*, 2021, 62(3):178–186.
3. Walters AM, O'Brien MA, Selmic LE, et al. Evaluation of the agreement between focused assessment with sonography for trauma (AFAST/TFAST) and computed tomography in dogs and cats with recent trauma. *J Vet Emerg Crit Care*, 2018, 28(5):429–435.
4. Boysen S, Chalhoub S, Romero A. Veterinary point-of-care ultrasound transducer orientation for detection of pleural effusion in dog cadavers by novice sonographers: A pilot study. *Ultrasound J*, 2020, 12(Suppl 1):45.
5. Boysen S, Chalhoub S, Gommeren K. Clinical applications of PLUS: Is there pleural effusion, yes/no? In: Boysen S, Chalhoub C, Gommeren K, eds. *The Essentials of Veterinary Point-of-Care Ultrasound: Pleural Space and Lung*. Edra publishing, Laval, Quebec; 2022, 36–62.

SECTION 3

VETERINARY CARDIAC POINT-OF-CARE ULTRASOUND

Section Editor Tereza Stastny

CHAPTER 6

INTRODUCTION TO VETERINARY CARDIAC POINT-OF-CARE ULTRASOUND

Tereza Stastny

Veterinary cardiac point-of-care ultrasound (Cardiac POCUS), while less sensitive and specific than comprehensive echocardiography, provides valuable noninvasive information about cardiac structure and function. It serves as a useful adjunct to physical examination findings and thoracic radiographs. C-POCUS is particularly helpful for assessing dogs and cats in respiratory distress, distinguishing between cardiac and non-cardiac causes of dyspnea. It also aids in evaluating conditions such as pericardial effusion, pleural effusion, arrhythmias, and hemodynamic instability. This chapter will cover image acquisition, transducer selection, and the various C-POCUS windows and views.

IMAGE ACQUISITION

Patients are typically scanned in sternal recumbency or standing for Cardiac POCUS. If the patient is stable enough and tolerates lateral recumbency, then the right lateral is optimal, as the heart lies closer to the thoracic wall due to gravity, which improves image quality and allows for standard cardiac views to be obtained.[1] To effectively scan the heart from this position, the patient should be placed near the edge of the table or on a cutout echocardiography table.

PATIENT PREPARATION AND TRANSDUCER SELECTION

Shaving patient fur is typically not required to obtain quality Cardiac POCUS images. Fur is parted, and an acoustic coupling agent is applied over the scanning site (see Chapter 3). Either a micro-convex (curvilinear) transducer or a phased-array transducer can be used for Cardiac POCUS. Micro-convex (curvilinear) transducer is the most readily available transducer in the emergency setting, but a phased-array transducer may be preferred when imaging the heart, due to better temporal resolution (the ability of ultrasound to distinguish between instantaneous events of rapidly moving structures, such as during the cardiac cycle) and a smaller footprint (the region the sound wave leaves and returns to the transducer).[2,3] Many portable ultrasound units have a preset cardiac mode which inverts the image left-to-right to achieve standard echocardiographic orientation used by cardiologists (**Figure 6.1**).[1] Using a micro-convex (curvilinear) transducer with standard preset thoracic mode, or a phased-array transducer with preset cardiac mode, are both viable options when performing Cardiac POCUS. It is important to recognize that the inversion associated with the cardiac mode will cause anatomic structures to appear on the opposite side of the screen (achieving standard echocardiographic orientation [**Figure 6.1**]).[1]

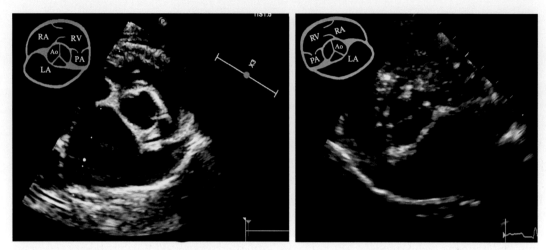

FIGURE 6.1 Right parasternal short axis view of the heart depicting the aorta (Ao), left atrium (LA), right atrium (RA), right ventricle (RV), and pulmonary artery (PA) as illustrated in the upper left diagram of each image. The image on the left depicts standard echocardiographic orientation. Note that the LA appears on the bottom left of the screen in the standard echocardiographic orientation and is inverted in the thoracic mode appearing on the bottom right of the screen.

ACOUSTIC WINDOWS AND VIEWS:

There are **two basic acoustic windows**; the right parasternal and subxiphoid windows, which allow standard views of the heart to be obtained for Cardiac POCUS.

A complete echocardiographic study encompasses imaging both the right and left sides of the thorax and involves more than 15 different views, using 2D echocardiography, M-mode echocardiography, color-wave Doppler, pulsed-wave Doppler, tissue Doppler, among other modes. While all views are useful to evaluate cardiac anatomy and function, **three views** (two short-axis views and one long-axis [**Figure 6.6**]) are most often used for the purposes of Cardiac POCUS on an emergency basis:[1,5]

1. **Right parasternal short-axis left ventricular ("mushroom") view (Figure 6.2):** Located on the right hemithorax, the transducer is positioned parasternal in the region of the 3rd-5th intercostal spaces. The transducer head is rotated clockwise to a 30 to 45-degree angle to horizontal, with the marker directed cranioventrally. This ensures the ultrasound beam is in short axis to the left ventricle. The "mushroom view" is then identified by sweeping the transducer to the level of the left midventricular-papillary muscle region just below the mitral valve. This view is commonly used to estimate volume status, contractility, hypertrophy, and pseudohypertrophy, and to assess for pericardial effusion.

2. **Right parasternal short-axis left atrium-to-aortic ratio (LA:Ao) view (Figure 6.3):** Located on the right hemithorax, the transducer is positioned parasternal at the location of the cardiac apex beat (where the heartbeat is palpated), in the region of the 3th-5th intercostal space. The "mushroom view" is first identified (see above) and then the transducer is swept dorsally in small increments until the mitral valve becomes more apparent ("fish-mouth" view). The transducer is then fanned dorsally toward the base of the heart by angulating the transducer tail

Introduction to Veterinary Cardiac Point-of-Care Ultrasound

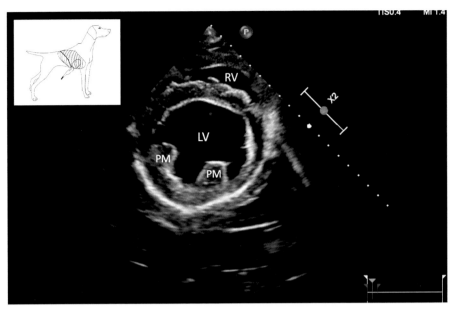

FIGURE 6.2 Right parasternal short axis left ventricular ("mushroom") view of the heart depicting the right ventricle (RV), left ventricle (LV), and papillary muscles (PM), using a phased-array transducer in cardiac mode. The transducer head is rotated clockwise to a 30 to 45-degree angle to horizontal, with the marker directed cranioventrally ensuring the ultrasound beam is in short axis to the left ventricle.

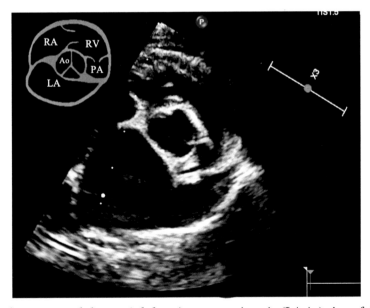

FIGURE 6.3 Right parasternal short-axis left-atrium-to-aortic ratio (LA:Ao) view of the heart depicting the aorta (Ao) and the left atrium (LA), using a phased-array transducer in cardiac mode. The transducer marker is facing cranioventrally ensuring the ultrasound beam is in short axis to the base of the heart (see schematic in Figure 2). This patient has significant heart disease and an enlarged LA:Ao ratio (>2:1).

downward until the left-atrium-to-aortic ratio is visualized. The correct LA:Ao view is obtained when the semilunar valves of the aorta form a "Mercedes" sign during diastole, and the left atrium creates the shape of a "whale" (**Figure 6.3**). LA:Ao reference intervals vary depending on the angle of insonation (the angle at which the ultrasound beam traverses the target structure), whether or not the pulmonary artery is included in the measurement, and operator's skill level. In both species, an LA:Ao ratio >2:1 is abnormal and indicates left atrial dilation.[1] Left atrial dilation suggests increased left atrial volume which is often associated with underlying cardiac disease, with or without left-sided congestive heart failure, or iatrogenic fluid overload.

3. **Right parasternal long-axis (4-chamber) view (Figure 6.4):** The transducer is first positioned at the right parasternal short-axis "mushroom" view. Then the head of the transducer is rotated 90 degrees to lie almost parallel with the ribs, with the marker directed craniodorsally. This ensures the ultrasound beam is in long axis to the left ventricle.[1,5,6] The transducer is fanned cranially and caudally until the ventricular lumen diameter is maximized, ensuring the ultrasound beam does not inadvertently oblique the ventricles. This view allows subjective assessment of the size of all cardiac chambers; left ventricle (LV), left atrium (LA), right ventricle (RV), and right atrium (RA). The RV internal chamber diameter should be approximately one-third that of the LV internal diameter.[5,6] The LV free wall and interventricular septum should be a similar thickness, while the RV free wall is approximately one-third to one-half the thickness of the LV free wall in health.

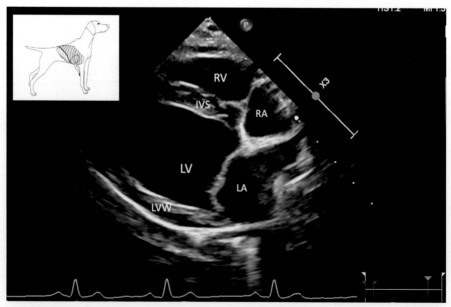

FIGURE 6.4 Right parasternal long-axis (4-chamber) view of the heart depicting the right ventricle (RV), left ventricle (LV), intraventricular septum (IVS), left ventricular wall (LVW), right atrium (RA), and left atrium (LA) using a phased-array transducer in cardiac mode. Then the head of the transducer is rotated 90 degrees to lie almost parallel with the ribs, with the marker directed craniodorsally.

The RA and the LA should have an approximate 1:1 ratio.[5,6] This view can be used to estimate volume status, contractility, changes in chamber size, and to assess for pericardial effusion.

The **subxiphoid window** is the second acoustic window that is most often used during Cardiac POCUS. This window is used to assess two views:

1. **Subxiphoid cardiac long-axis view (Figure 6.6):** To obtain this view, the transducer is placed at a 45-degree angle caudal to the xiphoid process in a short or long-axis plane. The liver should be visible in the near field, and the apex and LV free wall in the far field, separated by the hyperechoic line of the diaphragm. If the heart is not initially visualized, slide and then rock the transducer craniodorsally, tucking the head of the transducer under the xiphoid process. This view can be used to assess for pericardial effusion, pleural effusion, chamber size, and the caudal lung regions.

2. **Subxiphoid caudal vena cava view:** The caudal vena cava can also be evaluated at this site (please refer to Chapter 6 for image acquisition and technique).

Pearls and pitfalls:
- Clinical context correlated with triage exam findings and Cardiac POCUS is important when assessing a patient in respiratory distress.
- In general, a normal canine LA:Ao ratio is less than 1.5:1[1], and a normal feline LA:Ao ratio is less than 1.6, although there is an

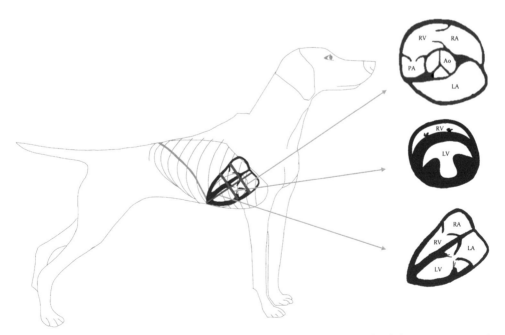

FIGURE 6.5 Schematic heart chamber diagram in right parasternal standard thoracic view mode. The bottom right cardiac diagram depicts the right parasternal long axis 4-chamber view. The middle image on the right depicts the right parasternal short axis left ventricular ("mushroom") view with the left ventricle on the bottom and right ventricle on top. Sliding the transducer dorsally and then fanning the transducer upwards identifies the right parasternal short-axis LA:Ao view, depicted in the top image on the right.

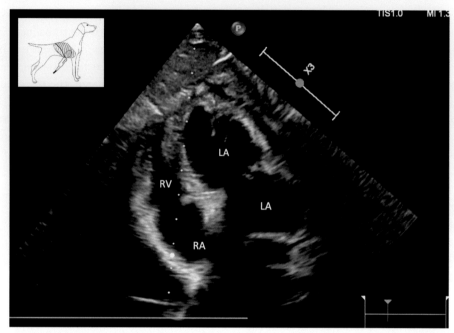

FIGURE 6.6 Subxiphoid cardiac long-axis view of the heart depicting the right ventricle (RV), left ventricle (LV), right atrium (RA), and left atrium (LA) using a phased-array transducer in cardiac mode with the indicator marker facing caudally.

equivocal area with values ranging up to 1.7 in cats.[6–8]
- Although objective LA:Ao measurements can be made, evidence suggests that subjective assessment of left atrial size is adequate for most Cardiac POCUS scans.[3,4,7,8]
- The LA:Ao view is obtained at the level of the heart base and must be perpendicular to the aortic semilunar valves to obtain an accurate LA:Ao ratio. If the right parasternal image is off-angle, errors in image interpretation can occur.[1]
- In cats, cardiac assessment from the subxiphoid site may be impeded by lung parenchyma that may occupy this space.[7]

REFERENCES

1. DeFrancesco TC, Ward JL. Focused Canine Cardiac Ultrasound. Vet Clin North Am Small Anim Pract. 2021;51(6):1203-1216.
2. Neelis D, Mattoon J, Nyland T. Thorax. In: Mattoon J, Nyland T, eds. *Small animal diagnostic ultrasound*. 3rd ed. St Louis: Elsevier, 2015;188–216.
3. Lisciandro GR. The Vet BLUE lung scan. In: Lisciandro GR, ed. *Focused ultrasound techniques for the small animal practitioner*. Ames, Iowa: Wiley-Blackwell, 2014;166–187.
4. Loughran KA, Rush JE, Rozanski EA, et al. The use of focused cardiac ultrasound to screen for occult heart disease in asymptomatic cats. J Vet Intern Med. 2019;33:1892–1901.
5. Ware, W. Chapter 4: Echocardiography. In: Cardiovascular Disease in Small Animal Medicine. 2nd ed. London: Manson publishing; 2018. p.71-130. Specific page(s):97-101.
6. Thomas, W. P., Gaber, C. E., Jacobs, G. J., Kaplan, P. M., Lombard, C. W., Moise, N. S., and Moses, B. L. Recommendations for

Standards in Transthoracic Two-Dimensional Echocardiography in the Dog and Cat. Echocardiography Committee of the Specialty of Cardiology, American College of Veterinary Internal Medicine. Journal of Veterinary Internal Medicine 1993;7:247-52.

7. Adin DB, Diley-Poston L. Papillary muscle measurements in cats with normal echocardiograms and cats with concentric left ventricular hypertrophy. J Vet Intern Med 2007;21(4):737–41.

8. Darnis E, Merveille AC, Desquilbet L, et al. Interobserver Agreement Between Non-Cardiologist Veterinarians and a Cardiologist After a 6-hour Training Course for Echographic Evaluation of Basic Echocardiographic Parameters and Caudal Vena Cava Diameter in 15 Healthy Beagles. J Vet Emerg Crit Care. 2019;29(5):495-504.

CHAPTER 7

CARDIAC POINT-OF-CARE ULTRASOUND CASES

CHAPTER 7 – CASE 1

CHIHUAHUA WITH A 3-DAY HISTORY OF HEART MURMUR, TACHYPNEA, AND A 24-HOUR HISTORY OF SEVERE RESPIRATORY DISTRESS

Tereza Stastny

HISTORY, TRIAGE, AND STABILIZATION

A 12-year-old male neutered Chihuahua presented with a 3-day history of tachypnea and a 24-h history of severe respiratory distress. The patient was known to have a historical 3/6 left apical systolic heart murmur, last noted 1 year ago on a routine wellness exam.

Triage exam findings (vitals):
- Mentation: dyspneic, alert, responsive
- Respiratory rate: 66 breaths per minute, significant inspiratory and expiratory dyspnea, increased bronchovesicular sounds bilaterally, with pulmonary crackles auscultated in the caudo-dorsal lung fields
- Heart rate: 196 beats per minute, 4/6 left apical systolic murmur
- Mucous membranes: pale pink, 2.5 sec capillary refill time
- Femoral and dorsal pedal arterial pulses: weak and regular
- Temperature: 36.8°C (98.3°F)
- Doppler blood pressure: 68 mmHg systolic

POCUS exam(s) to perform:
- PLUS
- Cardiac POCUS

Additional point-of-care diagnostics and initial management:
- Supplemental oxygen, butorphanol (0.2 mg/kg IV), and furosemide (2 mg/kg IV).

Abnormal POCUS exam results:
See **Figures 7.1.1** and **7.1.2**.

QUESTIONS AND ANSWERS

1. What are the differentials to rule in or out with POCUS based on history and physical exam?
2. What are the sonographic findings?
3. What is the sonographic diagnosis?
4. If necessary, what additional sonographic examination or findings would help rule in or out the differential diagnoses?

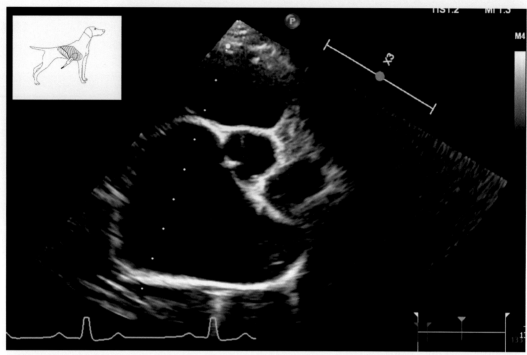

FIGURE 7.1.1 Right parasternal short-axis left atrial to aortic ratio (LA:Ao) view of the heart, using a phased-array probe in cardiac mode, 8 MHz, and 6 cm depth. The transducer marker is facing cranioventrally ensuring that the ultrasound beam is in short axis to the base of the heart. The fur is parted and 70% isopropyl alcohol is used as the coupling agent.

1. **Differential diagnosis to rule in or out with POCUS:**
   Left-sided congestive heart failure (most likely secondary to myxomatous mitral valve disease [MMVD]), aspiration pneumonia, pulmonary hypertension, pulmonary thromboembolism, less likely neoplasia, non-cardiogenic pulmonary edema, other.
2. **Describe your sonographic findings:**
   - Enlarged left atrial to aortic ratio (LA:Ao) (**Figure 7.1.3**).
   - Coalescing perihilar B-lines (**Figure 7.1.4**).
3. **What is your sonographic diagnosis?**
   Enlarged LA:Ao with coalescing perihilar B-lines secondary to left-sided congestive heart (L-CHF).
4. **What additional sonographic examination or finding would help you rule in or rule out your differential diagnoses (if necessary)?**
   - Formal echocardiogram.

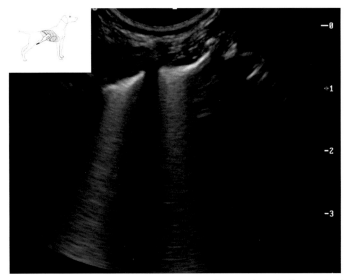

FIGURE 7.1.2 Left perihilar short-axis view of the lungs obtained with the probe located two-thirds of the way up the thorax using a microconvex curvilinear probe with the depth set at 4 in and the frequency set at 12 Hz. The fur is parted and 70% isopropyl alcohol is used as the coupling agent.

CLINICAL INTEGRATION

Suspected diagnosis based on ultrasound findings:

Based on the history, presenting signs, and clinical findings, MMVD with secondary L-CHF is suspected.

Sonographic interventions, monitoring, and outcome:

Given the unstable nature of the patient, supplemental oxygen was provided, and PLUS and cardiac POCUS scans were applied at the time of triage to assess for the presence of B-lines and an enlarged LA:Ao ratio. The presence of coalescing B-lines and a markedly enlarged LA:Ao (>2.5:1) further corroborated a diagnosis of L-CHF. Serial PLUS was performed to monitor the number, severity, and resolution of B-lines and pulmonary edema. The dog was initially treated with a furosemide constant rate infusion and pimobendan, and was subsequently switched to intermittent dosing of furosemide. An echocardiogram was performed on the second day of hospitalization. The dog was ultimately diagnosed with Stage C MMVD and discharged with a long-term follow-up plan through the cardiology service.

Important concepts regarding the sonographic diagnosis of MMVD and L-CHF on cardiac POCUS:

- When assessed together, two key sonographic findings support the diagnosis of L-CHF:
  1. Severe LA dilation.
  2. Numerous diffuse bilateral positive B-line sites on lung ultrasound.
- Generally, dogs with L-CHF have an LA:Ao ratio >2:1 (canine normal LA:Ao ratio is less than 1.5:1).[1]
- Dogs with respiratory distress due to L-CHF typically have bilateral, diffuse, and coalescing B-lines on lung ultrasound. A recent study showed that lung ultrasound had good diagnostic accuracy in identifying cardiogenic pulmonary edema. Specifically, dogs in Stage C MMVD with radiographic signs of pulmonary edema had numerous

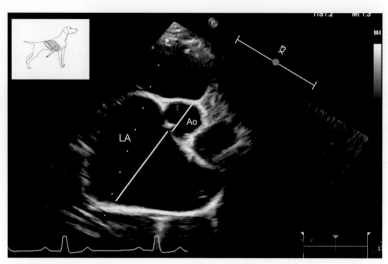

FIGURE 7.1.3 Right parasternal short-axis left atrial to aortic ratio (LA:Ao) view of the heart in cardiac mode, depicting the aorta (Ao) and the left atrium (LA). The white lines in the diagram indicate the measurements to obtain for LA:Ao determination.

coalescing B-lines in 18 of 20 cases (90%), and lung ultrasound detected pulmonary edema with a sensitivity of 90%, a specificity of 93%, and positive and negative predictive values of 85.7% and 95.2%, respectively.[3]

- MMVD is most common in older small-breed dogs and is associated with a loud left apical systolic heart murmur on physical exam. Echocardiogram findings include a thickened mitral valve, mitral regurgitation, severe LA enlargement, and left ventricle (LV) volume overload with compensatory eccentric hypertrophy, and normal to hyperdynamic LV systolic function which can be seen and calculated in M-mode.[4]
- If a patient has significant respiratory distress, a left apical systolic murmur, severe LA dilation, and positive B-lines detected on PLUS, stabilization should be initiated with oxygen, anxiolytics, and diuretic therapy. Comprehensive POCUS scans and diagnostic imaging should occur at a later point once the patient is stable.
- Congestive heart failure secondary to MMVD is a type of volume overload and must be distinguished from iatrogenic fluid volume overload. In both clinical scenarios, the patients will have an enlarged LA:Ao ratio and hyperkinetic left ventricle enlargement. Clinical context must be considered to differentiate the two, i.e., an older small-breed dog with a historically loud left apical systolic heart murmur that develops respiratory distress while at home is more suggestive of MMVD, whereas a patient hospitalized for pancreatitis receiving a large amount of crystalloids that acutely develops dyspnea is more suggestive of iatrogenic volume overload.[5]

Techniques and tips to identify the appropriate ultrasound images/views:

- Both short-axis and long-axis views of the heart can be obtained at the right parasternal acoustic window. This acoustic window is located on the right hemithorax at the cardiac apex beat (where the heartbeat is palpated).[1]
- LA:Ao view is obtained at the level of the heart base, and must be perpendicular to the heart with the aortic semilunar valves visualized (creating the appearance of a Mercedes sign) to obtain an accurate LA:Ao ratio. If the right parasternal image is off-angle, errors in image interpretation can occur.[5]

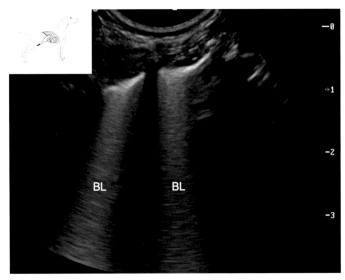

FIGURE 7.1.4 **Coalescing B-lines (BL).** Vertical white lines originating from the visceral surface of the pleural line and extending to the far field.

- The right parasternal image acoustic window is small, and therefore requires slight movements of the transducer (rotation, sliding, fanning, and rocking) to optimize image quality.[1]
- Dogs with L-CHF typically have an LA:Ao ratio >2:1. The exception to this rule is the scenario of a ruptured chordae tendineae, where sudden massive worsening of mitral regurgitation causes pulmonary edema before the LA has time to dilate and remodel.[1]

Take-home messages
- Clinical context is important when assessing a patient in respiratory distress secondary to MMVD with L-CHF. MMVD is most common in older small-breed dogs with an associated left apical systolic heart murmur identified on auscultation.
- The two key sonographic findings that when assessed together support a diagnosis of L-CHF secondary to MMVD are an enlarged LA:Ao ratio (>2:1) and bilateral B-lines on lung ultrasound.

REFERENCES

1. DeFrancesco TC, Ward JL. Focused Canine Cardiac Ultrasound. *Vet Clin Small Anim* 2021;51(6):1203–1216.
2. Lisciandro GR, Lisciandro SC. Global FAST for Patient Monitoring and Staging in Dogs and Cats. *Vet Clin Small Anim* 2021;51(6):1315–1333.
3. Vezzosi T, Mannucci T, Pistoresi A et al. Assessment of Lung Ultrasound B-Lines in Dogs with Different Stages of Chronic Valvular Heart Disease. *J Vet Intern Med* 2017;31(3):700–704.
4. Keene BW, Atkins CE, Bonagura JD, et al. ACVIM Consensus Guidelines for the Diagnosis and Treatment of Myxomatous Mitral Valve Disease in Dogs. *J Vet Intern Med* 2019;33(3):1127–1140.
5. Chalhoub S, Boysen SR. Veterinary Point of Care Ultrasound (POCUS): Abdomen, Pleural Space, Lung, Heart and Vascular Systems. *UCVM Clinical Skills Laboratory Manual* 2021:1–88.
6. Lisciandro GR. *Focused Ultrasound Techniques for the Small Animal Practitioner*, 2nd ed. John Wiley & Sons, Inc; 2014.

CHAPTER 7 – CASE 2

GOLDEN RETRIEVER WITH A 2-DAY HISTORY OF COUGHING AND EXERCISE INTOLERANCE FOLLOWING AN EPISODE OF VOMITING

Chiara Debie and Christopher Kennedy

HISTORY, TRIAGE, AND STABILIZATION

A 7-year-old neutered male Golden Retriever presented with a 2-day history of coughing and exercise intolerance. The dog vomited once prior to presentation.

Triage exam findings (vitals):
- Mentation: quiet, alert, and responsive
- Respiratory rate: 36 breaths per minute
- Heart rate: 164 beats per minute
- Mucous membranes: pale pink and moist, <2 sec capillary refill time
- Femoral arterial pulses: palpable and mostly synchronous with the heartbeat, with occasional dropped beats
- Temperature: 38.5°C (101.3°F)
- Cardiopulmonary auscultation: muffled heart sounds

POCUS exam(s) to perform:
- Cardiac POCUS
- PLUS
- Abdominal POCUS

Abnormal POCUS exam results:
See **Figures 7.2.1, 7.2.2, and 7.2.3**.

Additional point-of-care diagnostics and initial management:
- Intravenous catheter placed.
- Packed cell volume/total solids: 24%/58 g/L.
- Glucose: 4.4 mmol/L.
- Lactate: 4.6 mmol/L.
- Coagulation times (prothrombin time/partial prothrombin time): within normal limits.
- Systolic blood pressure (Doppler): 140 mmHg.
- Lead II bedside ECG: occasional, monoform, ventricular premature complexes.
- Cardiac troponin I: 92 ng/L.

QUESTIONS AND ANSWERS

1. What are the differentials to rule in or out with POCUS based on history and physical exam?
2. What are the sonographic findings?
3. What is the sonographic diagnosis?
4. If necessary, what additional sonographic examination or findings would help rule in or out the differential diagnoses?

DOI: 10.1201/9781003436690-33

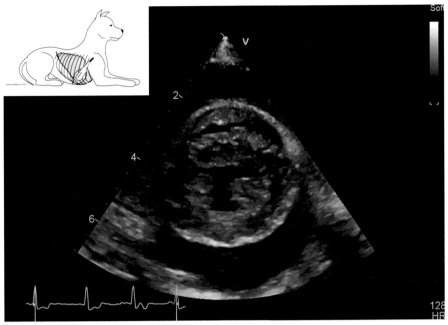

FIGURE 7.2.1 Right parasternal short-axis view at the level of the papillary muscles using a phased-array probe, with dog in sternal. Cardiac mode was used (image indicator to the right of the ultrasound screen) with the marker directed away from the right shoulder. Depth of the probe is 7 cm and the frequency is set at 8 Hz. The fur is parted and 70% isopropyl alcohol is used as the coupling agent.

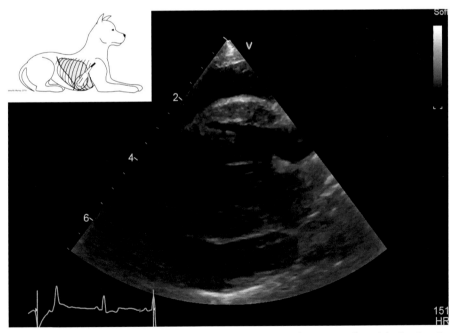

FIGURE 7.2.2 Right parasternal long-axis image using a phased-array probe with the depth at 7 cm and the frequency set at 8 Hz. Cardiac mode was used (image indicator to the right of the ultrasound screen) with the probe indicator marker directed caudally and ventrally (in line with the long axis of the heart). The fur is parted and 70% isopropyl alcohol is used as the coupling agent.

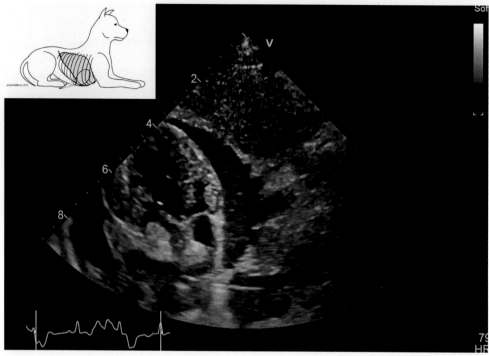

FIGURE 7.2.3 Subxiphoid view using a phased-array probe with the depth at 9 cm and the frequency set at 8 Hz. Cardiac mode was used (image indicator to the right of the ultrasound screen). The probe is placed at the xiphoid and pointed cranially. The fur is parted and 70% isopropyl alcohol is used as the coupling agent.

1. **Differential diagnosis to rule in or rule out with POCUS:**
   - Pleural, pericardial, and/or peritoneal effusion, with hemorrhage being the most likely type of effusion in all three cases; decreased cardiac systolic function (dilated cardiomyopathy-phenotype) +/− left-sided congestive heart failure.
2. **Describe your sonographic findings:**
   - Anechoic fluid surrounding the heart in all images (**Figures 7.2.4, 7.2.5, and 7.2.6**).
3. **What is your sonographic diagnosis?**
   - Pericardial effusion (PE).
4. **What additional sonographic examination or finding would help you rule in or rule out your differential diagnoses (if necessary)?**
   - Pericardiodiaphragmatic window: the absence of the mediastinal triangle or the mediastinal triangle replaced by fluid is an indication of pleural effusion. The mediastinal triangle is the triangle formed between the diaphragm, pericardium, and pleural wall.[1]
   - PLUS: the absence of B-lines or pleural effusion will help to rule out additional causes for bicavitary effusions and alveolar interstitial syndrome.
   - Caudal vena cava (CVC): the dilation of the CVC is supportive of, but not necessary for, the diagnosis of pericardial effusion. If present, it is suggestive of pericardial effusion with tamponade.
   - Abdominal POCUS: peritoneal effusion may be encountered with pericardial effusion, though it is not necessary for diagnosis. If present, it is suggestive of pericardial effusion with tamponade.

History of Coughing and Exercise Intolerance

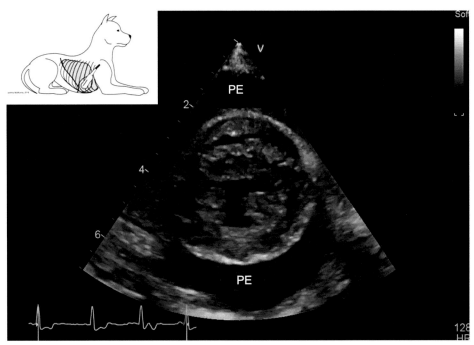

FIGURE 7.2.4 Right parasternal short-axis view depicting a band of anechoic fluid surrounding the heart indicating pericardial effusion (PE).

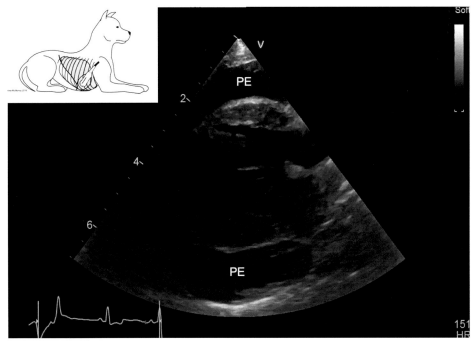

FIGURE 7.2.5 Right parasternal long-axis view depicting a band of anechoic fluid surrounding the heart indicating pericardial effusion (PE).

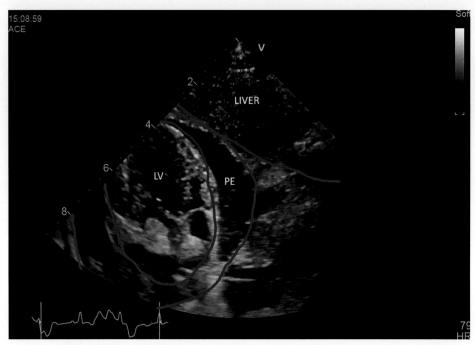

FIGURE 7.2.6 Subxiphoid view of the heart depicting anechoic effusion separating the left ventricular (LV) free wall from the diaphragm and liver. The effusion is contained within a sac contouring the heart, indicating pericardial effusion (PE). The inner blue line outlines the epicardium, the outer blue line the pericardial sac, and the single blue line the diaphragm.

CLINICAL INTEGRATION

Suspected diagnosis based on ultrasound findings:

Pericardial effusion. Without further examination via echocardiography, CT scan, fluid analysis, and/or pericardial histopathology, no further characterization of PE etiology can be made.

Sonographic interventions, monitoring, and outcome:

A pericardiocentesis was performed with ultrasound guidance. An echocardiogram and pericardial fluid analysis did not reveal an underlying cause for the effusion. Serial POCUS scans were performed over a 24-h period to assess for re-effusion. No recurrence. The dog was discharged the following day. Six months later, the pericardial effusion remained absent (assessed by POCUS) and the dog was clinically normal per the owners.

Important concepts regarding the sonographic diagnosis of idiopathic pericardial effusion:

- Cardiac ultrasound is quick and sensitive for detecting pericardial effusion.[1,4]
- Ventriculomegaly, an enlarged left atrium, or pleural effusion can be mistaken for pericardial effusion. Utilizing multiple cardiac views (right parasternal long and short axis and subxiphoid) can help differentiate pericardial effusion.[1,2,3]
- Cardiac ultrasound cannot determine the origin or type of effusion. However,

- an overt cardiac mass is suggestive of a neoplastic process. Idiopathic pericardial effusion is a diagnosis of exclusion.
- Cardiac tamponade can be deduced from a combination of pericardial effusion, tachycardia, and signs of poor perfusion. Right atrial inversion is a sensitive but not specific indicator of cardiac tamponade physiology.[2,4]

Techniques and tips to identify the appropriate ultrasound images/views:
- Start with a right-sided parasternal long-axis view (obtained by aligning the probe along the long axis of the heart, running apex to base, with the marker pointing toward the head). This gives a repeatable overview of the whole heart. Right atrial inversion is best appreciated via this view.
- Rotate 90° to obtain the right parasternal short-axis views. Pericardial effusion is often better appreciated at the level of the papillary muscles (the "mushroom view") rather than at the heart base (the "LA:Ao view").
- The subxiphoid view may be used to confirm pericardial effusion.
- All these images can be obtained in sternal, lateral decubitus, or while standing.
- Right parasternal long-axis view is the best view to detect cardiac tamponade.

Take-home messages
- With a good history and clinical examination, pericardial effusion can be suspected.
- Sonography is the preferred imaging modality for diagnosis and should be prioritized over radiography.
- Idiopathic pericardial effusion is a diagnosis by exclusion.

REFERENCES

1. Boysen, S., Gommeren, K., & Chalhoub, S. (2022) *The Essentials of Veterinary Point of Care Ultrasound*. Grupo Editorial Patria.
2. Boysen, S. R., & Lisciandro, G. R. (2013). The Use of Ultrasound for Dogs and Cats in the Emergency Room: AFAST and TFAST. *Veterinary Clinics of North America: Small Animal Practice*, 43(4), 773–797. https://doi.org/10.1016/j.cvsm.2013.03.011
3. Lisciandro, G. R. (2015). The Use of the Diaphragmatico-Hepatic (DH) Views of the Abdominal and Thoracic Focused Assessment with Sonography for Triage (AFAST/TFAST) Examinations for the Detection of Pericardial Effusion in 24 Dogs (2011–2012). *Journal of Veterinary Emergency and Critical Care*, 26(1), 125–131. https://doi.org/10.1111/vec.12374
4. Silverstein, D., & Hopper, K. (2014). Pericardial Diseases. Ware W, Small Animal Critical Care Medicine (2nd ed., pp. 239–246). Saunders.

CHAPTER 7 – CASE 3

ACUTE COLLAPSE IN A CAVALIER KING CHARLES SPANIEL

Shari Raheb and Xiu Ting Yiew

HISTORY, TRIAGE, AND STABILIZATION

An 11-year-old male neutered Cavalier King Charles Spaniel collapsed while chasing a squirrel. He abruptly stopped in the middle of the run and collapsed without losing consciousness. Several minutes later, he stood up and was ambulatory but thereafter remained less energetic.

Triage exam findings (vitals):
- Mentation: quiet, alert, responsive
- Heart rate and rhythm: 177 beats per minute, femoral arterial pulses fair with occasional pulse deficits, muffled heart sounds, Grade IV/VI holosystolic heart murmur
- Mucous membranes: pale pink mucous membranes, 2 sec capillary refill time
- Respiratory rate: 40 breaths per minute, slightly muffled lung sounds bilaterally
- Temperature: 38.3°C (100.9°F)
- Electrocardiogram: sinus tachycardia with occasional supraventricular premature complexes

POCUS exam(s) to perform:
- Cardiac POCUS
- PLUS

Abnormal POCUS exam results:
See **Figures 7.3.1** and **7.3.2** as well as **Video 7.3.1**.

Additional point-of-care diagnostics and initial management:
Point-of-care blood work was unremarkable, except for mild anemia and hyperlactatemia (3.1 mmol/L). Clotting times and oscillometric blood pressures were normal.

QUESTIONS AND ANSWERS

1. What are the differentials to rule in or out with POCUS based on history and physical exam?
2. What are the sonographic findings?
3. What is the sonographic diagnosis?
4. If necessary, what additional sonographic examination or findings would help rule in or out the differential diagnoses?

DOI: 10.1201/9781003436690-34

Acute Collapse In A Cavalier King Charles Spaniel

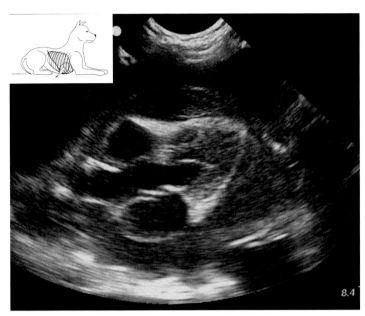

FIGURE 7.3.1 AND VIDEO 7.3.1 Right parasternal long-axis left ventricular outflow view, using a microconvex array probe, 8 MHz, 8 cm depth, with patient in sternal, and the probe placed in the right third to fifth intercostal space with the reference mark directed toward the head and shoulder blade. The fur is parted and 70% isopropyl alcohol is used as the coupling agent. Placing the probe directly on the point of maximal intensity (PMI) can help locate the heart, especially in larger animals. Once the heart is found, depth is adjusted until the entire heart is visualized within the ultrasound field of view. Without sliding the probe, slowly rotate the stationary probe clockwise or counter-clockwise until the right parasternal long-axis view is identified. Fine, discreet probe movements are essential during cardiac POCUS; avoid drastic probe movements. Once the right parasternal long-axis view is identified, fan the probe in either direction to visualize the atria and auricles from multiple angles.

1. **Differential diagnosis to rule in or rule out with POCUS:**
   Based on the history of collapse and muffled heart/lung sounds on auscultation, pericardial effusion ± cardiac tamponade with pleural effusion are the top differentials to rule out.
   Given the breed and heart murmur, left atrial (LA) rupture secondary to severe myxomatous mitral valve disease (MMVD) should be considered as the most likely cause of pericardial effusion. This can be ruled in by assessing for LA enlargement ± the presence of a heterogenous structure suggestive of a thrombus. Other differential diagnoses include a cardiac-associated mass (chemodectoma, hemangiosarcoma), mesothelioma, infectious pericarditis, and idiopathic pericardial effusion. Common locations for cardiac-associated masses include the right atrium/auricle or heart base, sometimes identifiable by cardiac POCUS.
   Differential diagnoses for pleural effusion include a transudate from heart failure or neoplastic effusion.

2. **Describe your sonographic findings:**
   - An organized echogenic structure extended fully around the heart

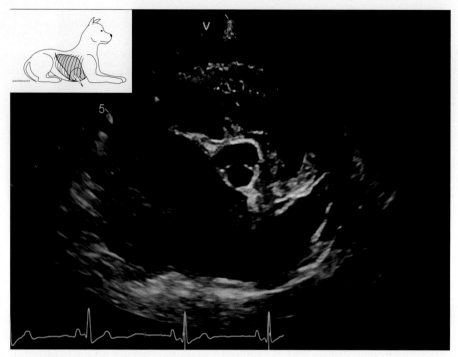

FIGURE 7.3.2 Right parasternal short-axis view of the heart base, optimized for left atrium and aortic valve view, using a 2.4–8.0 MHz phased-array probe in cardiac mode, 8 MHz, and 8 cm depth. Patient is in sternal and the probe is placed at the right third to fifth intercostal space, rotated 90° from Figure 7.3.1, and fanned slowly toward the head until the heart base is visualized. The fur is parted and 70% isopropyl alcohol is used as the coupling agent.

inside the pericardial space, confined by the hyperechoic pericardial sac (**Figure 7.3.3**).
- Free-formed anechoic fluid is present within the pleural space (**Figure 7.3.3**).
- Small volume pericardial effusion (**Figure 7.3.4**).
- Markedly enlarged left atrial to aortic (LA:Ao) ratio of 2.4:1 (**Figure 7.3.4**).

3. **What is your sonographic diagnosis?**
   - Pericardial effusion and pericardial thrombus.
   - Pleural effusion.
   - LA enlargement.
4. **What additional sonographic examination or finding would help you rule in or rule out your differential diagnoses (if necessary)?**
   - A complete echocardiogram is ideal to confirm the suspicion of LA rupture with a pericardial thrombus.
   - Two-dimensional color Doppler echocardiography can confirm the presence of MMVD with mitral regurgitation (MR).
   - Often the suspected thrombus can be identified adjacent to the LA wall.
   - Contrast-enhanced ultrasonography can determine active intra-pericardial hemorrhage from LA ruptures[1] and potentially guide decisions for pericardiocentesis in stable patients.

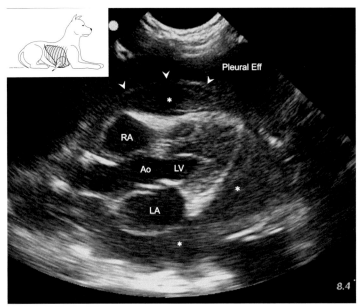

FIGURE 7.3.3 AND VIDEO 7.3.3 An organized echogenic structure (asterisks) extends fully around the heart inside the pericardial space indicating a pericardial thrombus, confined within the hyperechoic pericardial sac (arrowheads). Free-formed anechoic fluid is present within the pleural space. RA: right atrium; LV: left ventricle; Ao: aorta; LA: left atrium; Eff: effusion.

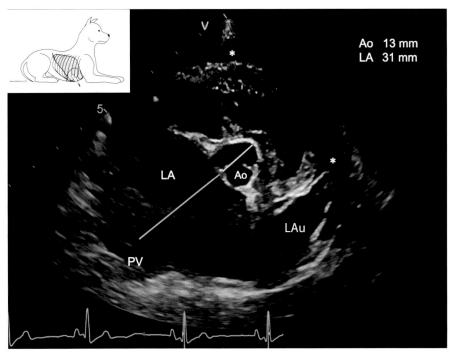

FIGURE 7.3.4 The gray lines represent the measurements required to determine the left atrial to aorta (LA:Ao) ratio. The left atrium is markedly dilated (LA:Ao 2.4:1), and a dilated pulmonary vein (PV) is seen. Pericardial effusion (asterisks) is surrounding the heart and the tip of the left auricle (LAu).

CLINICAL INTEGRATION

Suspected diagnosis based on ultrasound findings:
LA rupture with large pericardial thrombus caused by severe chronic MR secondary to MMVD.

Sonographic interventions, monitoring, and outcome:
- As the patient was relatively stable with a large pericardial thrombus, pericardiocentesis was not performed, and oxygen was provided.
- The LA rupture likely caused an acute reduction in LA pressure and avoided pulmonary edema formation at this time.
- Two-dimensional color Doppler echocardiography confirmed MMVD causing severe mitral regurgitation and a thrombus originating from the enlarged LA (**Figures 7.3.5** and **7.3.6**, **Videos 7.3.2** and **7.3.3**).
- If significant cardiovascular instability secondary to cardiac tamponade is present, pericardiocentesis may be necessary.
- However, relief of pericardial pressure may destabilize the thrombus and worsen ongoing hemorrhage necessitating blood transfusions or emergency thoracotomy.[1-3]
- With minimal pericardial effusion for drainage (**Figure 7.3.5**), thoracotomy may be required to remove the large pericardial thrombus restricting cardiac filling.
- Surgical repair[4] of the LA or transseptal puncture creating left-to-right shunting to relieve LA pressures may be attempted.[2]

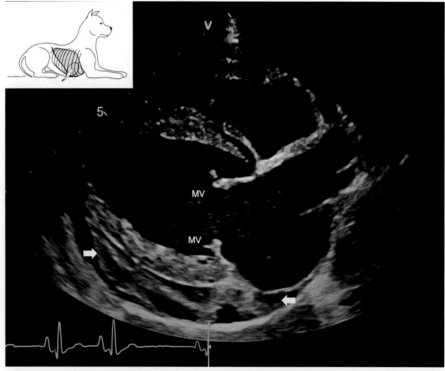

FIGURE 7.3.5 AND VIDEO 7.3.2 Right parasternal long-axis echocardiogram with small volume pericardial effusion and echogenic material suggestive of a thrombus within the pericardial space (depicted between the two arrows). The left atrium and left ventricle are dilated with preserved systolic function (contractility) of the left ventricle. The mitral valve (MV) leaflets are thickened and demonstrate prolapse.

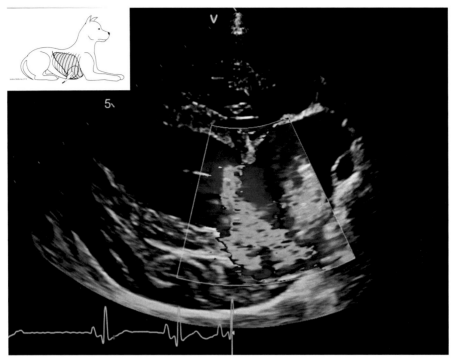

FIGURE 7.3.6 AND VIDEO 7.3.3 **Right parasternal long-axis echocardiogram with color Doppler interrogation of the mitral valve showing severe mitral regurgitation directed toward the left atrial free wall and curling around within the left atrium. Blue color represents blood flow away from the probe; red represents blood flow toward the probe; and green represents turbulent blood flow. As in Video 7.3.2, the left atrium and left ventricle are dilated and there is small volume pericardial effusion along with echogenic material suggestive of a thrombus in the area of the left atrium within the pericardial space.**

- Furosemide, pimobendan, benazepril, and spironolactone were initiated for severe MMVD and imminent heart failure, and to reduce LA pressures.[5]
- Serial cardiac POCUS and PLUS showed a stable pericardial thrombus with resolving pleural effusion over 48 h.
- The long-term outcome for LA rupture is poor to grave[2] with recent evidence of improved median survival (203 days, range: 0–760 days) depending on previous CHF history.[6]

Important concepts regarding the sonographic diagnosis of left atrial rupture (with clot):

- LA rupture is an uncommon cause of pericardial effusion (2%)[2] but should be suspected in older, small to medium-sized dogs presenting for collapse, coughing, and/or dyspnea with a left apical systolic heart murmur consistent with MR.[2,6,7]
- Chronic severe MR leads to LA enlargement and elevated pressure. Stretching of the LA and trauma of high-velocity MR to the LA wall are thought to contribute to LA tearing.[4,7]
- LA rupture is suggested when pericardial effusion is detected with LA enlargement,[2,6] especially if a suspicious echogenic structure is visualized in the pericardial space.[3,4,7,8]
- The absence of a pericardial thrombus cannot rule out LA rupture as large blood clots may not be visible with gradual hemopericardium due to defibrination.[2]

- A mural LA thrombus may also form on the lateral endocardium instead of the epicardium.[1,9]

Techniques and tips to identify the appropriate ultrasound images/views:

- Fine, discreet probe movements are essential during cardiac POCUS; avoid drastic probe movements.
- Long- and short-axis views of the heart can be obtained from the right parasternal window in the third to fifth intercostal space, and the probe rotated 90° between the long and short axis (see **Figures 7.3.1, 7.3.2, 7.3.3,** and **7.3.4**) - these are the POCUS images.
- Pericardial effusion can be differentiated from pleural effusion by the presence of anechoic fluid surrounding the auricular tips (see **Figure 7.3.4**); when both pericardial and pleural effusion are present, the hyperechoic pericardial sac is highlighted in the near field (see **Figure 7.3.3**).

Take-home messages

- Dogs with pericardial effusion secondary to LA rupture can be identified by cardiac POCUS through the detection of LA enlargement ± an echogenic structure within the pericardial space, supported by compatible patient signalment (older, small-to-medium breed dogs), history (collapse, coughing, dyspnea, history of MMVD), and physical examination findings (presence of heart murmur).
- Pericardiocentesis may worsen hemorrhage and is only performed in severe hemodynamically compromised patients.

REFERENCES

1. Caivano D, Marchesi MC, Birettoni F, Lepri E, Porciello F. Left atrial mural thrombosis and hemopericardium in a dog with myxomatous mitral valve disease. *Vet Sci* 2021;8(6):112. doi:10.3390/vetsci8060112
2. Reineke EL, Burkett DE, Drobatz KJ. Left atrial rupture in dogs: 14 cases (1990–2005). *J Vet Emerg Crit Care* 2008;18(2):158–164. doi:10.1111/j.1476-4431.2008.00289.x
3. Prošek R, Sisson DD, Oyama MA. What is your diagnosis? Pericardial effusion with a clot in the pericardial space likely caused by left atrial rupture secondary to mitral regurgitation. *J Am Vet Med Assoc* 2003;222(4):441–442. doi:10.2460/javma.2003.222.441
4. Sadanaga KK, MacDonald MJ, Buchanan JW. Echocardiography and surgery in a dog with left atrial rupture and hemopericardium. *J Vet Intern Med* 1990;4(4):216–221. doi:10.1111/j.1939-1676.1990.tb00900.x
5. Suzuki S, Fukushima R, Ishikawa T, et al. The effect of pimobendan on left atrial pressure in dogs with mitral valve regurgitation. *J Vet Intern Med* 2011;25(6):1328–1333. doi:10.1111/j.1939-1676.2011.00800.x
6. Nakamura RK, Tompkins E, Russell NJ, et al. Left atrial rupture secondary to myxomatous mitral valve disease in 11 dogs. *J Am Anim Hosp Assoc* 2014;50(6):405–408. doi:10.5326/JAAHA-MS-6084
7. Buchanan JW, Kelly AM. Endocardial splitting of the left atrium in the dog with hemorrhage and hemopericardium. *Vet Radiol Ultrasound* 1964;5(1):28–38. doi:10.1111/j.1740-8261.1964.tb01302.x
8. Pelle NG, Testa F, Romito G. Echocardiographic evidence of acute left atrial rupture in a dog with myxomatous mitral valve disease. *J Small Anim Pract* 2021;62(10):931–931. doi:10.1111/jsap.13359
9. Sleeper MM, Maczuzak ME, Bender SJ. Myocardial infarct associated with a partial thickness left atrial tear in a dog with mitral insufficiency. *J Vet Cardiol* 2015;17(3):229–236. doi:10.1016/j.jvc.2015.04.003

CHAPTER 7 – CASE 4

GOLDEN RETRIEVER WITH COLLAPSE, LETHARGY, ATAXIA, AND MILD RESPIRATORY DISTRESS

Julien Guillaumin

HISTORY, TRIAGE, AND STABILIZATION

A 7-year-old female spayed Golden Retriever presented to the emergency room for a "heart attack." She was playing with another dog when she collapsed. She did not lose consciousness and the owner does not believe it was a seizure. Since then, she has been lethargic, stumbling, and breathing heavily.

Triage exam findings (vitals):
- Mentation: quiet, alert, responsive
- Heart rate and rhythm: 180 beats per minute, muffled heart sounds, femoral pulses weak with some dyssynchrony
- Mucous membranes: pink pale, 4 sec capillary refill time
- Respiratory rate: 60 breaths per minute, mild dyspnea, normal breath sounds
- Abdomen: mild abdominal distension, non-painful, organs difficult to palpate
- Temperature: 38.3°C (100.9°F)

POCUS exam(s) to perform:
- PLUS
- Cardiac POCUS
- Abdominal POCUS

Abnormal POCUS exam results:
- See **Figures 7.4.1** and **7.4.2**.
- Abdominal POCUS showed mild to moderate abdominal effusion (images not available).

Additional point-of-care diagnostics and initial management:
An intravenous catheter was placed, and a minimal database was performed (venous blood gas and electrolytes and packed cell volume and total proteins), as well as an oscillometric blood pressure and an electrocardiogram (ECG). Abnormalities included sinus tachycardia, electrical alternans, lactate 6.2 mmol/L, and a mixed metabolic acidosis with a compensatory respiratory alkalosis. The dog was provided flow-by oxygen and was resuscitated with 1000 mL of a balanced isotonic crystalloid while being prepared for pericardiocentesis.

QUESTIONS AND ANSWERS

1. What are the differentials to rule in or out with POCUS based on history and physical exam?
2. What are the sonographic findings?

DOI: 10.1201/9781003436690-35

3. What is the sonographic diagnosis?
4. If necessary, what additional sonographic examination or findings would help rule in or out the differential diagnoses?

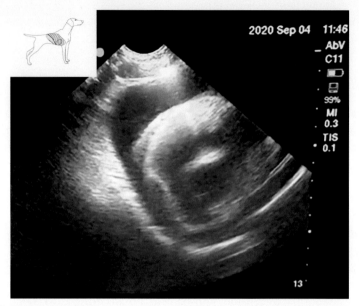

FIGURE 7.4.1 Right parasternal short-axis view of the heart using a microconvex curvilinear probe with depth set at 13 in (frequency not recorded). Patient is standing and the fur is parted and 70% isopropyl alcohol is used as the coupling agent.

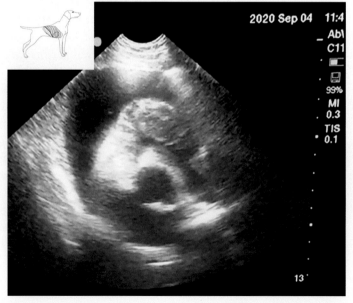

FIGURE 7.4.2 Right parasternal short-axis view of the heart using a microconvex curvilinear probe with depth set at 13 in (frequency not recorded). The fur is parted and 70% isopropyl alcohol is used as the coupling agent.

Collapse, Lethargy, Ataxia, and Mild Respiratory Distress

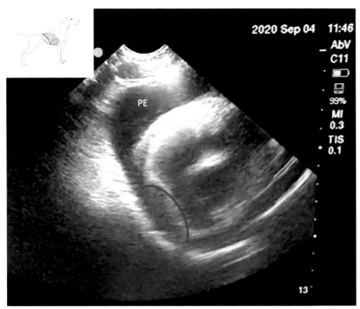

FIGURE 7.4.3 Right parasternal short-axis view of the heart with anechoic fluid surrounding the heart, illustrating pericardial effusion (PE). The wall of the right atrium (red line) is collapsed inward during diastole, consistent with cardiac tamponade.

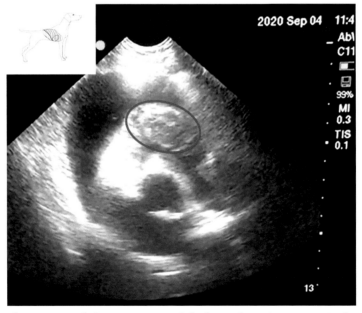

FIGURE 7.4.4 Right parasternal short-axis view of the heart depicting a mass in the right atrio-ventricular groove (encircled by red oval).

1. **Differential diagnosis to rule in or rule out with POCUS:**

   Pericardial effusion (PE) with cardiac tamponade and right atrial mass, associated with abdominal effusion (most likely from right-sided heart failure or less likely from an intra-abdominal bleed) (**Figure 7.4.1**).

2. **Describe your sonographic findings:**
   - Pericardial effusion is characterized by anechoic fluid surrounding the heart (**Figure 7.4.3**).
   - Cardiac tamponade, defined as diastolic right atrial collapse, can be seen as an inward motion of the right atrium (**Figure 7.4.3**).
   - Right atrial mass is shown as a heterogenous or mottled echotexture in the right atrio-ventricular groove (**Figure 7.4.4**).

3. **What is your sonographic diagnosis?**
   Right atrial mass with associated pericardial effusion leading to cardiac tamponade and right-sided heart failure.

4. **What additional sonographic examination or finding would help you rule in or rule out your differential diagnoses (if necessary)?**
   None.

CLINICAL INTEGRATION

Suspected diagnosis based on ultrasound findings:

Based on history, presenting complaint, and initial clinical findings, a right atrial neoplasia such as a hemangiosarcoma is suspected, with pericardial effusion and secondary right-sided heart failure.

Sonographic interventions, monitoring, and outcome:

- Due to the high clinical suspicion of pericardial effusion from the history (older large-breed dog, collapse) and initial physical examination (shock, muffled heart sounds), a cardiac POCUS was rapidly performed.
- Pericardial effusion was identified, and pericardiocentesis was performed with the dog in left lateral recumbency under sedation (butorphanol 0.2 mg/kg IV).
- After stabilization, the dog was administered an oral dose of aminocaproic acid and was hospitalized with continuous ECG monitoring and serial POCUS examinations.
- The patient was then referred to a multispecialty practice where surgery and chemotherapy occurred.
- The patient ultimately passed away 86 days after diagnosis.

Important concepts regarding the sonographic diagnosis of pericardial effusion with cardiac tamponade:

- Pericardial effusion is a simple diagnosis using POCUS. The main sonographic finding is anechoic fluid surrounding the heart contained within the pericardial sac. Pericardial effusion with cardiac tamponade can be assessed via the right parasternal short- or long-axis view. PE with tamponade is identified by
  - Right atrial collapse in systole.
  - Right ventricular collapse in diastole.
  - Distended caudal vena cava (subxiphoid view).
- A mass located on the right atrium can be identified in the right parasternal short-axis view, as seen in **Figures 7.4.2** and **7.4.4**. The right parasternal long-axis view can also be used.
- Pericardial effusion is common in older large-breed dogs. The mean age at diagnosis is approximately 8 years old, and the most common breeds are Golden Retriever and German Shepherd.[1] A retrospective study showed that 30% of dogs with pericardial effusion have a mass seen on echocardiogram. Dogs with a history of collapse are more likely to present with

an identifiable mass on echocardiogram.[1] The median survival time was 1068 days for echo-negative dogs and 26 days for echo-positive dogs.[1] Idiopathic pericardial effusion is the primary differential for pericardial effusion that is not caused by a mass.[2]

- It is estimated that echocardiography performed by a board certified cardiologist or cardiologist in training has a sensitivity of 82% and a specificity of 99% for the detection of a right atrial mass.[3] The same data is not available for clinicians trained in POCUS.
- Most right atrial masses are hemangiosarcomas, but other possible diagnoses include neuroendocrine tumors, ectopic thyroid gland tissue, mesothelioma, lymphosarcoma, or sarcoma.[3]
- Pericardiocentesis is paramount to the treatment of pericardial effusion with tamponade.[4] Fatal complications are rare and most common adverse events during or after pericardiocentesis are dysrhythmias.[5] Pericardial fluid effusion cytology can be performed, but its diagnostic utility is variable.[6]
- Stabilization prior to pericardiocentesis with rapid volume expansion is known to improve hemodynamics and cardiac output in cases of pericardial tamponade.[7,8]
- Aminocaproic acid, with or without Yunnan Baiyao, is commonly used, although its clinical efficacy is unclear.[9,10]

Techniques and tips to identify the appropriate ultrasound images/views:

- Traditional echocardiogram views may be more challenging to acquire with pericardial effusion present. Fluid surrounding the heart makes a mass easier to visualize.
- The sub-xiphoid view may help distinguish pleural versus pericardial effusion.[11]
- Although the pericardial window is on the right side, in cardiovascularly unstable dogs in lateral recumbency, the cardiac POCUS can also be performed from the left side. The probe may be slid underneath the dog. A rolled towel can be especially helpful in large-breed dogs.

Take-home messages

- Pericardial effusion is common in older large-breed dogs such as the Golden Retriever or German Shepherd with a history of collapse and muffled heart sounds.
- Diagnosis of pericardial effusion is simple using cardiac POCUS.
- Rapid volume expansion and pericardiocentesis are key elements for resuscitation.

REFERENCES

1. Stafford Johnson M, Martin M, Binns S, Day MJ. A retrospective study of clinical findings, treatment and outcome in 143 dogs with pericardial effusion. *J Small Anim Pract* 2004;45(11):546–552. doi:10.1111/j.1748-5827.2004.tb00202.x
2. Shaw SP, Rush JE. Canine pericardial effusion: Pathophysiology and cause. *Compend Contin Educ Vet* 2007;29(7):400–403; quiz 404.
3. MacDonald KA, Cagney O, Magne ML. Echocardiographic and clinicopathologic characterization of pericardial effusion in dogs: 107 cases (1985–2006). *J Am Vet Med Assoc* 2009;235(12):1456–1461. doi:10.2460/javma.235.12.1456
4. Kirkland LL, Taylor RW. Pericardiocentesis. *Crit Care Clin* 1992;8(4):699–712.
5. Humm KR, Keenaghan-Clark EA, Boag AK. Adverse events associated with pericardiocentesis in dogs: 85 cases (1999–2006). *J Vet Emerg Crit Care (San Antonio)* 2009;19(4):352–356. doi:10.1111/j.1476-4431.2009.00436.x
6. Cagle LA, Epstein SE, Owens SD, Mellema MS, Hopper K, Burton AG. Diagnostic yield of cytologic analysis of pericardial effusion in dogs.

J Vet Intern Med 2014;28(1):66–71. doi:10.1111/jvim.12253

7. Singh V, Dwivedi SK, Chandra S, et al. Optimal fluid amount for haemodynamic benefit in cardiac tamponade. *Eur Heart J Acute Cardiovasc Care* 2014;3(2):158–164. doi:10.1177/2048872613516017

8. Sagristà-Sauleda J, Angel J, Sambola A, Permanyer-Miralda G. Hemodynamic effects of volume expansion in patients with cardiac tamponade. *Circulation* 2008;117(12):1545–1549. doi:10.1161/CIRCULATIONAHA.107.737841

9. Davis M, Bracker K. Retrospective study of 122 dogs that were treated with the antifibrinolytic drug aminocaproic acid: 2010–2012. *J Am Anim Hosp Assoc* 2016;52(3):144–148. doi:10.5326/JAAHA-MS-6298

10. Murphy LA, Panek CM, Bianco D, Nakamura RK. Use of Yunnan Baiyao and epsilon aminocaproic acid in dogs with right atrial masses and pericardial effusion. *J Vet Emerg Crit Care (San Antonio)* 2017;27(1):121–126. doi:10.1111/vec.12529

11. Lisciandro GR. The use of the diaphragmaticohepatic (DH) views of the abdominal and thoracic focused assessment with sonography for triage (AFAST/TFAST) examinations for the detection of pericardial effusion in 24 dogs (2011–2012). *J Vet Emerg Crit Care (San Antonio)* 2016;26(1):125–131. doi:10.1111/vec.12374

CHAPTER 7 – CASE 5

HEARTWORM-POSITIVE GERMAN SHEPHERD MIX WITH SUDDEN ONSET LETHARGY, WEAKNESS, AND DARK RED–COLORED URINE

LM Bacek and Kendon Wu Kuo

HISTORY, TRIAGE, AND STABILIZATION

A 4-year-old male intact German Shepherd mix presented with sudden onset lethargy, weakness, and dark red–colored urine. The dog is heartworm positive.

Triage exam findings (vitals):
- Mentation: quiet, alert, responsive
- Respiratory rate: 40 breaths per minute
- Heart rate: 152 beats per minute
- Mucous membranes: pale pink, 3 sec capillary refill time
- Femoral and dorsal pedal arterial pulses: weak, thready
- Temperature: 38.4°C (101.2°F)

POCUS exam(s) to perform:
- Cardiac POCUS
- Abdominal POCUS

Abnormal POCUS exam results:
- See **Figures 7.5.1** and **7.5.2**.
- Abdominal POCUS: Moderate free fluid found with an abdominal fluid score of 2/4.

Additional point-of-care diagnostics and initial management:
Packed cell volume/total protein (PCV/TP) of 22% and 7.8 g/dL. Urinalysis showed hemoglobinuria

QUESTIONS AND ANSWERS

1. What are the differentials to rule in or out with POCUS based on history and physical exam?
2. What are the sonographic findings?
3. What is the sonographic diagnosis?
4. If necessary, what additional sonographic examination or findings would help rule in or out the differential diagnoses?

DOI: 10.1201/9781003436690-36

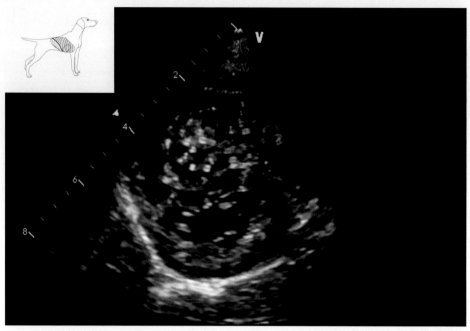

FIGURE 7.5.1 Right parasternal short-axis left ventricular ("mushroom") view of the heart using a phased-array probe, 5 MHz, with the depth set at 8 cm. Patient is standing and the probe is placed just behind the right forelimb and marker directed toward the elbow. The fur is parted and 70% isopropyl alcohol is used as the coupling agent.

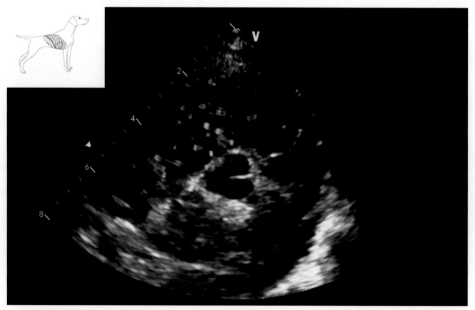

FIGURE 7.5.2 Right parasternal short-axis left atrial-to-aortic ratio (LA:Ao) view of the heart using a phased-array probe, 5 MHz, with the depth set at 8 cm. Patient is standing and the fur is parted and 70% isopropyl alcohol is used as the coupling agent.

1. **Differential diagnosis to rule in or rule out with POCUS:**
   Heartworm disease. The short hyperechoic parallel lines or "equal signs" are definitive for adult heartworms as the imaging plane cuts across segments of heartworms (**Figure 7.5.1** and **Figure 7.5.2**).

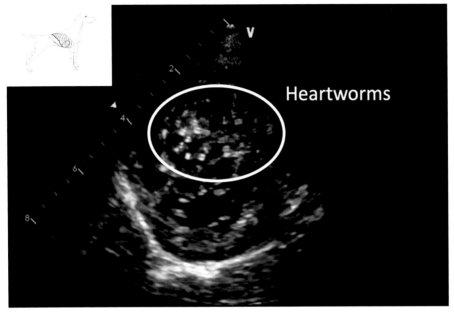

FIGURE 7.5.3 Mass of short hyperechoic parallel lines, or "equal signs," within the white circle depict adult heartworms in the right ventricle crossing the tricuspid valve into the right atrium. An enlarged and thickened right ventricular can be appreciated.

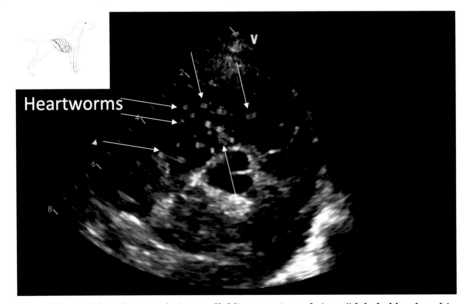

FIGURE 7.5.4 Mass of short hyperechoic parallel lines, or "equal signs," labeled by the white arrows, depict adult heartworms in the right atrium. An enlarged and thickened right atrium can be appreciated.

2. **Describe your sonographic findings:**
   - Mass of heartworms found in the right ventricle and crossing the tricuspid valve into the right atrium (**Figures 7.5.3** and **7.5.4**).
   - Enlarged and thickened right atrium and right ventricular (**Figures 7.5.3** and **7.5.4**).

3. **What is your sonographic diagnosis?**
   Heartworm disease.

4. **What additional sonographic examination or finding would help you rule in or rule out your differential diagnoses (if necessary)?**
   None.

CLINICAL INTEGRATION

Suspected diagnosis based on ultrasound findings:
Caval syndrome due to the acute onset of signs, hemoglobinuria, and a large mass of heartworms in the right ventricle crossing the tricuspid valve into the right atrium.[3]

Sonographic interventions, monitoring, and outcome:
- Worm extraction was performed utilizing ultrasound guidance to visualize the retrieval device and worms. A total of 36 worms were removed. No worms were visualized in the right ventricle or right atrium at the end of the procedure.
- An abdominocentesis was performed and consistent with a high protein transudate due to right-sided heart failure. Serial abdominal fluid scores were performed and showed a decreasing volume.
- The patient was discharged after 48 h of monitoring and supportive care with pimobendan, sildenafil, doxycycline, prednisone, and codeine.

Important concepts regarding the sonographic diagnosis of heartworm and caval syndrome:
- Cardiac POCUS is not a sensitive test for heartworm disease in lightly infected dogs because worms are often found in the peripheral branches of the pulmonary artery.[3]
- As the worm burden increases and hemodynamic changes such as acute worsening of pulmonary hypertension and reduced cardiac output occur, worms can move backward into the main pulmonary artery, right ventricle, right atrium, and vena cava, and are more easily found via ultrasound.[4]
- The presence of intracardiac heartworms is not definitive for caval syndrome. In addition to the visualization of heartworms within the tricuspid valve, clinical signs such as sudden onset severe lethargy and weakness plus hemoglobinemia and hemoglobinuria are required for a more definitive diagnosis.[3]
- The long-axis four-chamber view is helpful in visualizing worms crossing the tricuspid valve and comparing the relative size of the right side of the heart vs the left side of the heart. The left ventricle is normally about twice the size of the right ventricle, but severe pulmonary hypertension from heartworm disease can lead to right atrial and right ventricular dilation and hypertrophy.[1]
- Flattening of the interventricular septum can also be seen on short-axis views.

Techniques and tips to identify the appropriate ultrasound images/views:
See Chapter 6: Cardiac POCUS.

Take-home message
- Heartworm-positive dogs with anemia, pigmenturia, and sudden onset of clinical signs should have a cardiac POCUS performed to visualize heartworms (hyperechoic "equal signs") crossing the tricuspid valve to confirm caval syndrome.

REFERENCES

1. Atkins CE, Keene BW, McGuirk SM. Pathophysiologic mechanism of cardiac dysfunction in experimentally induced heartworm caval syndrome in dogs: An echocardiographic study. *Am J Vet Res* 1988; 49(3): 403–410.
2. Bove CM, Gordon SG, Saunders AB, et al. Outcome of minimally invasive surgical treatment of heartworm caval syndrome in dogs: 42 cases (1999–2007). *J Am Vet Med Assoc* 2010; 236(2): 187–192.
3. American Heartworm Society. Current canine guidelines for the prevention, diagnosis, and management of heartworm (*Dirofilaria immitis*) infection in dogs. 2020. https://www.heartwormsociety.org/veterinary-resources/american-heartworm-society-guidelines.
4. Romano AE, Saunders AB, Gordon SG, Wesselowski S. Intracardiac heartworms in dogs: Clinical and echocardiographic characteristics in 72 cases (2010–2019). *J Vet Intern Med* 2021; 35(1): 88–97.

CHAPTER 7 – CASE 6

MAIN COON WITH ACUTE ONSET OF LABORED BREATHING

Liz-Valérie S Guieu and Charles T Talbot

HISTORY, TRIAGE, AND STABILIZATION:

A 3-year-old male castrated Main Coon presented for acute onset of labored breathing. The cat had no other previous medical history.

Triage exam findings (vitals):
- Mentation: quiet, alert, responsive
- Respiratory rate: 70 breaths per minute, crackles throughout all lung fields
- Heart rate: 140 beats per minute, gallop rhythm auscultated
- Mucous membranes: pink pale, <3 sec capillary refill time
- Femoral and dorsal pedal arterial pulses: weak and regular
- Temperature: 97.5°F (36.4°C)
- Doppler: 100 mmHg

POCUS exam(s) to perform:
- PLUS
- Cardiac POCUS

Abnormal POCUS exam results:
See **Figures 7.6.1** and **7.6.2**.

Additional point-of-care diagnostics and initial management:
- IV catheter placement.
- Oxygen cage.
- Butorphanol (0.3 mg/kg IV).
- Furosemide (1 mg/kg IV).
- Electrocardiogram – sinus bradycardia.

QUESTIONS AND ANSWERS

1. What are the differentials to rule in or out with POCUS based on history and physical exam?
2. What are the sonographic findings?
3. What is the sonographic diagnosis?
4. If necessary, what additional sonographic examination or findings would help rule in or out the differential diagnoses?

Main Coon with Acute Onset of Labored Breathing

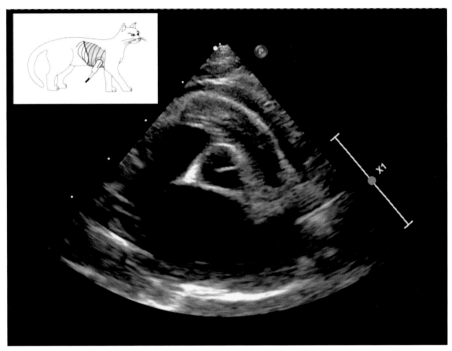

FIGURE 7.6.1 Right parasternal short-axis left ventricular ("mushroom") view the heart using a microconvex curvilinear probe with a depth of 5 cm and a frequency of 9 Hz. The cat is sternal and the fur is parted and 70% isopropyl alcohol is used as the coupling agent.

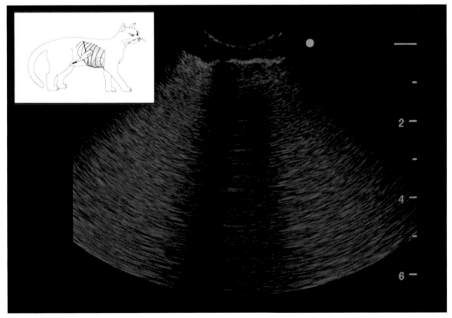

FIGURE 7.6.2 Right perihilar view of the lungs obtained with a microconvex curvilinear probe with the depth set at 5 cm and the frequency set at 12 Hz. The probe is placed mid-thorax. The fur is parted and 70% isopropyl alcohol is used as the coupling agent.

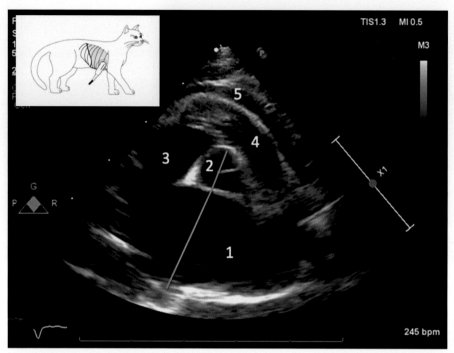

FIGURE 7.6.3 Right parasternal short-axis left atrium-to-aorta view of the heart depicting the left atrium (LA) (1), the aorta (Ao) (2), the right atrium (3), the right ventricle (4), and scant pericardial effusion (5). The blue line transecting the aorta and the left atrium can be used to subjectively assess the LA:Ao ratio of this patient. The LA:Ao in this instance is approximately 3, indicating a marked enlargement of the left atrium.

1. **Differential diagnosis to rule in or rule out with POCUS:**
   Left-sided congestive heart failure, infiltrative lung disease, pulmonary hypertension, pulmonary thromboembolism, and non-cardiogenic pulmonary edema (**Figure 7.6.1**).
2. **Describe your sonographic findings:**
   - Enlarged left atrial to aorta (LA:Ao) ratio (**Figure 7.6.3**).
   - Pericardial effusion (**Figure 7.6.3**).
   - Bilateral coalescing B-lines, ∞ B-lines, white lung (**Figure 7.6.2**).
3. **What is your sonographic diagnosis?**
   Enlarged LA:Ao with pericardial effusion and B-lines suggestive of left-sided congestive heart failure (L-CHF) (**Figure 7.6.4**).
4. **What additional sonographic examination or finding would help you rule in or rule out your differential diagnoses (if necessary)?**
   None.

CLINICAL INTEGRATION

Suspected diagnosis based on ultrasound findings:

Based on the history, presenting signs, clinical and ultrasonographic findings, cardiomyopathy with secondary left-sided congestive heart failure (L-CHF) is suspected.

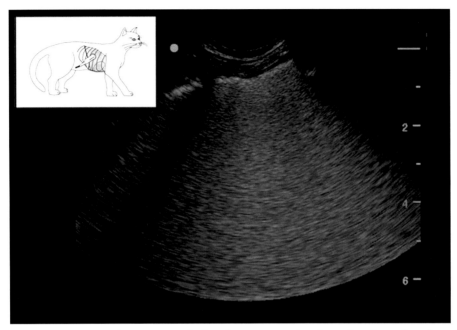

FIGURE 7.6.4 Coalescing B-lines. Vertical white lines originating from the visceral surface of the pleural line and extending to the far field, indicating the presence of alveolo-interstitial disease (i.e., pulmonary edema in this case).

Sonographic interventions, monitoring, and outcome:

- Identification of a gallop rhythm and hypothermia along with an enlarged LA:Ao, pericardial effusion, and B-lines enabled a bedside diagnosis of L-CHF.
- Due to cardio-respiratory instability, the cat was maintained in an oxygen cage on respiratory watch with limited handling. Furosemide was initially given at 1 mg/kg IV q6 for the first 24 h and switched to twice a day due to the normalization of the respiratory rate and effort on day two of hospitalization.
- Serial PLUS scans were performed with the cat maintained in the oxygen cage to monitor the progressive resolution of B-lines.
- The cat was ultimately diagnosed with hypertrophic cardiomyopathy (HCM) by the cardiology service on the second day, and started on and started on clopidogrel and pimobendan.

Important concepts regarding the sonographic diagnosis of hypertrophic cardiomyopathy:

- Cats presenting for respiratory distress with respiratory rates >80 breaths/min, gallop sounds, rectal temperatures <37.5°C (99.5°F), or heart rates >200 bpm are more likely to have CHF than other causes of dyspnea.[1]
- POCUS findings are highly suggestive of feline CHF:
  - Subjective LA:Ao enlargement >2 on right parasternal short-axis view is 97.0% sensitive and 100% specific[2] for feline CHF.
  - Normal LA:Ao 0.8–1.6.[3]
  - Enlarged LA:Ao >1.6, suspicious for CHF.[4]
  - Enlarged LA:Ao >2, highly suspicious for CHF.[5]
  - Presence of pericardial effusion, 100% specific and 60.6% sensitive for feline CHF.[2]

- Presence of more than one site strongly positive for B-lines (greater than three B-lines per site), 83.3% specific and 78.8% sensitive for feline CHF.[2]
- Pleural effusion combined with an enlarged LA:Ao.[6]
- Spontaneous echogenic contrast, left atrial "smoke."[7]
- Echocardiographic changes associated with hypovolemia, including a small left ventricular chamber size and secondary increased wall thickness (pseudohypertrophy), can be misleading. Clinical context and left atrial size should aid in differentiating pseudohypertrophy from HCM.[7]
- Lower doses of furosemide are generally initially prescribed in cats (1 mg/kg) compared to dogs (2 mg/kg) and should be titrated to maintain a resting or sleeping respiratory rate at home of <30 breaths/min. Pimobendan can be considered in cats without any clinically relevant left ventricular outflow obstruction.[6]
- A cat with an enlarged left atrium, left atrial spontaneous echogenic contrast, and/or an observed intracardiac thrombus should receive prophylactic treatment with clopidogrel to prevent aortic thromboembolism.[6]

Techniques and tips to identify the appropriate ultrasound images/views:

- Lateral recumbency is generally preferred for cardiac ultrasound due to limited air interference between the thoracic wall and heart. Cats with CHF, however, might not tolerate this position. The cardiac POCUS is best performed in sternal recumbency in dyspneic cats, allowing for improved lung ventilation and less resistance.[7] Rolled-up towels can be placed under the cranial aspect of the thorax, allowing for ease of probe manipulation and fanning of the heart.
- The mushroom view allows a subjective evaluation of cardiac contractility, left ventricle (LV) chamber dimension, and wall thickness. Wall thickness is generally evaluated at the end of diastole. HCM is characterized by a left ventricle wall thickness >6 mm.
- Once the mushroom view is visualized, the probe is fanned toward the base of the heart until the view required to evaluate the left atrial to aorta ratio is obtained.
- In cases with coalescent B-lines, it can be difficult to visualize the LA:Ao adequately. Moving the probe cranially by one intercostal space can aid in obtaining a diagnostic image.

Take-home messages

- Cats presenting in respiratory distress with hypothermia and a gallop rhythm have a high likelihood of being in CHF.
- Identification of an LA:Ao >2, B-lines, and/or pleural effusion on POCUS is indicative of CHF in cats presenting in respiratory distress.
- Cardiomyopathy is the primary cause of pericardial effusion in cats. Medical management for CHF generally resolves pericardial effusion. Pericardiocentesis is rarely indicated.

REFERENCES

1. Dickson D, Little CJL, Harris J, Rishniw M. Rapid assessment with physical examination in dyspnoeic cats: The RAPID CAT study. *J Small Anim Pract* 2018;59(2):75–84.

2. Ward JL et al. Evaluation of point-of-care thoracic ultrasound and NT-proBNP for the diagnosis of congestive heart failure in cats with respiratory distress. *J Vet Intern Med* 2018;32(5):1530–1540.

3. Adin DB, Diley-Poston L. Papillary muscle measurements in cats with normal echocardiograms and cats with concentric left ventricular hypertrophy. *J Vet Intern Med* 2007;21(4):737–741..
4. DeFrancesco T. Chapter 11. Focused or COAST[3] – Echo (heart). In: *Focused Ultrasound Techniques for the Small Animal Practitioner*, 2nd ed. John Wiley & Sons, Inc; 2014, pp. 189–205.
5. Campbell FE, Kittleson MD. The effect of hydration status on the echocardiographic measurements of normal cats. *J Vet Intern Med* 2007;21(5):1008–1015.
6. Luis Fuentes V, Abbott J, Chetboul V, et al. ACVIM consensus statement guidelines for the classification, diagnosis, and management of cardiomyopathies in cats. *J Vet Intern Med* 2020;34(3):1062–1077.
7. Loughran K. Focused cardiac ultrasonography in cats. *Vet Clin Small Anim* 2021;51(6):1183–1202.

CHAPTER 7 – CASE 7

GREAT DANE WITH A 5-DAY HISTORY OF WEAKNESS, SYNCOPE, DIFFICULTY BREATHING, OCCASIONAL COUGHING, AND ABDOMINAL DISTENSION

Priscilla Burnotte

HISTORY, TRIAGE, AND STABILIZATION

A 5-year-old male neutered Great Dane presented with a 5-day history of weakness, difficulty breathing, occasional coughing, and abdominal distension. The owners reported exercise intolerance for 2 months and two episodes of syncope.

Triage exam findings (vitals):
- Mentation: quiet, alert, responsive
- Respiratory rate: 48 breaths per minute, mild expiratory dyspnea
- Heart rate: 200 beats per minute with an irregular rhythm
- Mucous membranes: pale pink, <2 sec capillary refill time
- Femoral and dorsal pedal arterial pulses: weak with frequent pulse deficits
- Temperature: 37.6°C (98.6°F)
- Marked abdominal distension (with a positive fluid wave)
- Jugular distension

POCUS exam(s) to perform:
- PLUS
- Cardiac POCUS
- Abdominal POCUS

Abnormal POCUS exam results:
See **Figures 7.7.1–7.7.4**, as well as **Videos 7.7.1–7.7.4**.

Additional point of care diagnostics and initial management:
- Supplemental oxygen.
- Butorphanol (0.2 mg/kg IV).
- Electrocardiogram (ECG) revealing atrial fibrillation.
- Doppler systolic blood pressure of 100 mmHg.

QUESTIONS AND ANSWERS

1. What are the differentials to rule in or out with POCUS based on history and physical exam?
2. What are the sonographic findings?

DOI: 10.1201/9781003436690-38

3. What is the sonographic diagnosis?
4. If necessary, what additional sonographic examination or findings would help rule in or out the differential diagnoses?

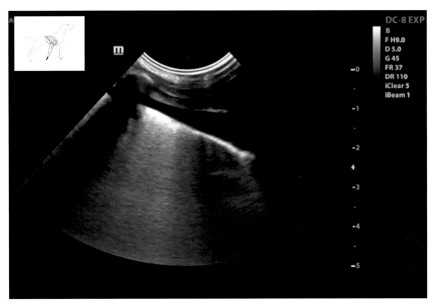

FIGURE 7.7.1 AND VIDEO 7.7.1 PLUS scan with the dog in standing position is performed using a microconvex curvilinear probe, depth set at 5 cm and frequency set at 7 Hz. The probe is oriented perpendicular to the ribs, located at the lowest portion of the thorax; 70% isopropyl alcohol is used as the coupling agent.

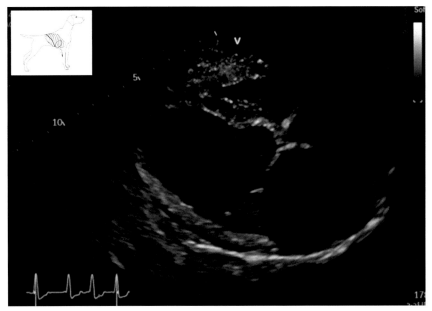

FIGURE 7.7.2 AND VIDEO 7.7.2 Right parasternal long-axis view of the heart in 2D. The dog is standing and the phased-array probe is oriented with the marker directed craniodorsally. The frequency and depth of the probe are 3 Hz and 14 cm, respectively; 70% isopropyl alcohol is used as the coupling agent.

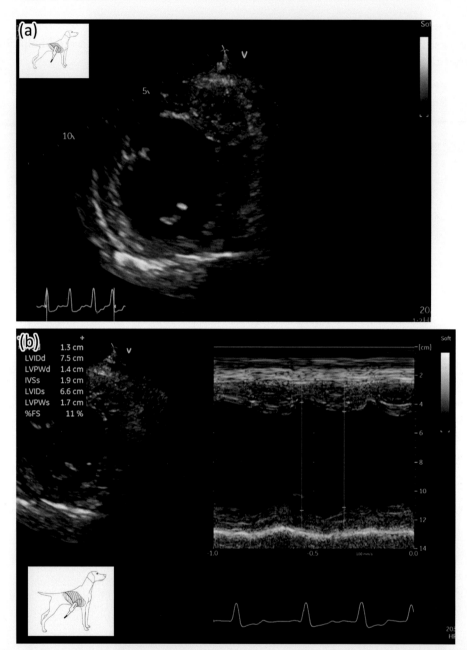

FIGURE 7.7.3 a,b AND VIDEO 7.7.3 Right parasternal short-axis view of the heart in (a) 2D and (b) M-mode to assess fractional shortening. The phased-array probe is rotated 90° clockwise from the right parasternal long-axis view with the marker directed cranioventrally. The frequency and depth of the probe are 3 Hz and 14 cm, respectively; 70% isopropyl alcohol is used as the coupling agent to obtain this view.

1. **Differential diagnosis to rule in or rule out with POCUS:**

   The most likely differential diagnosis based on signalment, history, and physical exam is right-sided congestive heart failure (R-CHF) secondary to dilated cardiomyopathy. Other differential diagnoses causing right-sided heart failure include severe pulmonary hypertension, right heart congenital anomalies, or pericardial effusion with associated tamponade. Due to the abnormal cardiopulmonary auscultation, PLUS is the first sonographic exam performed to investigate the presence of B-lines and pleural effusion. Due to the presence of an arrhythmia, jugular pulses, and ascites, the heart is evaluated for cardiac tamponade, systolic dysfunction, and significant right-sided heart remodeling. Abdominal POCUS at the subxiphoid view is performed to diagnose the presence of ascites and a distended non-compliant vena cava, supporting the diagnosis of right-sided congestive heart failure.

2. **Describe your sonographic findings:**
   - Presence of mild free fluid seen on PLUS (**Figure 7.7.5**).
   - Enlarged left and right heart chambers with thinned ventricular walls and cardiac hypo-contractility, seen on the parasternal long- and short-axis views (**Figures 7.7.6** and **7.7.7**).
   - Distended non-compliant caudal vena cava (CVC) in the subxiphoid view (**Figure 7.7.8**).
   - Presence of free fluid seen in the subxiphoid view (**Figure 7.7.8**).

3. **What is your sonographic diagnosis?**

   Generalized heart chamber enlargement with severe systolic dysfunction

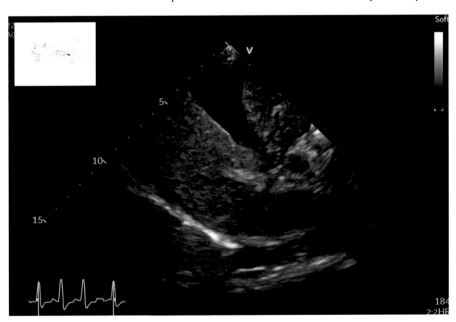

FIGURE 7.7.4 AND VIDEO 7.7.4 Abdominal POCUS centered on the subxiphoid view, specifically assessing the caudal vena cava (CVC) with the dog in right lateral recumbency. The phased-array probe is placed longitudinally under the subxiphoid process to obtain this view. If the CVC is not seen immediately, the probe can be fanned left and right from the midline until the CVC crossing the diaphragm is clearly visualized. Pressure on the probe should be minimal while performing these manipulations to avoid collapse of the CVC. The frequency and depth of the probe are 3 Hz and 15 cm, respectively; 70% isopropyl alcohol is used as the coupling agent to obtain this view.

associated with a mild amount of pleural effusion, ascites, and a distended non-compliant caudal vena cava secondary to right-sided congestive heart failure.

4. **What additional sonographic examination or finding would help you rule in or rule out your differential diagnoses (if necessary)?**
   None.

CLINICAL INTEGRATION

Suspected diagnosis based on ultrasound findings:
Based on the history, presenting signs, and sonographic findings, primary or secondary dilated cardiomyopathy (DCM) with secondary R-CHF is suspected.

Sonographic interventions, monitoring, and outcome:
- Supplemental oxygen and butorphanol were provided while the PLUS and Cardiac POCUS were performed.
- The four heart chambers were enlarged with a thinned and hypo-contractile left ventricular wall.
- R-CHF was confirmed with the presence of a mild amount of pleural effusion, ascites, and a distended non-compliant caudal vena cava.
- A thoracocentesis and abdominocentesis were performed, revealing a modified transudate. An echocardiogram confirmed the diagnosis of DCM associated with atrial fibrillation.
- The dog was discharged with furosemide, pimobendan, spironolactone, diltiazem, and digoxin.

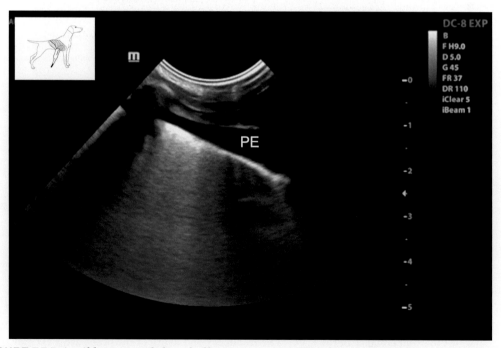

FIGURE 7.7.5 **A mild amount of pleural effusion is present (PE: pleural effusion).**

Important concepts regarding the sonographic diagnosis of DCM and R-CHF on POCUS:

- DCM is the most common cardiac disease in adult large or medium-sized dogs and is associated with systolic dysfunction characterized by reduced fractional shortening (FS) of the left ventricle below 20% (normal FS: 25%–45%).[1,2]
- Tachyarrhythmias are often present and may be primary or secondary to DCM.[1] Systolic function should be interpreted carefully in the presence of tachyarrhythmia as severe tachycardia reduces left ventricular preload and subsequently decreases FS.[2] Dogs with decompensated DCM can present with either L-CHF, R-CHF, or both.
- The normal ratio between the right ventricle and the left ventricle in this patient was mostly preserved (the normal ratio is 1:3).[1] Therefore, to diagnose R-CHF in this dog, it was important to identify enlarged non-compliant caudal vena cava and cavitary effusion without evidence of tamponade.[1,3]

Techniques and tips to identify the appropriate ultrasound images/views:

- Both short-axis and long-axis views of the heart can be obtained at the right parasternal acoustic window, located on the right hemithorax at the cardiac apex beat.[1]
- The left ventricle in short-axis view should appear round with symmetrical papillary muscles, not ovoid.[4]
- Left ventricular function should be assessed just below the mitral valve in the short-axis view.[1]
- To measure FS, the cursor should be placed perpendicular to the interventricular septum, in the middle of the papillary muscles.

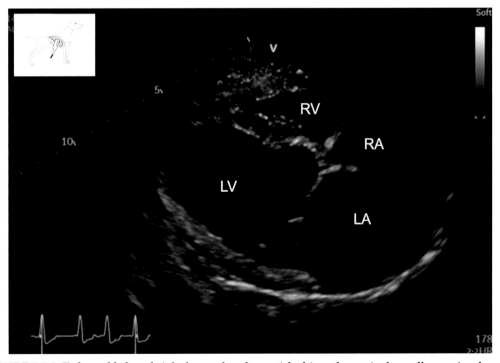

FIGURE 7.7.6 Enlarged left and right heart chambers with thinned ventricular walls are visualized on parasternal long-axis view (LV: left ventricle; RV: right ventricle; LA: left atrium; RA: right atrium).

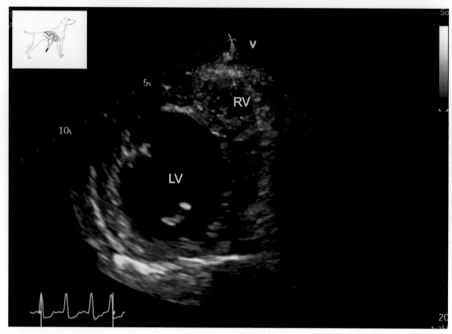

FIGURE 7.7.7 Enlarged left ventricle with thinned ventricular walls is visualized on parasternal short-axis view (LV: left ventricle; RV: right ventricle).

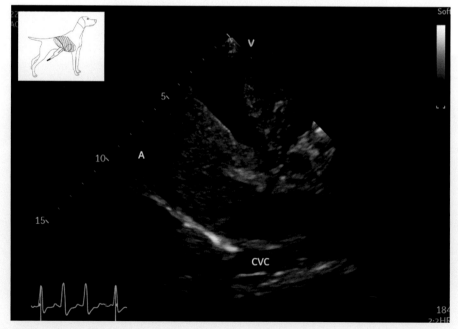

FIGURE 7.7.8 A distended CVC is visualized, as well as free fluid seen caudal to the diaphragm, cranial to the liver (A: ascites).

An average of five to ten measurements should be taken in dogs with tachyarrhythmias as preload and contractility will be more variable.[5]

Take-home message
- The sonographic findings that support a diagnosis of DCM and R-CHF are decreased systolic function, distended non-compliant caudal vena cava, ascites, and less often pleural or pericardial effusion.

REFERENCES

1. DeFrancesco TC, Ward JL. Focused canine cardiac ultrasound. *Vet Clin Small Anim* 51(6) (2021): 1203–1216.
2. Bonagura JD, Visser MSC. Echocardiographic assessment of dilated cardiomyopathy in dogs. *J Vet Cardiol* 40 (2022): 15–50.
3. Chalhoub S, Boysen SR. Veterinary point of care ultrasound (POCUS): Abdomen, pleural space, lung, heart and vascular systems. *UCVM Clinical Skills Laboratory Manual* (2021): 1–88.
4. Thomas WP, Gaber CE, Jacobs GJ, et al. Recommendations for standards in transthoracic two-dimensional echocardiography in the dog and cat. *J Vet Intern Med* 7(4) (1993): 247–252.
5. Dukes-McEwan J, Borgarelli M, Tidholm A, et al. Proposed guidelines for the diagnosis of canine idiopathic dilated cardiomyopathy. *J Vet Cardiol* 5(2) (2003): 7–19.

CHAPTER 7 – CASE 8

BOXER WITH ACUTE VOMITING, COUGH WITH TERMINAL WRETCH, TACHYPNEA, WEAKNESS, WEIGHT LOSS, AND LETHARGY

Dana Caldwell and Amanda Liggett

HISTORY, TRIAGE, AND STABILIZATION

A 7-year-old male neutered Boxer from Arizona presented on emergency with acute vomiting, cough with terminal wretch, tachypnea, weakness, and 1-month history of weight loss and lethargy.

Triage exam findings (vitals):
- Mentation: obtunded
- Respiratory rate: tachypnea, mild dyspnea, 50 breaths per minute
- Temperature: 39.3°C (102.8° F)
- Heart rate: 150 beats per minute
- Doppler blood pressure: 90 mmHg
- Mucous membranes: pale pink, tacky, 3 sec capillary refill time
- Femoral and dorsal pedal pulses: weak, thready
- Thoracic auscultation: muffled heart sounds, muffled lung sounds right ventral thorax
- Jugular venous distension with pulsation (Kussmaul's sign)
- Palpable fluid wave on abdominal palpation

POCUS exam(s) to perform:
- Cardiac POCUS
- Abdominal POCUS
- Abdominal POCUS with gallbladder assessment

Abnormal POCUS exam results:
See **Figures 7.8.1–7.8.3**.

Additional point-of-care diagnostics and initial management:
- Initial point-of-care testing included electrocardiogram (ECG) with sinus tachycardia, Doppler blood pressure, packed cell volume/total solids (PCV/TS) of 50% and 7.5 g/dL, venous blood gas showed lactic acidosis (pH 7.15, lactate 8.0 mmol/L).
- Initial management included flow-by oxygen, crystalloid fluid bolus, and emergent POCUS-guided pericardiocentesis and thoracocentesis. Blood was drawn for later complete blood count, chemistry, and serology for the infectious fungal disease coccidioidomycosis.

QUESTIONS AND ANSWERS

1. What are the differentials to rule in or out with POCUS based on history and physical exam?
2. What are the sonographic findings?

DOI: 10.1201/9781003436690-39

3. What is the sonographic diagnosis?
4. If necessary, what additional sonographic examination or findings would help rule in or out the differential diagnoses?

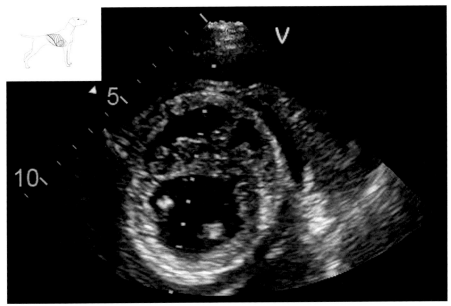

FIGURE 7.8.1 Right parasternal short-axis left ventricular papillary view of the heart at the fourth to sixth intercostal spaces using GE Cardiac 5S probe, 5 MHz frequency, and 10 cm depth. Dog is sternal and the fur is parted and 70% isopropyl alcohol is used as the coupling agent.

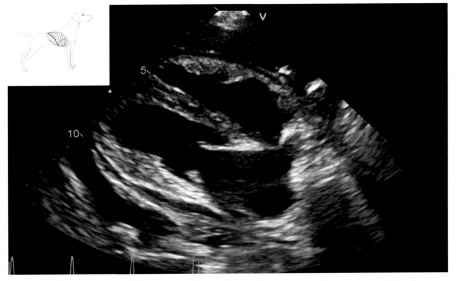

FIGURE 7.8.2 Right parasternal long-axis view of the left atrium and ventricle of the heart using GE Cardiac 5S probe, 5 MHz frequency, and 15 cm depth. Dog is sternal and the fur is parted and 70% isopropyl alcohol is used as the coupling agent.

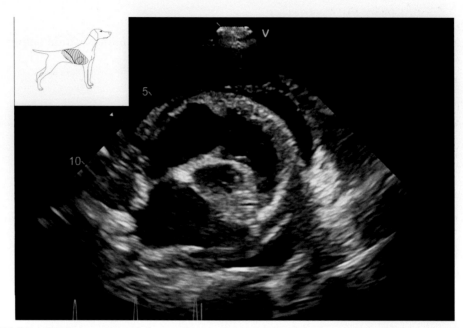

FIGURE 7.8.3 Right parasternal short-axis transaortic view of the heart using GE Cardiac 5S probe, 5 MHz frequency, and 15 cm depth. To facilitate this view, the transducer is fanned toward the base of the heart or it can be moved cranially by one intercostal space. The fur is parted and 70% isopropyl alcohol is used as the coupling agent.

1. **Differential diagnosis to rule in or rule out with POCUS:**

    The most probable diagnosis is pericarditis with pericardial effusion likely secondary to fungal, bacterial, foreign body, or neoplasia[1] Additional differentials included hemangiosarcoma, chemodectoma, coagulopathy, idiopathic pericardial effusion, heart failure, and left atrial tear. Most likely differential for pleural effusion includes right-sided congestive heart failure secondary to pericardial tamponade, chylous effusion, neoplasia, fungal disease, hemorrhage, neoplasia, and pyothorax.[2]

2. **Describe your sonographic findings:**
    - Anechoic pericardial effusion with pericardial thickening. The pericardium does not appear adhered to the epicardium. There is also moderate pleural effusion (**Figures 7.8.4, 7.8.5,** and **7.8.6**) and left atrial enlargement (**Figure 7.8.6**).

3. **What is your sonographic diagnosis?**
    Pericardial effusion with pericarditis of a constrictive nature and right-sided congestive heart failure.

4. **What additional sonographic examination or finding would help you rule in or rule out your differential diagnoses (if necessary)?**
    - Abdominal POCUS and subxiphoid views (images not shown) assist with the evaluation of right-sided congestive heart failure sequela: peritoneal effusion, caudal vena cava and hepatic venous distension, and gallbladder wall edema. The subxiphoid view can mitigate air interference when imaging through the acoustic window of the liver and gallbladder while evaluating for pericardial effusion.
    - Cardiac POCUS assists with the identification of fibrin tags on the epicardial surface and within the pericardial sac commonly seen with pericarditis.

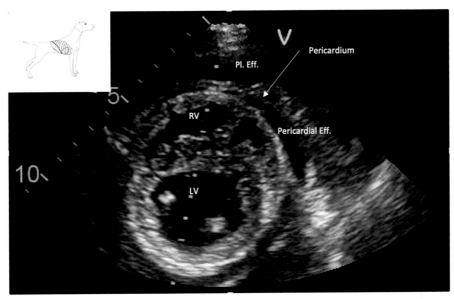

FIGURE 7.8.4 Right parasternal short-axis left ventricular papillary view of the heart using GE Cardiac 5S probe, 5 MHz frequency, and 10 cm depth. Mild pericardial effusion with thickening of the pericardium. Independent motion of the pericardium from the epicardium was visualized indicating that the pericardium is likely not adhered to the epicardium in the obtained views. LV: left ventricle; RV: right ventricle; Pl Eff: pleural effusion.

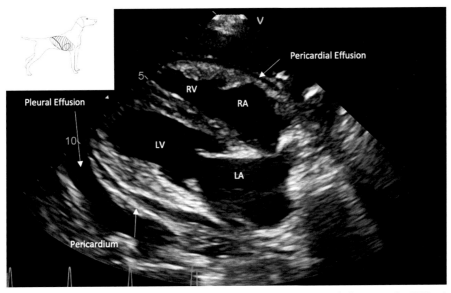

FIGURE 7.8.5 Right parasternal long-axis view of the left atrium (LA), left ventricle (LV), right atrium (RA), and right ventricle (RV). Mild pleural and pericardial effusion is seen with thickening of the pericardium. Note that there is no significant right heart enlargement, despite evidence of right-sided congestive heart failure. This is due to the limited capacity and constriction caused by the thickened and fibrosed pericardium.[1]

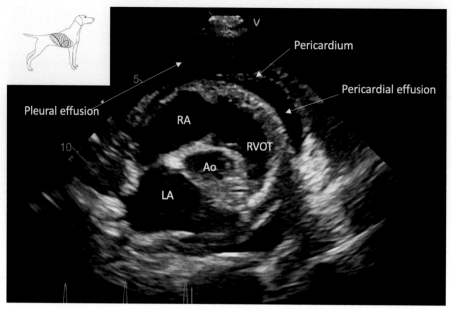

FIGURE 7.8.6 Right parasternal short-axis transaortic view. Structures viewed include aorta (Ao), left atrium (LA), and right ventricular outflow tract (RVOT). There is left atrial enlargement with normal right heart size despite the presence of right heart failure signs. This is due to the constriction of the heart by the pericardium and epicardium in this case. An enlarged left atrial to aortic ratio (LA:Ao) of 2:1 is associated with mitral regurgitation.

Distortion of the right ventricle may also be noted if the pericardium has adhered to the epicardium. Occasionally, pericardial granulomas are visualized. In addition, cardiac POCUS can evaluate for right atrial and ventricular diastolic collapse associated with cardiac tamponade.[2,3]

CLINICAL INTEGRATION

Suspected diagnosis based on ultrasound findings:

Based on history, physical exam, and clinical findings, pericarditis with pericardial effusion and signs of right-sided congestive heart failure. This case of pericarditis is suspected to be associated with *Coccidioides immitis* given the enzootic region of this patient and 1:32 positive *C. immitis* titer.[4]

Sonographic interventions, monitoring, and outcome:

- An ultrasound-guided therapeutic thoracocentesis and diagnostic right-sided pericardiocentesis was performed.
- Serial POCUS monitored for the progression or resolution of pleural and pericardial effusions.
- Cytologic evaluation of the pericardial effusion was consistent with a modified transudate and pyogranulomatous inflammation. Protein concentration of 3.5 mg/dL and PCV of 15%. Fungal and bacterial cultures of the effusion were negative for growth; however, serology for *C. immitis* by autoimmune gastrointestinal dysmotility (AGID) revealed a positive titer of 1:32. Evidence has shown that the diagnostic yield for cytology is greater for pericardial

effusion samples with hematocrit of less than 10%.[5]
- Surgical treatment with subtotal pericardiectomy relieved right-sided heart failure signs. Post-operative echocardiogram showed successful mitigation of pericardial constriction.
- Pericardium histopathology showed fibrosing pyogranulomatous pericarditis.
- Recommended medical management included antifungal agents, anti-inflammatory steroids, pimobendan, and furosemide.
- An echocardiogram with a board-certified cardiologist is recommended to confirm constrictive pericarditis. In addition to pericardial thickening and effusion, findings of hepatic venous distention, mild vena cava plethora (indicating high right atrial pressure and a less compliant venous system), a restrictive filling pattern on left ventricular inflow velocities with respiratory variation, increased E' velocity, and a septal bounce (indicating ventricular interdependence due to the constriction) were all consistent with constrictive pericarditis.[4] With classic constrictive pericarditis, an unusual septal bounce occurs from the interaction of the left and right ventricular diastolic pressures and is accompanied by an inspiratory septal shift. The presence of mitral regurgitation is common in patients with constrictive pericarditis. It is uncommon for this mitral insufficiency to progress to severe regurgitation with left heart failure.

Important concepts regarding the sonographic diagnosis of pericarditis:
- Several sonographic features can support the diagnosis of pericarditis:
  - Thickening of the pericardium.
  - Pericardial effusion can be present.
  - Constrictive pericarditis restricts diastolic filling of the heart and is commonly associated with POCUS findings consistent with right-sided congestive heart failure including peritoneal effusion, hepatic venous and caudal vena cava distension, pleural effusion, and gallbladder wall edema, despite an unremarkable right heart.
  - Fluid analysis including cytology, culture, and serology are important for the diagnostic investigation of pericarditis.[6]

Technique and tips to identify the appropriate ultrasound images/views:
- Subxiphoid cardiac long-axis view or apical transducer positioning can help evaluate for the presence of pericardial thickening and pericardial effusion which can be best observed low toward the apex of the heart.[3]
- Pericardial effusion can be identified as an anechoic or hypoechoic space evident between the two layers of the pericardial sac (visceral epicardial layer and fibrous parietal layer).
- In people, normal pericardial thickness is generally considered less than or equal to 2 mm while a thickened pericardium is >2 mm.[7]
- Generally, right parasternal long-axis view is ideal for assessing tamponade.
- POCUS evidence of cardiac tamponade can include diastolic collapse of the right ventricle and right atrium and inferior vena cava plethora.
- When evaluating for pericardial disorders, ensure the entire heart is visualized or fills at least two-thirds of the screen.[3]
- The lack of pericardial elasticity associated with pericarditis can lead to increased ventricular interdependence, which is characterized by septal flattening and paradoxical septal motion.

Take-home messages
- Constrictive pericarditis is suspected when a thickened pericardium, pericardial effusion, and signs supportive of right-sided congestive heart failure are present.
- Cardiac POCUS and subxiphoid views are helpful for the evaluation of pericardial effusion and pericarditis. These views allow the assessment of right-sided cardiac failure sequelae such as hepatic venous and caudal vena cava distension, gallbladder wall edema, peritoneal effusion, and pleural effusion.
- *C. immitis* infection can be associated with pericarditis, epicarditis, or myocarditis in dogs. Coccidioidomycosis is considered a differential in dogs from enzootic areas with pericardial thickening with effusion.[6]

REFERENCES

1. Heinritz CK, Gilson SD, Soderstrom MJ, Robertson TA, Gorman SC, Boston RC. Subtotal pericardectomy and epicardial excision for treatment of coccidioidomycosis induced effusive constrictive pericarditis in dogs: 17 cases (1999–2003). *J Am Vet Med Assoc* 2005; **227**(3): 435–440.
2. MacDonald KA, Cagney O, Magne M. Echocardiographic and clinicopathologic characterization of pericardial effusion in dogs: 107 cases (1985–2006). *J Am Vet Med Assoc* 2009; **235**(12): 1456–1461.
3. DeFrancesco TC, Ward JL. Focused canine cardiac ultrasound. *Vet Clin North Am Small Anim Pract* 2021; **51**(6): 1203–1216.
4. Oh JK, Kane GC. *The Echo Manual*, 4th ed. Wolters Kluwer Health; 2018, pp. 282–304.
5. Cagle LA, Epstein SE, Owens SD, Mellema MS, Hopper K, Burton AG. Diagnostic yield of cytologic analysis of pericardial effusion in dogs. *J Vet Intern Med* 2013; **28**(1): 66–71.
6. Shubitz LF, Matz ME, Noon TH, Reggiardo CC, Bradley GA. Constrictive pericarditis secondary to Coccidioides immitis infection in a dog. *J Am Vet Med Assoc* 2001; **218**(4): 537–540.
7. Talreja DR, Edwards WD, Danelson GK, Schaff HV, Tajik AJ, Tazelaar HD, Breen JF, Oh JK. Constrictive pericarditis in 26 patients with histologically normal pericardial thickness. *Circulation* 2003; **108**(15): 1852–1857.

CHAPTER 7 – CASE 9

TERRIER WITH A 2-WEEK HISTORY OF COUGH, EXERCISE INTOLERANCE, PROGRESSIVE TACHYPNEA, AND SUSPECTED SYNCOPE

Dave Beeston and Kris Gommeren

HISTORY, TRIAGE, AND STABILIZATION

A 9-year-old female spayed Maltese Terrier presented with a 2-week history of cough, exercise intolerance, and progressive tachypnea. In the week prior to presentation there were three episodes of collapse, suspected syncope.

Triage exam findings (vitals):
- Mentation: appropriate
- Respiratory rate: 50 breaths per minute with increased inspiratory and expiratory effort. Bronchovesicular sounds: harsh but no crackles
- Heart rate: 170 beats per minute with a Grade II/VI left apical systolic murmur
- Mucous membranes: pink and moist, 2 sec capillary refill time
- Femoral and dorsal pedal arterial pulses: good quality and synchronous
- Temperature: 37.9°C (100.2°F)

POCUS exam(s) to perform:
- Cardiac POCUS
- PLUS
- Caudal vena cava (CaVC) assessment

Abnormal POCUS exam results:
See **Figures** 7.9.1–7.9.3, as well as **Videos** 7.9.1–7.9.3.

Additional point-of-care diagnostics and initial management:
Non-invasive Doppler blood pressure was within normal limits (110 mmHg). Arterial blood gas analysis documented hypoxemia (56 mmHg). The patient was placed in an oxygen kennel and received 0.3 mg/kg butorphanol intramuscularly.

QUESTIONS AND ANSWERS

1. What are the differentials to rule in or out with POCUS based on history and physical exam?
2. What are the sonographic findings?

DOI: 10.1201/9781003436690-40

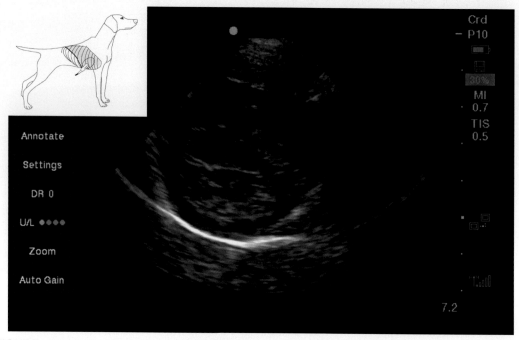

FIGURE 7.9.1 AND VIDEO 7.9.1 Cardiac POCUS with the dog standing, at the right parasternal short axis (mushroom view), using a phased-array probe, 8–4 MHz at 7.2 cm depth, with the probe marker cranial and toward the spine. The fur was not clipped but parted and alcohol/gel/combo was used as the coupling agent.

3. What is the sonographic diagnosis?
4. If necessary, what additional sonographic examination or findings would help rule in or out the differential diagnoses?

1. **Differential diagnosis to rule in or rule out with POCUS:**
   Pulmonary hypertension (PH), left-sided congestive heart failure (L-CHF), pulmonary thromboembolism, less likely negative pressure pulmonary edema, and aspiration pneumonia.
2. **Describe your sonographic findings:**
   - Flattened interventricular septum (IVS) and right ventricular (RV) dilation and hypertrophy (**Figure 7.9.4**).
   - Enlarged pulmonary artery to aortic ratio (**Figure 7.9.5**).
   - Distended, non-compliant CaVC and peritoneal effusion (**Figure 7.9.6**).
3. **What is your sonographic diagnosis?**
   Evidence of PH and right-sided congestive heart failure (R-CHF).
4. **What additional sonographic examination or finding would help you rule in or rule out your differential diagnoses (if necessary)?**
   - Formal echocardiography for assessment of the severity of PH.
   - Complementary examinations to identify the underlying etiology.

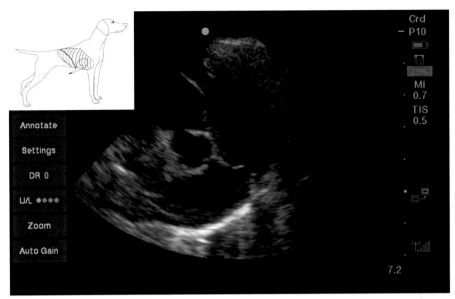

FIGURE 7.9.2 AND VIDEO 7.9.2 Cardiac POCUS using 8–4 MHz cardiac probe at 7.2 cm depth. Image taken with probe placed in transverse orientation with marker directed toward the spine. The standard right parasternal short-axis left atrial:aortic view is initially obtained. The probe is then rocked toward the head (tail of the probe lifted toward ceiling) to optimize for the main pulmonary trunk and pulmonary artery (PA).

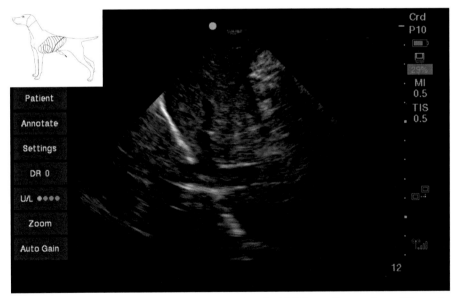

FIGURE 7.9.3 AND VIDEO 7.9.3 CaVC assessment using 8–4 MHz cardiac probe at 7.2 cm depth. Image taken with probe placed in longitudinal orientation with marker directed toward the head at the subxiphoid location. The CaVC is found crossing the diaphragm in this view as it passes through the liver. The CaVC can be challenging to find, especially when "normal" and in deep-chested dogs. Finding the gallbladder slightly right of midline at the subxiphoid location and fanning ventrally can help identify the CaVC as it crosses the diaphragm. See Video 7.9.3 for an example obtained at this site.

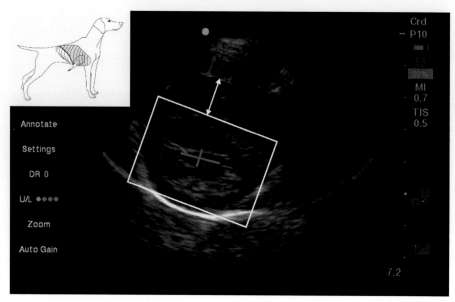

FIGURE 7.9.4 (Figure 7.9.1 labeled) The left ventricle has become flattened and has changed from a circular to a rectangular shape (white rectangle). The right ventricle is dilated (white arrow) and the right ventricular free wall is thickened (red arrow). The eccentricity index (EI) can be calculated by taking the ratio of the lateral-lateral and cranio-caudal left ventricular dimensions (1.47/0.64 = 2.3). EI in dogs without pulmonary hypertension in one recent abstract was 1.12 [0.94 – 1.3].[1]

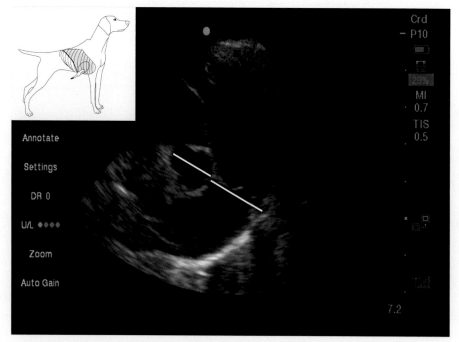

FIGURE 7.9.5 The white line indicates the direction of measurement to compare the main pulmonary artery trunk to the aorta. The above example was a PA:Ao ratio of 1.23:1 (<1 being a normal ratio).[2]

CLINICAL INTEGRATION

Suspected diagnosis based on ultrasound findings:
Pre-capillary PH.

Sonographic interventions, monitoring, and outcome:
Formal echocardiography documented severe PH (pulmonary arterial pressure [PAP] derived from the simplified Bernoulli equation >75 mmHg). Computed tomography identified pulmonary thromboembolism secondary to a hypercoagulable state in association with a transitional cell carcinoma of the bladder wall. The patient responded favorably to treatment with clopidogrel and a phosphodiesterase V inhibitor (sildenafil), surviving to discharge.

Important concepts regarding the sonographic diagnosis of pulmonary hypertension:

- PH is defined by abnormally increased pressure within the pulmonary vasculature caused by a variety of respiratory, cardiovascular, and systemic diseases.[3] Patients can present with signs of syncope, respiratory distress at rest, exercise intolerance, and signs of R-CHF.[3]
- The gold standard for PH diagnosis (right-heart catheterization) is rarely available or practiced in veterinary medicine.[3] The assessment of PH often relies on the estimation of pulmonary arterial pressure and right ventricular changes via Doppler echocardiographic assessment.[4,5] PH can be loosely categorized as pre- or post-capillary based on the absence or presence of increased pulmonary arterial wedge pressure (PAWP) and elevated left atrial (LA) pressures, respectively.[3] Patients can present with severe respiratory distress, so ruling out severe PH in respiratory compromise is an important step in emergency room evaluation and stabilization.
- Recently, a 10-point scoring system for pre-capillary PH adopted a gray-zone approach to rule in (scores ≥ 5) or rule out (scores <2) severe PH with >90% accuracy.[6] The

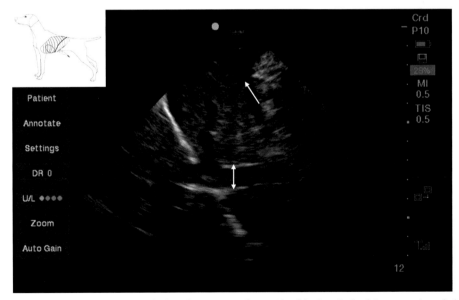

FIGURE 7.9.6 The CaVC is distended and non-compliant (double-headed white arrow) and there is a small volume of peritoneal effusion (single-headed white arrow).

scoring system assessed right atrial (RA) and right ventricular enlargement, RV hypertrophy, flattening of the interventricular septum, PA enlargement, and signs of R-CHF (presence of abdominal effusion and a distended, non-compliant CaVC).[6]
Videos 7.9.1–7.9.3 show a bedside assessment of these sites that were suggestive of severe PH (score of ≥7/10). Additionally, the eccentricity index (EI), assessed in the right parasternal short-axis view, evaluates the shape of the left ventricle and IVS in either systole or diastole, providing rapid quantitative assessment of PH, as it has been shown to correlate with the severity of PH.[1] The eccentricity index is measured at the mid-ventricular level from a parasternal short-axis view as the ratio of the latero-lateral and cranio-caudal left ventricular cavity dimensions. Dogs without evidence of pulmonary hypertension were shown to have a median EI of 1.12.[1]

Techniques and tips to identify the appropriate ultrasound images/views:

- Echocardiographic examination of PH can be challenging even when performed by board-certified cardiologists, so POCUS evaluation should form part of a comprehensive workup.
- It is easy to over-interpret right ventricular dilation in the right parasternal short-axis view depending on transducer angle. If suspicious of right ventricular dilation, the right parasternal long-axis view can also help support your findings. The right ventricle should be smaller than the left, although published veterinary ratios of right and left ventricular size are lacking.
- The pulmonary artery to aortic ratio view can be challenging to obtain and measurements should be assessed with both the aortic and pulmonary valves in view. To achieve this view, it is likely easiest to start at the right parasternal short-axis left atrial:aortic view and fan the tail of the probe toward the ceiling to move the PA into view.
- The CaVC can also be challenging to obtain and ease of assessment via the subxiphoid, right hepatic, or sublumbar views may vary between patients (e.g., right hepatic view may be easier in deep-chested dogs).[7]
- The 10-point scoring system can be used to aid decision-making in emergency cases, with >90% certainty to rule in (scores ≥5) or rule out (scores <2) severe PH.

Take-home messages

- Severe PH can be readily identified using POCUS in patients with respiratory compromise and is an important differential to rule out, owing to its specific treatment options.
- Routinely incorporating the pulmonary artery view on cardiac POCUS can help recognition of normal and subsequently abnormal pulmonary artery diameters.

REFERENCES

1. Lekane M, Merveille AC, Gommeren K, et al. Left ventricular eccentricity index for Assessment of precapillary pulmonary hypertension in dogs. In: *ECVIM-CA Online Congress*, 2021.
2. Visser LC, Im MK, Johnson LR, Stern JA. Diagnostic value of right pulmonary artery distensibility index in dogs with pulmonary hypertension: Comparison with Doppler echocardiographic estimates of pulmonary arterial pressure. *J Vet Intern Med* 2016; 30(2): 543–552.
3. Reinero C, Visser LC, Kellihan HB, et al. ACVIM consensus statement guidelines for the

diagnosis, classification, treatment, and monitoring of pulmonary hypertension in dogs. *J Vet Intern Med* 2020; 34(2): 549–573.
4. Vezzosi T, Domenech O, Costa G, et al. Echocardiographic evaluation of the right ventricular dimension and systolic function in dogs with pulmonary hypertension. *J Vet Intern Med* 2018; 32(5): 1541–1548.
5. Visser LC, Wood JE, Johnson LR. Survival characteristics and prognostic importance of echocardiographic measurements of right heart size and function in dogs with pulmonary hypertension. *J Vet Intern Med* 2020; 34(4): 1379–1388.
6. Lyssens A, Lekane M, Gommeren K, Merveille AC. Focused cardiac ultrasound to detect pre-capillary pulmonary hypertension. *Front Vet Sci* 2022; 9: 80.
7. Darnis E, Boysen S, Merveille AC, et al. Establishment of reference values of the caudal vena cava by fast-ultrasonography through different views in healthy dogs. *J Vet Intern Med* 2018; 32(4): 1308–1318.

CHAPTER 7–CASE 10

DOMESTIC SHORT HAIR CAT REFERRED FOR CARDIOVASCULAR COLLAPSE FOLLOWING SUSPECTED TRAUMA

Steffi Jalava and Ivayla Yozova

HISTORY, TRIAGE, AND STABILIZATION

AN 8-year-old male neutered domestic short-hair cat was referred by the primary care veterinarian for cardiovascular collapse following suspected trauma.

Triage exam findings (vitals):
- Mentation: dull, able to stand unassisted
- Heart rate: 152 beats per minute, heart sounds muffled right-hand side
- Femoral pulses: weak
- Mucous membranes: pale, 1–2 sec capillary refill time
- Temperature: 33.4°C (92.1°F)
- Respiratory rate: 64 breaths per minute, increased inspiratory effort and noise
- Other findings: severe scuffing of multiple nails on all four paws

POCUS exam(s) to perform:
- Cardiac POCUS
- PLUS
- Abdominal POCUS

Abnormal POCUS exam results:
See **Figures 7.10.1– 7.10.4**, as well as **Videos 7.10.1 and 7.10.2**.

Additional point-of-care diagnostics and initial management:
- Oxygen supplementation was provided by mask, butorphanol at 0.2 mg/kg IM was administered as a sedative, and active warming via Bair Hugger® was initiated.
- Packed cell volume and total proteins (PCV/TP) 33%/6.2 g/dL.
- Venous blood gas; metabolic acidosis (pH 7.19), with hyperlactatemia (8.4 mmol/L).
- Additional PLUS findings:
- A small volume of pleural effusion (not shown).
- Abdominal POCUS:
  - Abdominal fluid score 2/4; small amounts of abdominal effusion detected craniodorsal to the liver, delineating the abdominal border of the thorax, and cranial to the bladder (not shown).

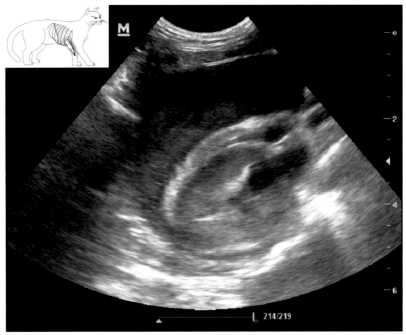

FIGURE 7.10.1 AND VIDEO 7.10.1 Cardiac POCUS of the right parasternal long-axis view of the heart using a micro convex curvilinear probe with depth set at 3cm frequency of 9 Hz. Marker caudal. The fur is parted, and 70% isopropyl alcohol is used as the coupling agent.

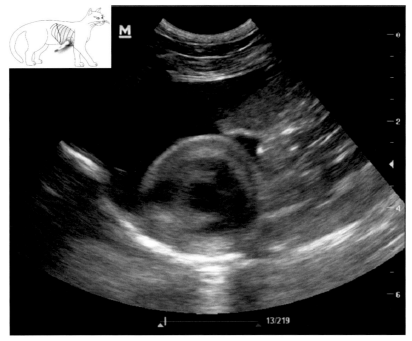

FIGURE 7.10.2 AND VIDEO 7.10.2 Cardiac POCUS of the right parasternal short-axis view of the heart using a micro convex curvilinear probe with depth set at 3 cm frequency of 9 Hz. Marker caudal. The fur is parted, and 70% isopropyl alcohol is used as the coupling agent.

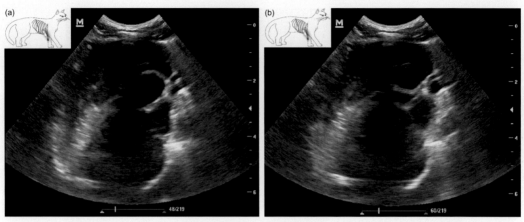

FIGURE 7.10.3A AND FIGURE 7.10.3B Cardiac POCUS of the right parasternal long-axis view during systole (7.10.3a) and diastole (7.10.3b), using a micro convex curvilinear probe with depth set at 3 cm and frequency of 9 Hz. (note; view is slightly oblique, possibly due to displacement of the heart in the pericardial sack). Marker caudal. The fur is parted, and 70% isopropyl alcohol is used as the coupling agent.

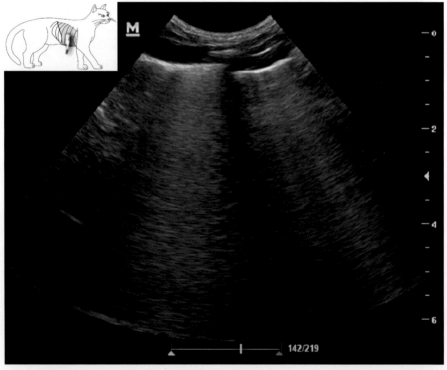

FIGURE 7.10.4 Right short axis of the lungs (note; similar findings throughout the right hemithorax,) using a micro convex curvilinear probe with depth set at 3 cm and frequency of 9 Hz. Marker caudal. The fur is parted, and 70% isopropyl alcohol is used as the coupling agent.

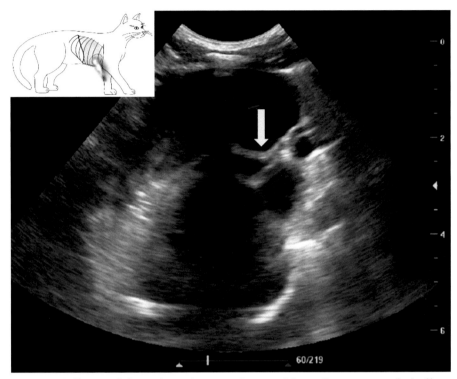

FIGURE 7.10.5 Collapse of the right atrium consistent with cardiac tamponade (yellow arrow).

QUESTIONS AND ANSWERS

1. What are the differentials to rule in or out with POCUS based on history and physical exam?
2. What are the sonographic findings?
3. What is the sonographic diagnosis?
4. If necessary, what additional sonographic examination or findings would help rule in or out the differential diagnoses?

1. **Differential diagnosis to rule in or rule out with POCUS:**
   - Pulmonary contusions, pneumothorax, hemothorax, diaphragmatic hernia, hemoabdomen, uroabdomen, urinary bladder hematoma, kidney avulsion, ureteral tear (both manifested as retroperitoneal effusion), abdominal wall herniation are all potential injuries that can be detected ultrasonographically in feline patients following trauma.

2. **Describe your sonographic findings:**
   - Severe pericardial effusion leading to cardiac tamponade (**Figures 7.10.1–7.10.5**).
   - Thickened left ventricular wall to lumen size (**Figure 7.10.1**) and a small LA:Ao ratio (**Figure 7.10.2**).
   - Dynamic collapse of the right atrium during the cardiac cycle (**Figure 7.10.3a and 7.10.3b**).
   - Unilateral diffuse to coalescing B-lines of the right hemithorax (**Figure 7.10.4**).

3. **What is your sonographic diagnosis?**
   - Obstructive shock (compounded by a degree of hypovolaemia) resulting in microcardia.
   - Pericardial effusion leading to cardiac tamponade, with collapse of the right atrium.
   - Diffuse unilateral pulmonary contusions of the right hemithorax.
4. **What additional sonographic examination or finding would help you rule in or rule out your differential diagnoses (if necessary)?**
   None.

CLINICAL INTEGRATION

Suspected diagnosis based on ultrasound findings:

The most likely differential based on history, physical exam, and ultrasonographic findings is pericardial effusion causing cardiac tamponade (unknown fluid type without further diagnostics), with right-sided pulmonary contusions and mild abdominal effusion (fluid type also unknown without further diagnostics) with a fluid score of 2/4.

Although a traumatic event was suspected in this case, whether the pericardial effusion was trauma induced (i.e. hemorrhagic pericardial effusion) or whether the pericardial effusion predisposed to the traumatic event occurring, was unknown following initial diagnostics.

Sonographic interventions, monitoring, and outcome:

- The cause of shock, in this case, was primarily cardiac tamponade-induced obstructive shock from the pericardial effusion. As a result, a pericardiocentesis was performed.
- Pericardial effusion: PCV/TP 30%/5.0 mg/dL and cytology-supported hemorrhage.
- An autotransfusion of the collected blood was performed.
- Ultrasound-guided abdominocentesis performed for diagnostic purposes; findings supporting hemorrhagic abdominal effusion.
- Ultrasound-guided thoracocentesis for diagnostic purposes performed prior to pericardiocentesis to avoid leakage of pericardial effusion and contamination of the pleural effusion; findings supportive of hemorrhagic pleural effusion.
- Serial POCUS examinations were performed to monitor the resolution of pericardial, pleural, and abdominal effusions, and pulmonary contusions.
- Blood results:
  - ALT severely elevated; indicated hepatocellular damage, supportive of suspected blunt force injury.
  - CK and AST were severely elevated; indicating muscle damage, also supportive of the suspected blunt-force trauma.
  - Severely elevated troponin levels; supportive of cardiac muscle damage.
- Following a comprehensive workup, a traumatic hemorrhagic pericardial effusion was diagnosed.
- Following the relief of the pericardial effusion and autotransfusion of the retrieved fluid, stabilization rapidly occurred, with improvement and eventual resolution of the clinical and abnormal sonographic findings.

Importance concepts regarding the sonographic diagnosis of pericardial effusion:

- Diagnosis of pericardial effusion is made by the visualization of a circumferential accumulation of either an anechoic fluid or in the case of more cellular effusions (hemorrhagic or septic), an increased echogenic fluid which may appear to be 'swirling, around the heart. Fibrin and pyogenic effusions can result in the presence of flocculation (hyperechoic strands moving within the fluid).[1-3]
- A hemorrhagic pericardial effusion can be caused by a cardiac chamber tear (more common on the low-pressure right side of the heart) or a pericardial tear from a traumatic event (suspected in this case), from an atrial tear secondary to mitral valve disease, or from a bleeding cardiac mass. Clots may be observed as tubular areas of increased echogenicity.[2,3]
- Cardiac tamponade arises from the pressure of the accumulated pericardial effusion causing cardiac chamber compression and preventing chamber filling. Collapse of the lower pressure right side of the heart occurs, with collapse of the right atrium and occasionally (in more advanced cases) the right ventricle occurring. The parasternal long-axis view is the best to visualize the right atrium and ventricle to assess for collapse.[1-3]
- Multiple cardiac views, including the parasternal short-axis, parasternal long-axis, and subxiphoid view, and all 4 cardiac chambers should be identified when diagnosing pericardial effusion to avoid mistaken identification of dilated cardiac chambers as pericardial effusion with consequential disastrous results if centesis is performed.[3,4]
- Pleural effusion can also be misdiagnosed as pericardial effusion. The distribution of pleural effusion forms 'sharp angles' or triangles especially between the heart and lungs and/or diaphragm, while pericardial effusion will form 'curves' around the heart as it is contained within the pericardial sac.[4]
- In obese cats, increased pericardial fat should not be mistakenly diagnosed as pericardial effusion.[5]

Techniques and tips to identify the appropriate ultrasound images/views:

- To ensure the heart is visualized in its entirety, identify the pericardium as a hyperechoic line in the deep field of view.[6]
- Perform all standard focused cardiac ultrasound views; including parasternal short axis, parasternal long axis, and subxiphoid views, and visualize all chambers of the heart when diagnosing pericardial effusion to avoid accidental misdiagnosis.[6]
- The pericardium attaches to the top of the atriums, therefore visualized pericardial effusion should reduce at the top of the atrium in the long axis view.

Take-home messages

- Pericardial effusion with cardiac tamponade is life-threatening, with focused cardiac ultrasound enabling swift diagnosis and treatment, to correct deranged hemodynamics, and prevent potential cardiac arrest.

- Multiple cardiac views should be performed, with all 4 cardiac chambers being identified to avoid misdiagnosis.
- Pericardial effusion is visualized as a circumferential accumulation of fluid around the heart, or forming 'curves' around the heart.

REFERENCES

1. Alerhand, S. and Carter, J. What echocardiographic findings suggest a pericardial effusion is causing tamponade? *American Journal of Emergency Medicine* 2019; 37: 321–326.
2. Goodman, A. et al. The role of bedside ultrasound in the diagnosis of pericardial effusion and cardiac tamponade. *Journal of Emergencies, Trauma, and Shock* 2012; 5(1): 72–75.
3. Huang, Y. et al. Traumatic pericardial effusion: Impact of diagnostic and surgical approaches. *Resuscitation* 2010; 81: 1682–1686
4. Lisciandro, G. Cageside Ultrasonography in the Emergency Room and Intensive Care Unit. *Veterinary Clinics: Small Animal Practice* 2020; 50: 1445–1467.
5. Lisciandro, G. TFAST Accurate Diagnosis of Pleural and Pericardial Effusion, Caudal Vena Cava in Dogs and Cats. *Veterinary Clinics: Small Animal Practice* 2021; 51(6): 1169–1182.
6. Point-of-Care Ultrasound Techniques for the Small Animal Practitioner. Chapter 21. POCUS: Heart–Pericardial Effusion and Pericardiocentesis. *Teresa DeFrancesco*. 417–424.

SECTION 4

VASCULAR APPLICATIONS OF VETERINARY POINT-OF-CARE ULTRASOUND

Section Editor Tereza Stastny

CHAPTER 8A
INTRODUCTION TO ULTRASOUND-GUIDED VASCULAR ACCESS

Tereza Stastny

In human medicine, ultrasound-guided vascular catheterization is frequently employed because it is faster than blind techniques, has a higher success rate, and a lower complication rate.[1,2] Ultrasound-guided catheterization is particularly useful in situations when vascular access is challenging due to small vessel size, thrombosis, subcutaneous edema, obesity, marked peripheral vasoconstriction, and vascular collapse.[1,2] Limited available evidence in veterinary medicine suggests that complication rates and time to place ultrasound-guided central lines in anesthetized dogs are comparable to blind techniques,[3] ultrasound-guided femoral arterial catheterization in anesthetized dogs has a high success rate,[4,5] and ultrasound-guided catheterization in canine cadavers in scenarios where landmarks are difficult to visualize has a high success rate.[6] Beyond cadaveric studies and studies performed on anesthetized dogs, there is a paucity of literature assessing ultrasound-guided catheterization in veterinary medicine. However, given that situations leading to difficult vascular access are common in the emergency and critical care setting, familiarity with ultrasound-guided vascular access techniques in these patients is valuable.

PATIENT PREPARATION AND TRANSDUCER SELECTION

Patients are positioned in sternal recumbency to access the cephalic veins, and lateral recumbency when accessing the jugular or saphenous veins. For ultrasound-guided central venous catheterization, the patient is placed in lateral recumbency with the neck extended and the limbs gently pulled caudally, fur is clipped widely, and the jugular area is prepared aseptically (refer to Chapter 9 for a full description of the preparation and ultrasound-guided jugular catheterization technique). For ultrasound-guided peripheral venous catheterization, the vein is prepared for routine catheter placement.

A high-frequency linear array transducer (8–13 MHz) is preferred when attempting vascular catheterization, but a microconvex (curvilinear) transducer may also be used if a linear transducer is unavailable. The head of the transducer should be cleaned with a low-level disinfectant such as hydrogen peroxide or diluted bleach and a sterile lubricant should be applied to the ultrasound site. For sterile procedures such as central or arterial line placement, a gel-filled sterile glove may be placed over the ultrasound transducer, and sterile lubricant applied to the ultrasound site.

ULTRASOUND-GUIDED VASCULAR CATHETERIZATION TECHNIQUES

Ultrasound-guided catheter placement can be performed using two techniques: the in-plane technique and the out-of-plane technique.

Each technique comes with certain advantages and disadvantages (see section titled "Pearls and Pitfalls").

DOI: 10.1201/9781003436690-43

IN-PLANE TECHNIQUE:

1. Prepare the patient, vessel, and transducer appropriately (see above).
2. Position the ultrasound machine in front of the sonographer and set the depth to maximize the vessel size without losing the vessel in the far field of the image.
3. Distinguish vein from artery by applying pressure to the transducer, collapsing the vein at a much lower pressure than the artery.[2]
4. Ask an assistant to hold off the vein to further prevent venous collapse.
5. Scan the vessel of interest to gauge its direction, diameter, and depth from the skin surface.
6. Select an appropriately sized catheter. A longer catheter is preferred in situations of subcutaneous edema, hematoma formation, or obese patients.
7. Orient the marker on the transducer to visualize the catheter entering from either the left or the right side of the ultrasound image depending on operator preference (**Figure 8.1a**).
8. With the *in-plane technique*, the head of the transducer is oriented parallel (long axis) to the vessel.
9. Insert the stylet tip at an angle of 30° to 45° to the transducer, just distal to the ultrasound transducer, and then visualize the catheter and stylet tip traverse the skin, subcutaneous tissues, and superficial wall of the vessel, into the lumen (**Figure 8.2a**).
10. Once in the lumen, the angle of the catheter is decreased to run more horizontal to the vessel, the catheter and stylet are advanced slightly within the vessel, and then once the catheter is well situated in the vessel, the catheter is advanced off the stylet and down the vessel lumen.
11. The entire procedure can be visualized using in-plane techniques (**Figure 8.3a**).

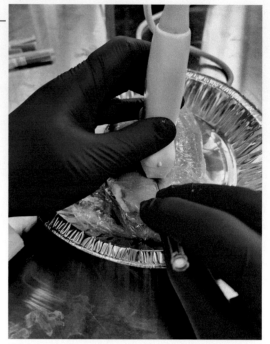

FIGURE 8.1A **In-plane ultrasound-guided vascular access using a high-frequency linear array transducer (8–13 MHz) demonstrating operator orientation using a chicken phantom. The marker on the transducer is faced cranially, and the catheter is visualized entering the chicken breast.** Image courtesy of Katrine Gillett with permission.

OUT-OF-PLANE TECHNIQUE:

1. Follow Steps 1–7 of the in-plane technique.
2. With the *out-of-plane* technique, the head of the transducer is oriented perpendicular (short axis) at a 90° angle to the vessel (**Figure 8.4a**).
3. Identify the vessel of interest, then focus attention on the *ultrasound transducer–skin interface* (not the ultrasound screen).

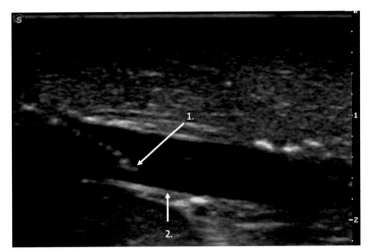

FIGURE 8.2A In-plane ultrasound-guided vascular access demonstrated using a high-frequency linear array transducer (8–13 MHz) with the depth set at 2 in. Ultrasound image of the vessel and catheter in long axis (longitudinal). (1) Stylet tip. (2) Distal vessel wall. Image courtesy of Katrine Gillett with permission.

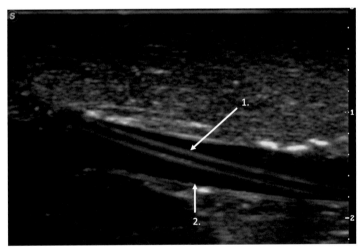

FIGURE 8.3A In-plane ultrasound-guided vascular access demonstrated using a high-frequency linear array transducer (8–13 MHz) with the depth set at 2 in. Ultrasound image of the vessel and catheter in long axis (longitudinal). (1) Catheter shaft. (2) Distal vessel wall. Image courtesy of Katrine Gillett with permission.

4. Insert the needle at an angle of 30° to 45° to the transducer, just distal to the ultrasound transducer.
5. Direct the bevel of the stylet up, toward the ultrasound transducer, to maximize the reflection of ultrasound waves and enhance stylet tip visualization.
6. Once the tip of the stylet has entered the skin, focus attention on the *ultrasound screen*.
7. Advance the stylet tip until it becomes visible within the ultrasound image as a white hyperechoic dot. Stop advancing the catheter as soon as the stylet tip

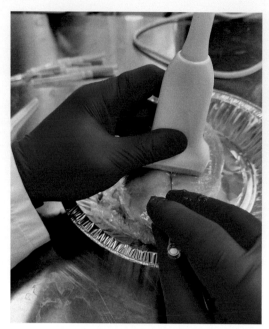

FIGURE 8.4A Out-of-plane ultrasound-guided vascular access using a high-frequency linear array transducer (8–13 MHz) demonstrating operator orientation using a chicken phantom. The head of the transducer is perpendicular to the vessel, and the catheter is visualized entering the chicken breast in the center of the transducer head. Image courtesy of Katrine Gillett with permission.

is seen on the ultrasound screen. Then, incrementally (millimeter by millimeter) sweep the transducer proximally along the vessel until the stylet tip is once again out-of-plane. Advance the stylet tip again until it becomes visible within the ultrasound image as a white hyperechoic dot. Then, incrementally sweep the transducer proximally along the vessel until the stylet tip is once again out-of-plane.

8. The process is repeated until the stylet tip can be visualized within the vessel lumen (seen as a hyperechoic dot [**Figure 8.5a**]).
9. Failure to sweep the transducer completely off the stylet tip (completely out of the plane of the ultrasound image) can result in the catheter tip being advanced beyond the ultrasound beam when the catheter is advanced again.
10. Once the catheter tip is visualized within the vessel, a flash of blood will generally occur within the stylet.
11. The remainder of the procedure proceeds as with the blind technique for vascular catheter placement, or ultrasound guidance can be continued to the entirety of the process. The angle of the catheter is

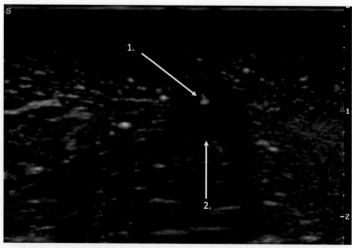

FIGURE 8.5A Out-of-plane ultrasound-guided vascular access demonstrated using a high-frequency linear array transducer (8–13 MHz) with the depth set at 2 in. Ultrasound image of the vessel and catheter in long axis (longitudinal). (1) Stylet tip. (2) Distal vessel wall. Image courtesy of Katrine Gillett with permission.

decreased to run more horizontal to the vessel, the catheter and stylet are advanced slightly within the vessel, and then the catheter is advanced off the stylet and down the vessel lumen.

ULTRASOUND PEARLS AND PITFALLS

In-plane technique:
- This technique requires practice to keep the catheter perfectly aligned in the plane of the ultrasound beam.
- Misalignment as small as 1° to 2° off center of the ultrasound transducer prevents catheter visualization.
- Linear array transducer's footprint has a narrow width (1–2 mm), making alignment particularly difficult, especially with smaller peripheral vessels.
- The visualization of surrounding structures during the in-plane technique is advantageous.

Out-of-plane technique:
- Failure to stop advancing the catheter as soon as the stylet tip appears as a hyperechoic dot on the ultrasound screen increases the risk of accidentally traversing the distal wall of the vessel.
- Advancing the stylet tip beyond the ultrasound beam results in uncertainty regarding catheter tip location.
- Advancing the stylet tip while simultaneously sweeping the transducer proximally requires more training and practice.

REFERENCES

1. Gottlieb M, Sundaram T, Holladay D, Nakitende D. Ultrasound-guided peripheral intravenous line placement: A narrative review of evidence-based best practices. *West J Emerg Med* 2017;18(6):1047–1054.
2. van Loon FHJ, Buise MP, Claassen JJF, Dierick-van Daele ATM, Bouwman ARA. Comparison of ultrasound guidance with palpation and direct visualization for peripheral vein cannulation in adult patients: A systematic review and meta-analysis. *Br J Anaesth* 2018;121(2):358–366.
3. Hundley D, Brooks A, Thomovsk J, et al. Comparison of ultrasound-guided and landmark-based techniques for central venous catheterization via the external jugular vein in healthy anesthetized dogs. *Am J Vet Res* 2018;79(6):628–636.
4. Pavlisko ND, Soares JHN, Henao-Guerrero NP, Williamson AJ. Ultrasound-guided catheterization of the femoral artery in a canine model of acute hemorrhagic shock. *J Vet Emerg Crit Care (San Antonio)* 2018;28(6):579–584. doi:10.1111/vec.12767.
5. Ringold SA, Kelmer E. Freehand ultrasound-guided femoral arterial catheterization in dogs. *J Vet Emerg Crit Care* 2008;18(3):306–311.
6. Chamberlin SC, Sullivan LA, Morley PS, Boscan P. Evaluation of ultrasound-guided vascular access in dogs. *J Vet Emerg Crit Care* 2013;23(5):498–503.

CHAPTER 8B
INTRODUCTION TO CAUDAL VENA CAVA COLLAPSIBILITY INDEX

Tereza Stastny

INTRODUCTION

Hypovolemia and hypervolemia are both associated with worse outcomes in people admitted to the ICU.[1] In human medicine, dynamic variables are proving more accurate in identifying fluid responsiveness than static variables such as central venous pressure, blood pressure, and heart rate.[2,3] Dynamic variables account for changes in cardiac output and venous return secondary to ventilation, which contributes to their superior accuracy.

Although controversial, the inferior vena cava collapsibility index (IVC-CI) is a dynamic variable that predicts fluid responsiveness in mechanically and spontaneously ventilating humans.[4,5] Currently, there is a paucity of literature in small animal veterinary medicine assessing the caudal vena cava collapsibility index (CVC-CI) in *spontaneously* breathing patients, aside from a small clinical study in spontaneously breathing dogs with compromised hemodynamics. This study suggests that CVC-CI can accurately predict fluid responsiveness, although the authors conclude that further research is necessary to assess the generalizability of their results.[6]

CVC-CI FORMULAS

The CVC-CI expresses the change (%) in the diameter of the caudal vena cava (CVCd) during the respiratory cycle. In spontaneously breathing patients, inspiration decreases transpulmonary/intrathoracic pressure, increasing preload and decreasing the diameter of the CVC at the site of the subxiphoid window. Conversely, on expiration, intrathoracic pressure increases, decreasing preload and subsequently increasing the diameter of the CVC at the site of the subxiphoid window. The opposite is true of mechanically ventilated patients.

CVC-CI = CVCd max − CVCd min/CVCd max

Spontaneous ventilation: CVC-CI = $CVCd_{end\text{-}expiration} - CVCd_{end\text{-}inspiration}/CVCd_{end\text{-}expiration}$

Mechanical ventilation: CVC-CI = $CVCd_{end\text{-}inspiration} - CVCd_{end\text{-}expiration}/CVCd_{end\text{-}inspiration}$

INTERPRETING VENTILATION-INDUCED CHANGES IN CVCD

Ventilation induces changes in intrathoracic pressure, which are reflected in the diameter of the caudal vena cava and depend on the central blood volume.[6] A collapsed or "flat" CVC with a CVC-CI of more than 50% (**Figure 8.1b**) is suggestive of hypovolemia based on human

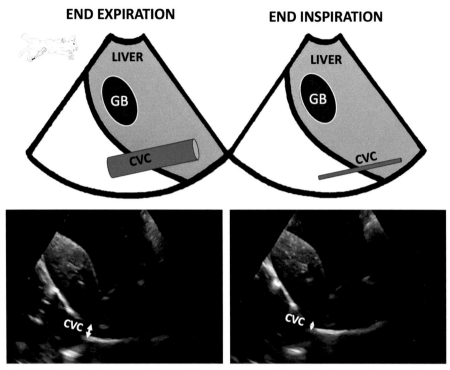

FIGURE 8.1B Schematic ultrasound image in a spontaneously ventilating canine at end-expiration and end-inspiration. The image depicts a volume-tolerant patient with a subjectively "flat" CVC on inspiration when compared to expiration. This patient's CVC-CI was calculated to be 80%.

literature.[3,4] A distended CVC with less than a 20% change in the CVC-CI suggests increased right atrial pressures and a fluid-intolerant patient (**Figure 8.2b**).[4,7] Differentials for a distended CVC include hypervolemia, cardiac tamponade, right-sided congestive heart disease, and pulmonary hypertension. Assessment of other point-of-care ultrasound (POCUS) examinations should be evaluated to differentiate causes of an enlarged CVC (see C-POCUS and ultrasound PLUS sections). There is a gray zone that exists between 20% and 50% CVC-CI. Given the overlap between the gray zone and the hypovolemic zone, some authors have moved toward using the terminology "fluid tolerant" and "fluid intolerant" rather than hypovolemic and hypervolemic.

IMAGE ACQUISITION

The *subxiphoid window* is the acoustic window most often used to assess the CVC-CI. The *caudal vena cava view* is assessed in the longitudinal plane at a 45° angle where it crosses the diaphragm. Slowly fan the probe to the right and left of the midline while looking for a "break" in the diaphragm. Once the CVC is located, the transducer is fanned off either side of the CVC to locate its widest diameter. To calculate the CVC-CI in spontaneously breathing patients, the widest diameter of the CVC is identified across several respiratory cycles. The image is then frozen, and measurements are taken of the CVC's maximum diameter

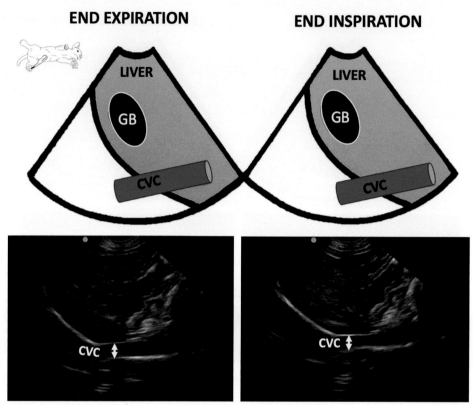

FIGURE 8.2B Schematic ultrasound image in a spontaneously ventilating canine at end-expiration and end-inspiration. The image depicts a volume-intolerant patient with a subjectively "flat" CVC on inspiration when compared to expiration. This patient's CVC-CI was calculated to be 10%.

during expiration and minimum diameter during inspiration. (see formula and explanation above). The CVC-CI can also be estimated rather than frozen and measured.

PEARLS AND PITFALLS

- Applying too much pressure to the transducer will create pressure artifact and result in a falsely collapsed CVC.
- In most patients, the CVC is identified just right of midline in the same plane that the gallbladder is visible.
- The caudal vena cava to aortic ratio (CVC:Ao ratio) is a second measurement that shows promise in evaluating intravascular volume status in dogs.[7] Further description is beyond the scope of this chapter.

REFERENCES

1. Joosten A, Alexander B, Cannesson M. Defining goals of resuscitation in the critically ill patient. *Crit Care Clin* 2015; 31(1): 113–132.
2. Cecconi M, De Backer D, Antonelli M, et al. Consensus on circulatory shock and hemodynamic monitoring: Task force of the European

society of intensive care medicine. *Intensive Care Med* 2014; 40(12): 1795–1815.
3. Monnet X, Marik PE, Teboul JL. Prediction of fluid responsiveness: An update. *Ann Intensive Care* 2016; 6(1): 111.
4. Via G, Hussain A, Wells M, et al. International liaison committee on focused cardiac ultrasound (ILC-FoCUS); international conference on focused cardiac ultrasound (IC-FoCUS). International evidence-based recommendations for focused cardiac ultrasound. *J Am Soc Echocardiogr* 2014; 27(7): 683.
5. Adler AC, Brown KA, Conlin FT, Thammasitboon S, Chandrakantan A. Cardiac and lung point-of-care ultrasound in pediatric anesthesia and critical care medicine: Uses, pitfalls, and future directions to optimize pediatric care. *Paediatr Anaesth* 2019; 29(8): 790–798.
6. Donati PA, Guevara JM, Ardiles V, et al. Caudal vena cava collapsibility index as a tool to predict fluid responsiveness in dogs. *J Vet Emerg Crit Care* 2020; 30(6): 677–686.
7. Kwak J, Yoon H, Kim J, Kim M, Eom K. Ultrasonographic measurement of caudal vena cava to aorta ratios for determination of volume depletion in normal beagle dogs. *Vet Radiol Ultrasound* 2018; 59(2): 203–211.

CHAPTER 9

VASCULAR POINT-OF-CARE CASES

CHAPTER 9 – CASE 1

BORZOI WITH ACUTE ONSET OF VOMITING, FOLLOWED BY SEVERE DYSPNEA AND HYPERTHERMIA

Laurentin Duriez and Kris Gommeren

HISTORY, TRIAGE, AND STABILIZATION

A 9-year-old male neutered Barzoï presented with acute onset of vomiting, followed by severe dyspnea and hyperthermia.

Triage exam findings (vitals):
- Mentation: quiet to obtunded, lateral recumbency
- Respiratory rate: severe inspiratory and expiratory dyspnea, tachypnea at 80 breaths/min
- Heart rate: 160 beats per minute
- Mucous membranes: pale, 3 sec capillary refill time
- Femoral arterial pulse: weak but regular
- Systolic arterial pressure: 80 mmHg
- Temperature: 40°C (104°F)

Initial Diagnostics
- Hematology: severe leukopenia $0.50 \times 10^9/L$ $(5.05 – 16.76 \times 10^9/L)$
- C-reactive protein (CRP): >100 mg/L (0–10)
- PLUS: bilateral cranioventral shred signs and coalescent B-lines

Diagnosis:
Aspiration pneumonia with secondary leukopenia and septic shock.

Initial management:
The patient had worsening dyspnea and work of breathing despite ampicillin/sulbactam (50 mg/kg IV q8), enrofloxacin (10 mg/kg IV q24) and high-flow oxygen therapy. Patient was therefore transitioned to mechanical ventilation. The nursing staff attempted to place a central venous catheter, but after several unsuccessful attempts transitioned to ultrasound-guided placement.

QUESTIONS AND ANSWERS

1. What is the advantage of having a central venous catheter (CVC) in this patient?
2. How do you differentiate the jugular vein from the carotid artery using ultrasound?
3. What are the complications associated with central line placement? Do point-of-care ultrasound (POCUS)-guided techniques reduce the risk of complications?

1. **What is the advantage of having a central venous catheter in this patient?**
   - Central veins are large vessels that are superficially located, making them accessible even in hypotensive patients. Reports in human medicine describe greater success rates placing central venous catheters with fewer complications when compared to peripheral catheters.[1]
   - Central venous catheters can have multiple access ports, allowing for simultaneous delivery of various fluids and medications, easy blood sampling, and measurement of central venous pressure to assess right heart function and fluid tolerance.
   - Central venous catheters are more biocompatible, allowing them to remain in place for longer periods in patients expected to have prolonged hospitalization.[2]
2. **How do you differentiate the jugular vein from the carotid artery using ultrasound?**
   - Jugular veins are not pulsatile, while the carotid arteries are pulsatile.[3] The pulsation of the carotid artery may, however, be translated to the jugular veins, which may confuse the clinician.
   - Veins are more compliant than arteries. By applying gentle probe pressure, jugular veins will collapse, while carotid arteries will not collapse.
3. **What are the complications associated with central line placement? Do POCUS-guided techniques reduce the risk of complications?**
   - According to human literature, the use of POCUS-guided catheterization during vascular cannulation greatly improves first-pass success and reduces complications such as:[3]
     - arterial puncture
     - hematoma formation
     - venous air embolism
     - nerve injury

CLINICAL INTEGRATION

What is required to perform POCUS-guided CVC placement technique (Figure 9.1.1):
- Clippers
- Surgical scrub
- Sterile gloves and drapes
- Central venous catheter kit
- #11 scalpel blade
- Suture material
- Scissors

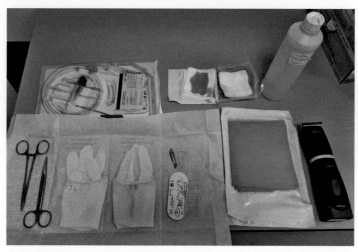

FIGURE 9.1.1 Material needed to place CVC.

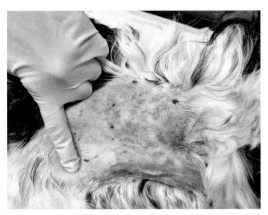

FIGURE 9.1.2 **Fur clipped widely and prepared aseptically.**

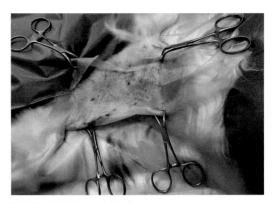

FIGURE 9.1.3 **Sterile drape covering the area.**

- Clamp
- Cleaned ultrasound linear probe
- Sterile ultrasound gel or alcohol

POCUS-guided CVC placement technique:

- Place the dog in lateral recumbency with the neck extended and the limbs gently pulled caudally.
- Clip fur widely and prepare the area aseptically (**Figure 9.1.2**).
- Clean the ultrasound (preferably) linear probe.
- Surgically scrub your hands and put on sterile gloves.
- Place sterile drapes to cover the area (**Figure 9.1.3**).
- Measure the distance from the insertion site to the third to fourth ICS (intercostal space). The central venous catheter should rest in the cranial vena cava just cranial to the right atrium.

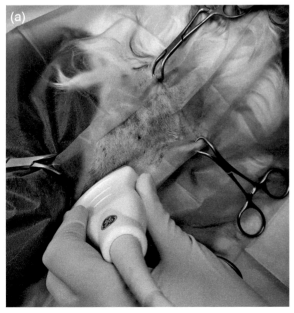
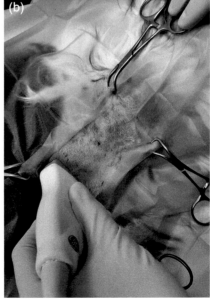

FIGURE 9.1.4 **(a) Probe is placed over and parallel to the vessel, depicting the in-plane technique; b) Probe is placed over and perpendicular to the vessel, depicting the out-of-plane technique.**

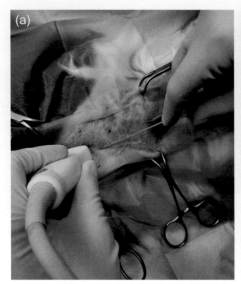
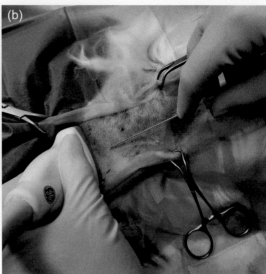

FIGURE 9.1.5 (a) Needle is in the same plane as the ultrasound probe depicting the in-plane technique, jugular vein is visualized, and over-the-needle catheter is advanced until it is visualized in the vein. (b) Needle is not in the ultrasound probe's plane, depicting the out-of-plane technique, jugular vein is visualized, and over-the-needle catheter is advanced until it is visualized in the vein.

- Place the probe caudal to the insertion site and look for the jugular vein.[2] The non-dominant hand should hold the ultrasound probe while the dominant hand controls the needle.[3]

- As soon as the jugular vein is identified, the clinician can choose between two different techniques: in-plane or out-of-plane views,[1] referring to the relationship between the needle of the catheter and the ultrasound probe.

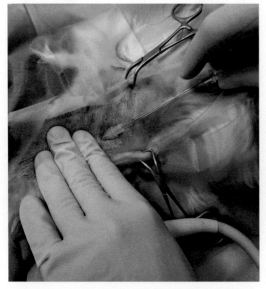

FIGURE 9.1.6 Catheter is advanced into the center of the vein and the stylet is withdrawn.

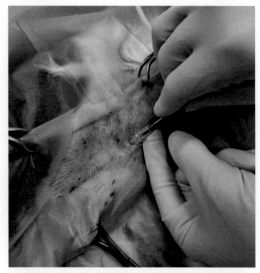

FIGURE 9.1.7 Guidewire is inserted into the vessel through the over-the-needle catheter to the pre-measured length.

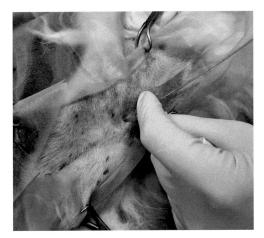

FIGURE 9.1.8 Catheter is removed.

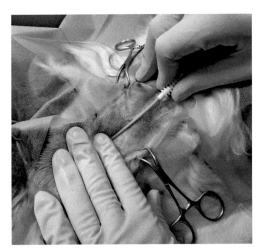

FIGURE 9.1.9 Plastic dilator is inserted into the vessel over the wire.

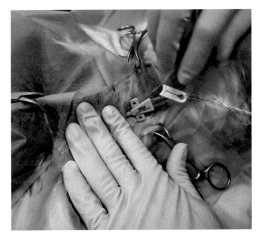

FIGURE 9.1.10 Catheter is inserted over the guidewire and into the vessel to the predetermined length.

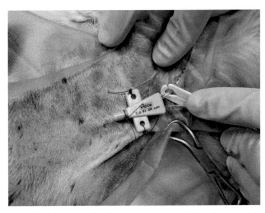

FIGURE 9.1.11 Catheter is secured in place with sutures.

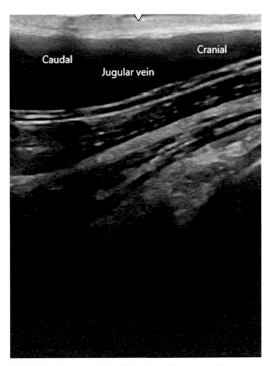

FIGURE 9.1.12 POCUS view of the jugular vein with the dog placed in lateral recumbency, using a linear probe, 12 MHz, 5 cm depth, and in-plane technique. Fur was clipped. Alcohol and gel in a sterile glove were used as coupling agents. The jugular vein is labeled.

- In-plane view: the needle is in the same plane as the ultrasound probe, visualizing the entire needle as it enters the vein (**Figure 9.1.4a**).

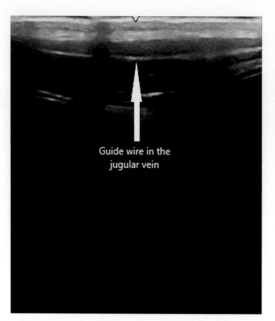

FIGURE 9.1.13 POCUS-guided catheterization with the dog placed in lateral recumbency, using a linear probe, 12 MHz, 5 cm depth, and in-plane technique. Fur was clipped. Alcohol and gel in a sterile glove were used as coupling agents. The guidewire is labeled within the jugular vein.

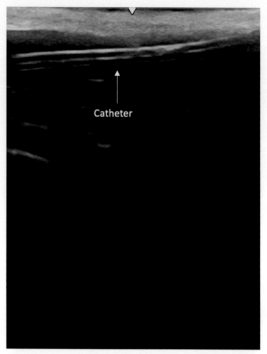

FIGURE 9.1.14 POCUS-guided catheterization with the dog placed in lateral recumbency, using a linear probe, 12 MHz, 5 cm depth, and in-plane technique. Fur was clipped. Alcohol and gel in a sterile glove were used as coupling agents. The catheter is depicted in the jugular vein.

- Out-of-plan view: the needle is not in the probe's plane, only visualizing the tip of the needle. This technique requires some training to follow the catheter tip with the probe (**Figure 9.1.4b**).
- Visualize the jugular vein, and advance the over-the-needle catheter until you see the catheter in the vein. Successful cannulation of the vessel is confirmed by direct visualization of the needle entering the vessel[3] (**Figure 9.1.5a** and **b**).
- Advance the catheter into the center of the vein, then withdraw the stylet (**Figure 9.1.6**).
- Insert the guidewire into the vessel through the over-the-needle catheter to the pre-measured length (**Figure 9.1.7**). Never let go of the guidewire at any point in the procedure.
- Remove the catheter (**Figure 9.1.8**).

- Insert the plastic dilator into the vessel by placing the wire through it and guiding it into the vessel (**Figure 9.1.9**). An incision into the skin using a #11 scalpel blade next to the wire may ease insertion. Then remove the dilator from the guidewire.
- Insert the catheter over the wire and guide it into the vessel to the predetermined length (**Figure 9.1.10**).
- Ensure the end of the wire is projecting out of the distal end of the catheter before advancement; the wire should be held as the catheter is advanced into the vessel and should exit the distal port of the catheter.
- Remove the wire and leave the catheter in place.

- Aspirate using a small syringe; blood should easily flow into the syringe. Flush the catheter with sterile saline. Repeat the procedure for each port if a multilumen catheter is used.
- Secure the catheter in place with sutures[2] (**Figure 9.1.11**).

Tips to identify ultrasound images:

- Another person applying compression caudally to the jugular vein induces congestion, increasing the diameter of the vein, making it easier to visualize and catheterize.
- Preferences between using the in-plane or out-of-plane technique is clinician dependent.

Take-home message

POCUS is a valuable tool to perform central venous catheterization, allowing for easy visualization of the jugular vein and decreasing the complication rate. The technique is easy to learn and can save critical patients.

REFERENCES

1. Chamberlin, S.C., Sullivan, L.A., Morley, P.S., Boscan, P. Evaluation of ultrasound-guided vascular access in dogs. *J. Vet. Emerg. Crit. Care* 2013 Sep–Oct; 23(5): 498–503. https://doi.org/10.1111/vec.12102
2. Campbell, M.T., Macintire, D.K. Catheterization of the venous compartment. *Adv. Monit. Proced. Small Anim. Emerg. Crit. Care* 2014: 49–68. https://doi.org/10.1002/9781118997246.ch4
3. Troianos, C.A., Hartman, G.S., Glas, K.E., Skubas, N.J., Eberhardt, R.T., Walker, J.D., Reeves, S.T. Guidelines for performing ultrasound guided vascular cannulation: Recommendations of the American Society of Echocardiography and the society of cardiovascular anesthesiologists. *J. Am. Soc. Echocardiogr.* 2011; 24(12): 1291–1318. https://doi.org/10.1016/j.echo.2011.09.021

CHAPTER 9 – CASE 2

GOLDEN RETRIEVER WITH A 24-HOUR HISTORY OF ACUTE VOMITING, DIARRHEA, AND SEVERE LETHARGY

Aurélie Jourdan and Kris Gommeren

HISTORY, TRIAGE, AND STABILIZATION

A 5-month-old male intact Golden Retriever presented with a 24-h history of acute vomiting, diarrhea, and severe lethargy. Not current on vaccinations. No known dietary indiscretion.

Triage exam findings (vitals):
- Mentation: lateral recumbency, stuporous
- Respiratory rate: 36 breaths per minute
- Heart rate: 150 beats per minute
- Mucous membranes: pale pink, 3 sec capillary refill time
- Femoral and dorsal pedal arterial pulses: weak but regular
- Temperature: 37.3°C (99.1°F)

POCUS exam(s) to perform:
- Abdominal POCUS
- Cardiac POCUS
- Caudal vena cava assessment

Abnormal POCUS exam results:
See **Figures 9.2.1–9.2.3**, as well as **Videos 9.2.1** and **9.2.2**.

Additional point-of-care diagnostics and initial management:
Initial point-of-care testing included a Doppler systolic blood pressure of 50 mmHg and a serum lactate of 4.4 mmol/L. The patient was given a 20 mL/kg crystalloid bolus over 15 min, which increased the systolic blood pressure to 90 mmHg. An antigen test for parvovirus showed a positive result, while hematology confirmed typical concomitant leukopenia. All complementary examinations were performed in an isolation room, wearing protective clothing and gloves when handling the patient.

QUESTIONS AND ANSWERS

1. What are the differentials to rule in or out with POCUS based on history and physical exam?
2. What are the sonographic findings?
3. What is the sonographic diagnosis?
4. If necessary, what additional sonographic examination or findings would help rule in or out the differential diagnoses?

DOI: 10.1201/9781003436690-47

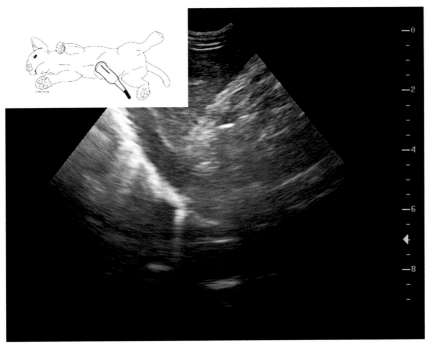

FIGURE 9.2.1 AND VIDEO 9.2.1 Caudal vena cava (CVC) view at the subxiphoid (diaphragmatic) site, using a microconvex curvilinear probe with the depth set at 8 cm and the frequency set at 7 MHz. Patient was placed in right lateral, the fur was not clipped and alcohol was used as the coupling agent.

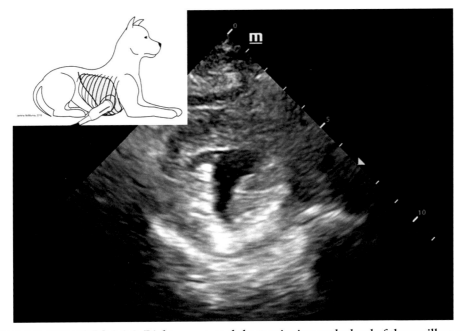

FIGURE 9.2.2 AND VIDEO 9.2.2 Right parasternal short-axis view at the level of the papillary muscles (the "mushroom view"), using a microconvex curvilinear probe with the depth set at 7 cm and the frequency set at 7 MHz. Patient was placed in sternal, the fur was not clipped and alcohol was used as the coupling agent.

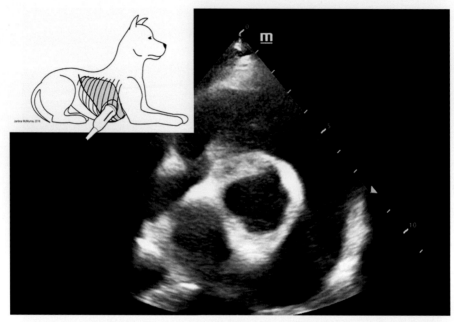

FIGURE 9.2.3 **Right parasternal short-axis view at the level of the heart base (the "LA:Ao view")** using a microconvex curvilinear probe with the depth set at 8 cm and the frequency set at 7 MHz. Patient was placed in sternal, the fur was not clipped and alcohol was used as the coupling agent.

1. **Differential diagnosis to rule in or rule out with POCUS:**
   - Hypovolemic shock (most likely differential) while assessing cardiac function and ruling out cardiogenic shock concurrently.
2. **What is your sonographic diagnosis?**
   - Hypovolemic shock as suggested by the observation of a flat and collapsing CVC, an underfilled ventricle with pseudohypertrophy of the ventricular walls (**Figure 9.2.5**), and a decreased left atrial to aortic (LA:Ao) ratio (**Figure 9.2.6**).
3. **What additional sonographic examination or finding would help you rule in or rule out your differential diagnoses (if necessary)?**
   - The CVC can be observed at the subxiphoid, paralumbar, and right hepatic sites. In humans, a flat/narrow inferior vena cava (IVC) with an increased inferior vena cava collapsibility index (IVC_{CI} >50%) is correlated with low central venous pressure and hypovolemia.
   - Despite a description of normal values[1] the appreciation of the CVC in veterinary medicine is often subjective, described as small or flat versus large or fat versus normal (**Figure 9.2.4**).
   - With hypovolemia, the CVC:Ao ratio at the sublumbar site will be less than 0.9, while values between 0.9 and 1.2 are described as normal (**Figure 9.2.7**).[2]
   - Recent veterinary studies assessed the accuracy of CVC:Ao and the collapsibility of CVC in predicting fluid responsiveness. The collapsibility index (CVC_{CI}) reflects the variation in the CVC diameter (CVC_d) during the cardiorespiratory cycle, and is calculated by the following formula[2]:

   $$CVC_{CI} = (CVC_d \text{ max} - CVC_d \text{ min})/CVC_d \text{ max}$$

   - The CVC_{CI} seems to be a promising marker of fluid responsiveness, but further research is needed before recommending its application in small animals. Changes are proportional to the severity of hypovolemia.

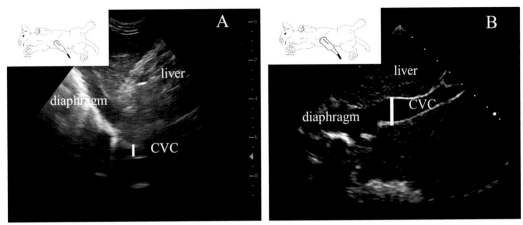

FIGURE 9.2.4 Caudal vena cava (CVC) view at the subxiphoid (diaphragmatic) site. The white line in the image indicates the CVC. With hypovolemia, the CVC appears as a small vessel and typically collapses during inspiration in spontaneously breathing patient (A). With hypervolemia or right-sided heart disease, the CVC appears to be large and does not change in size during the respiratory cycle (B).

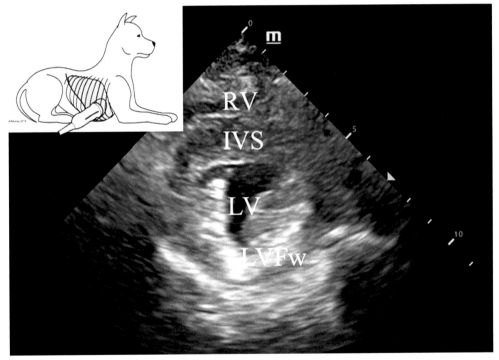

FIGURE 9.2.5 Right parasternal short-axis view at the level of the papillary muscles. Right ventricle (RV), interventricular septum (IVS), left ventricle (LV), and left ventricular free wall (LVFw) are indicated. With hypovolemia, LV is typically small, LVFw is thickened ("pseudohypertrophy"), and ventricular lumen disappears during systole ("kissing papillary muscles") at the ventricular level (absent here). Hyperdynamic cardiac contractions can be seen.

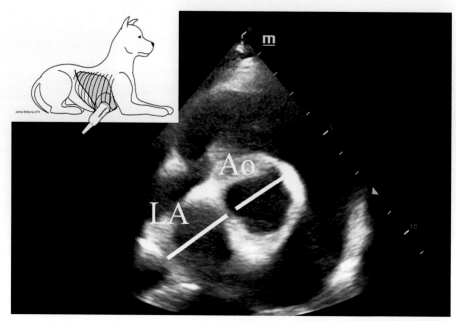

FIGURE 9.2.6 Right parasternal short-axis view at the level of the heart base (the "LA:Ao view"). The aorta (Ao) and left atrium (LA) are indicated. The white line in the image indicates the measurements to obtain for LA:Ao determination. A decreased LA:Ao ratio ≤1 is calculated, indicative of hypovolemia.

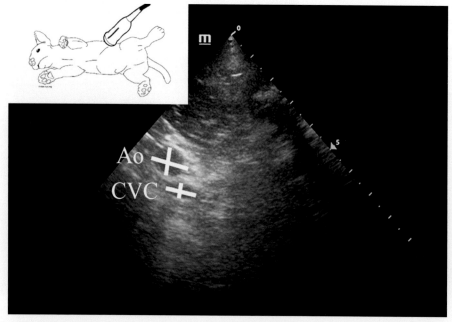

FIGURE 9.2.7 Flat caudal vena cava diameter at the sublumbar site, using a microconvex curvilinear probe with the depth set at 5 cm and the frequency set at 7 MHz. Fur was not clipped and alcohol was used as the coupling agent. The white line in the image indicates the measurements to obtain for CVC:Ao determination. Diameters are measured using lines perpendicular to both vessel walls. Means of both diameters for each vessel can be calculated.

CLINICAL INTEGRATION

Suspected diagnosis based on ultrasound findings:
Hypovolemic shock is diagnosed based on a small CVC that collapses during inspiration, underfilled ventricle, pseudohypertrophy of the ventricular walls, and a decreased LA:Ao ratio.

Sonographic interventions, monitoring, and outcome:
- A second cardiovascular POCUS scan (LA:Ao ratio, ventricular filling, ventricular wall thickness, CVC size and collapsibility) was performed following fluid resuscitation and revealed immediate clinical improvement and normalization of all parameters.
- The dog was treated medically, monitored closely, and eventually discharged after five days of hospital care.

Important concepts regarding the sonographic diagnosis of a hypovolemia:
- Estimation of hypovolemic status with cardiac POCUS is based on the subjective assessment of the LA:Ao ratio (<1), a small ventricular and atrial lumen size, thickened left ventricular walls (pseudohypertrophy), and subjective hyperdynamic contractility.
- A flat CVC with a CVC:Ao ratio <0.9 and a CVC collapsing during inspiration are indicative of hypovolemia.[2]
- The ultrasonographic assessment of CVC and CVC:Ao ratio has some limitations.[3] The CVC diameter can be influenced by
  - Cardiac function (right heart failure, cardiac tamponade).[3]
  - Respiratory effort (pneumothorax, pleural effusion, pulmonary thromboembolism).
  - Intra-abdominal pressure (presence of abdominal fluid or masses).
- Pressure artifact (excessive pressure applied to the skin by the operator).

Techniques and tips to identify the appropriate ultrasound images/views:
- Obtaining standard views for cardiac measurements is crucial for good interpretation.
- The short-axis right parasternal view is used to assess the left atrial size. This acoustic window is located on the right hemithorax at the cardiac apex beat.[4]
- The LA:Ao view is obtained at the level of the heart base, and must be perpendicular to the heart with the aortic semilunar valves visualized ("Mercedes sign").[5]
- Left ventricle function is assessed just below the mitral valves ("mushroom view"). The cross-sectional left ventricle must have a round shape (not an ovoid or "egg" shape).[6]
- The CVC can be observed at the subxiphoid, hepatic, or sublumbar view. The subxiphoid view is the standard view used in human medicine and allows for better assessment of collapsibility due to its close proximity to the thoracic cage. That said, the subxiphoid view has a higher inter-rater variability compared to the other views,[1] and high inter-rater variability has been described in cats too regardless of the view.[7] See Chapter 8b for the acquisition of standard CVC views.
- The CVC:Ao ratio can be obtained at the sublumbar view with the dog positioned in right lateral recumbency. After identifying the left kidney, the probe is moved caudally and then oriented dorsally to identify the CVC and Ao, with the caudal pole of the left kidney remaining visible. The probe is rotated 90° to obtain a cross-sectional circular view of both the CVC and Ao.[7]

Take-home message
- The key sonographic findings that support a diagnosis of hypovolemia are a flat and collapsing CVC, underfilled left ventricle, thickened left ventricular wall, small LA:Ao ratio, and hyperdynamic contractility.

REFERENCES

1. Darnis E, Boysen S, Merveille AC, Desquilbet L, Chalhoub S, Gommeren K. Establishment of reference values of the caudal vena cava by fast-ultrasonography through different views in healthy dogs. *J Vet Intern Med*. 2018 July.
2. EVECC CONGRESS (Ghent, 2022). *Abdominal POCUS: Basics in the Unstable Patient*. 16p.
3. Boysen SR, Gommeren K. Assessment of volume status and fluid responsiveness in small animals. *Front Vet Sci*. 2021 May 28.
4. Giraud L, Gommeren K, Merveille AC. Point of care ultrasound of the caudal vena cava in canine DMVD. Abstract. *J Vet Int Med*. 2020.
5. DeFrancesco TC, Ward JL. Focused canine cardiac ultrasound. *Vet Clin North Am Small Anim Pract*. 2021 Nov.
6. Chalhoub S, Boysen SR. Veterinary point of care ultrasound (POCUS): Abdomen, pleural space, lung, heart and vascular systems. *UCVM Clinical Skills Laboratory Manual* 2021:1–88.
7. Sänger F, Dorsch R, Hartmann K, Dörfelt R. Ultrasonographic assessment of the caudal vena cava diameter in cats during blood donation. *J Feline Med Surg*. 2022 Jun;24(6):484–492.
8. Cambournac M, Goy-Thollot I, Violé A, Boisvineau C, Pouzot-Nevoret C, Barthélemy A. Sonographic assessment of volaemia: Development and validation of a new method in dogs. *J Small Anim Pract*. 2018 Mar.

CHAPTER 9 – CASE 3

BEAUCERON WITH STATUS EPILEPTICUS SECONDARY TO HEAT STROKE, FEVER (40°C [104°F]), AND SEVERE PHLEBITIS

Laurentin Duriez and Kris Gommeren

HISTORY, TRIAGE, AND STABILIZATION

A 7-year-old male castrated Beauceron presented for status epilepticus and heat stroke. The dog had been hospitalized for several days and had received numerous anti-epileptic treatments (phenobarbital, levetiracetam, midazolam, and propofol boluses and constant rate infusion). On the fifth day, he developed a fever (40°C [104°F]) and severe phlebitis.

Triage exam findings (vitals):
- Mentation: quiet to obtunded (secondary to the seizures and anti-epileptics)
- Respiratory rate: 32 breaths per minute with normal bronchovesicular sounds
- Heart rate: 120 beats per minute
- Mucous membranes: pink and moist with a 2 sec capillary refill time
- Femoral and dorsal pedal arterial pulses: good quality and synchronous
- Systolic arterial pressure: 120 mmHg
- Temperature: 40°C (104°F)
- Phlebitis on the left anterior leg
- Lameness of the left anterior leg

POCUS exam(s) to perform:
- Left anterior limb POCUS
- Abdominal POCUS
- PLUS

Additional point-of-care diagnostics and initial management:
- No abnormality on urine analysis.
- Mild neutrophilic leukocytosis on blood analysis.

Abnormal POCUS exam results:
See **Figures 9.3.1–9.3.3**.

QUESTIONS AND ANSWERS

1. What are the differentials to rule in or out with POCUS based on history and physical exam?
2. What are the sonographic findings?
3. What is the sonographic diagnosis?
4. If necessary, what additional sonographic examination or findings would help rule in or out the differential diagnoses?

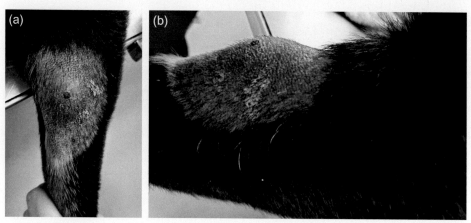

FIGURE 9.3.1 Phlebitis on the left anterior limb in (a) and (b). view

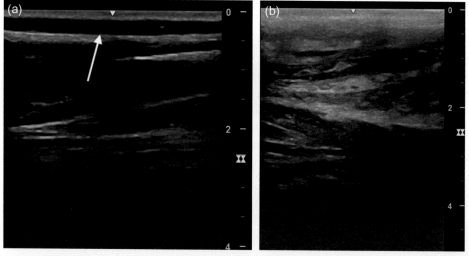

FIGURE 9.3.2 POCUS of a healthy dog using a linear probe, in-plane technique, with the depth set at 5 cm and the frequency set at 12 MHz. Fur was clipped and gel was used as the coupling agent. The vein is indicated by an (a). POCUS of the left anterior limb using a linear probe, in-plane technique, with the depth set at 5 cm and the frequency set at 12 MHz. Fur was clipped and gel was used as the coupling (b).

1. **Differential diagnosis to rule in or rule out with POCUS:**
   - Phlebitis/thrombophlebitis secondary to catheterization.
   - Aspiration bronchopneumonia secondary to status epilepticus.
2. **Describe your sonographic findings:**
   On limb POCUS, the subcutaneous tissue demonstrates perivascular edema associated with a hyperechoic zone. The cephalic vein has a heterogenous aspect of the vessel wall, making it difficult to distinguish it from the perivascular tissue. Finally, a 4 cm long thrombus can be seen within the lumen of the vein (indicated by an arrow on **Figure 9.3.4a** and **b**). The vein is not compliant when pressure is applied.
3. **What is the sonographic diagnosis?**
   Thrombophlebitis associated with cellulitis.

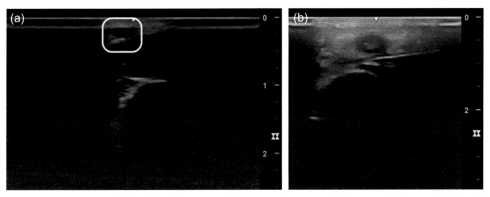

FIGURE 9.3.3 POCUS of a healthy dog using a linear probe, out-of-plane technique, with the depth set at 5 cm and the frequency set at 12 MHz. Fur was clipped and gel was used as the coupling agent. The vein is indicated within the (a). POCUS of the left anterior limb using a linear probe, out-of-plane technique, with the depth set at 5 cm and the frequency set at 12 MHz. Fur was clipped and gel was used as the coupling (b).

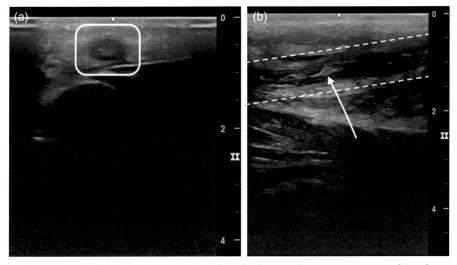

FIGURE 9.3.4 Perivascular edema is appreciated at the hyperechoic zone surrounding the rectangle. The cephalic vein, denoted within the rectangle, has a heterogenous vessel wall, making it difficult to distinguish it from the perivascular (a). A 4 cm long thrombus can be seen within the lumen of the vein (indicated by an arrow). The vein is outlined by the hashed (b).

CLINICAL INTEGRATION

Suspected diagnosis based on ultrasound findings:

Thrombophlebitis associated with cellulitis.

Sonographic interventions, monitoring, and outcome:

- No interventions.
- Daily POCUS monitoring can be performed to monitor the resolution of phlebitis and the thrombus.

Important concepts regarding the sonographic diagnosis of thrombophlebitis:

- A normal vein should have a smooth and thin wall, be easily collapsed when pressure is applied, and should demonstrate unidirectional, non-pulsatile flow without turbulence when assessed with Doppler.
- In Lodzinska et al.'s prospective study assessing phlebitis in 18 dogs, the main ultrasonographic changes to the veins noted were:
  - Wall thickening (83%)
  - Decreased compressibility (55%)
  - Filling defects consistent with intraluminal thrombus (55%)
  - Vessel wall hyperechogenicity (44%)
  - Abnormal color Doppler flow (39%)[1]
- A venous thrombus can be anechoic or hyperechoic depending on the organization of the thrombus.[2]

Techniques and tips to identify the appropriate ultrasound images/views:

- Color Doppler flow can be used to visualize abnormal flow created by a thrombus. The thrombus can be partially or totally obstructive.
- Veins can be difficult to discern when cellulitis and severe thrombophlebitis are present. The out-of-plane technique may help to visualize the vessel surrounded by inflammatory tissue.

Take-home message

- Phlebitis is a common complication of catheterization, which can lead to septicemia, pulmonary emboli, and endocarditis. Early detection of phlebitis, possibly by catheter monitoring, is crucial.

REFERENCES

1. Lodzinska, J., Leigh, H., Parys, M., Liuti, T., 2019. Vascular ultrasonographic findings in canine patients with clinically diagnosed phlebitis. *Vet. Radiol. Ultrasound* 60, 745–752. https://doi.org/10.1111/vru.12805

2. Adhikari, S., 2013. Point-of-care ultrasound diagnosis of peripheral vein septic thrombophlebitis in the emergency department. *J. Emerg. Med.* 44, 183–184. https://doi.org/10.1016/j.jemermed.2011.08.014

CHAPTER 9 – CASE 4

OLIGURIC DOMESTIC SHORTHAIR ON IV FLUIDS FOR SUSPECTED LILY INTOXICATION THAT HAS GAINED WEIGHT WHILE HOSPITALIZED

Tove M Hultman and Ivayla Yozova

HISTORY, TRIAGE, AND STABILIZATION

A 1-year-old male neutered domestic shorthair cat was referred for suspected lily intoxication. He had a 2-day history of inappetence, vomiting, and lethargy. Blood work at the referring veterinarian revealed moderately increased creatinine and blood urea nitrogen (BUN). The cat was treated for 48 hours with IV fluid therapy and supportive care. After 48 hours, the cat was suspected to be oliguric and had gained 800 g, which led to its referral.

Triage exam findings:
- Mentation: able to stand unassisted, responsive, but dull
- Heart rate and rhythm: 200 beats per minute, normal rhythm, normal femoral and dorsal pedal arterial pulses
- Mucous membranes: pink, 1–2 sec capillary refill time
- Temperature: 38.1°C (100.6°F)
- Respiratory rate: 48 breaths per minute, moderately increased respiratory effort and significantly increased bronchovesicular sounds bilaterally

POCUS exam(s) to perform:
- PLUS
- Caudal vena cava (CVC) measurements and assessment of caudal vena cava collapsibility (CVC-CI)
- Cardiac POCUS
- Abdominal POCUS

Abnormal POCUS exam results:
See **Figures 9.4.1–9.4.3** as well as **Videos 9.4.1–9.4.3**.

Additional point-of-care diagnostics and initial management:
Blood work showed severe azotemia with moderate hyperphosphatemia. The cat was initially treated with oxygen therapy and butorphanol.

QUESTIONS AND ANSWERS

1. What are the differentials to rule in or out with POCUS based on history and physical exam?
2. What are the sonographic findings?
3. What is the sonographic diagnosis?
4. If necessary, what additional sonographic examination or findings would help rule in or out the differential diagnoses?

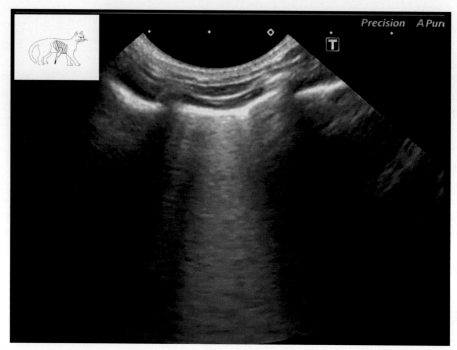

FIGURE 9.4.1 Pleural space and lung ultrasound using a microconvex curvilinear probe with the depth set at 5 cm and the frequency set at 7 MHz. Fur not clipped and alcohol used as the coupling agent. Image obtained with probe located one-third of the way up the thorax. The probe is oriented perpendicular to the ribs and the marker directed cranially.

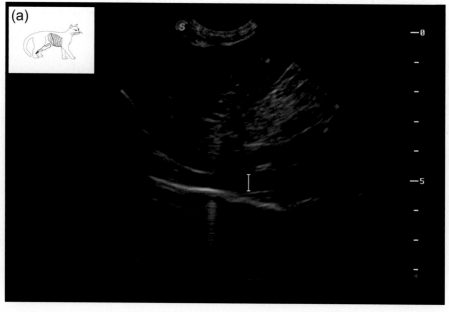

FIGURE 9.4.2A

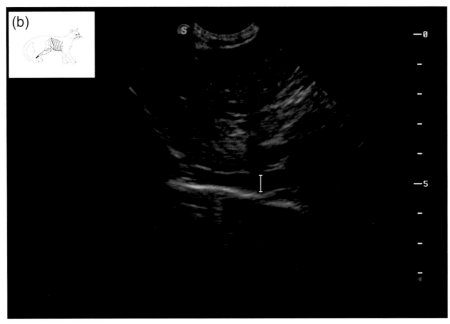

FIGURE 9.4.2B Caudal vena cava (CVC) view at the subxiphoid (diaphragmatic) site, on expiration (Figure 9.4.2a) and inspiration (Figure 9.4.2b), using a microconvex curvilinear probe with the depth set at 8 cm and the frequency set at 7 MHz. Fur not clipped and alcohol used as the coupling agent.

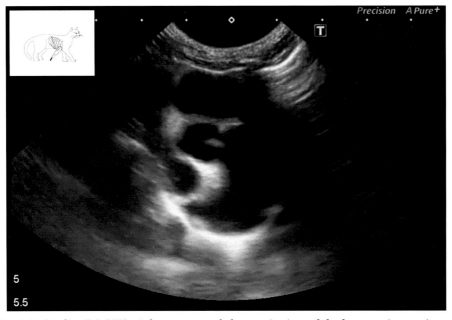

FIGURE 9.4.3 Cardiac POCUS, right parasternal short-axis view of the heart, using a microconvex curvilinear probe with the depth set at 5 cm and the frequency set at 7 MHz. Fur not clipped and alcohol used as the coupling agent.

1. **Differential diagnosis to rule in or rule out with POCUS:**
   - The most likely differential diagnosis based on history and physical exam is volume overload due to oliguria and overzealous fluid therapy. Other differentials to be considered are congestive heart failure with cardiogenic pulmonary edema, and less likely non-cardiogenic pulmonary edema and pneumonia.
2. **Describe your sonographic findings:**
   - B-lines observed in multiple locations on both sides of the hemithorax (**Figure 9.4.4, Video 9.4.1**).
   - Low (<20%) caudal vena cava collapsibility index (**Figure 9.4.5, Video 9.4.2**).
   - Increased left atrial to aortic (LA:Ao) ratio (**Figure 9.4.6, Video 9.4.3**).
3. **What is your sonographic diagnosis?**
   - Enlarged LA:Ao ratio with coalescing B-lines and decreased CVC-CI.
4. **What additional sonographic examination or finding would help you rule in or rule out your differential diagnoses (if necessary)?**
   - Echocardiogram to rule out occult cardiomyopathy and secondary congestive heart failure.

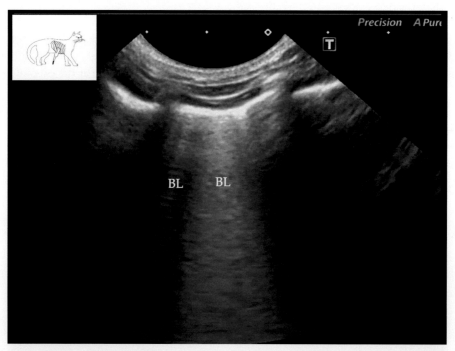

FIGURE 9.4.4 Coalescing B-lines (BL). A vertical white lines originating from the visceral surface of the pleural line and extending to the far field.

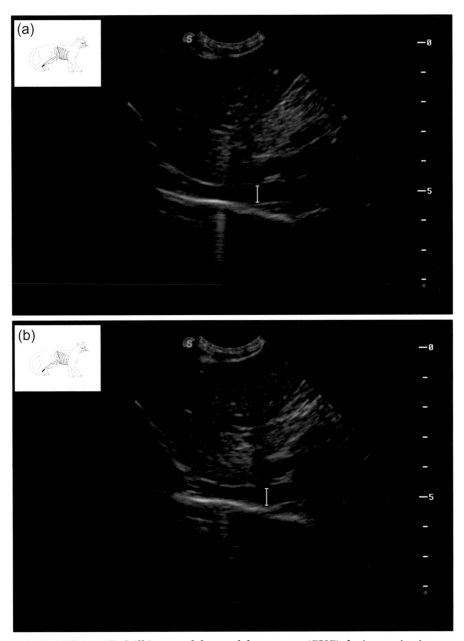

FIGURE 9.4.5A AND 9.4.5B Still image of the caudal vena cava (CVC) during expiration (Figure 9.4.5a) and inspiration (Figure 9.4.5b) depicting less than 20% collapse in diameter between inspiration and expiration. The line illustrates the diameter of the CVC.

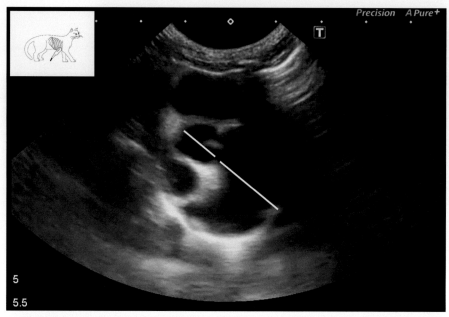

FIGURE 9.4.6 Right parasternal short-axis view of the heart depicting an enlarged LA:Ao ratio, consistent with volume overload. The white lines in the diagram indicate the measurements to obtain for LA:Ao determination.

CLINICAL INTEGRATION

Suspected diagnosis based on ultrasound findings:
- Volume overload and pulmonary edema.

Sonographic interventions, monitoring, and outcome:
- The patient was given oxygen therapy and furosemide (two doses in a 2-h interval) and an indwelling urinary catheter was placed for monitoring of urine production.
- Body weight, urine production, and respiratory rate were closely monitored.
- A second PLUS was performed to monitor the number and resolution of B-lines and pulmonary edema.
- Alternative to an indwelling urinary catheter, urinary output could have been estimated using POCUS and the following calculation: bladder length × height × width × 0.52 = urinary bladder volume in milliliters.[1]

- Four hours after presentation, a mild amount of pleural effusion could be detected. Due to financial constraints and poor prognosis the clients opted for euthanasia the next day.

Important concepts regarding the sonographic diagnosis of volume overload in cats:
- There are multiple sonographic findings that when used together support the diagnosis of volume overload in cats:
  - Decreased caudal vena cava collapsibility index (**Video 9.4.2**).
  - Abnormal amount of B-lines (**Video 9.4.1**).
  - Pleural effusion.
  - Enlarged left atrial size (**Video 9.4.3**).
  - Enlarged left ventricle lumen size.
  - Gallbladder wall edema (halo sign).
  - Ascites.

- Calculating or subjectively assessing the caudal vena cava collapsibility is one of the techniques that can help estimate volume status. The CVC-CI is calculated by measuring the *minimum diameter* of the CVC during *inspiration* and the *maximum diameter* during *expiration*. The CVC-CI is then expressed as a percentage by using the following calculation: CVC-CI = $(CVC_{max} - CVC_{min})/CVC_{max}$. A high CVC-CI is indicative of hypovolemia whereas a low CVC-CI is indicative of volume overload.[2]
- CVC-CI has been evaluated in a large group of cats (pending publication), suggesting a median CVC-CI of 38.3% with a range of 29%–54%. A CVC-CI < 29%, together with appropriate history and clinical findings (including Cardiac POCUS and lung ultrasound) can therefore be suggestive of volume overload. The CVC-CI can be measured in both standing position and lateral recumbency.[3]
- Intravascular volume can also be estimated using Cardiac POCUS, evaluating parameters such as atrial lumen size, ventricular wall thickness, and cardiac contractility. The use of cardiac POCUS to assess volume overload is most often performed using a two-dimensional short-axis right parasternal view of the heart where the LA:Ao ratio is measured (or subjectively estimated). If the LA:Ao ratio is above 1.5 or if the area of the left atrium is 2.5 times the area of the aorta, the left atrium should be considered enlarged which could be indicative of volume overload or left-sided congestive heart failure.
- Increased extravascular lung water (EVLW) accumulates during volume overload and manifests ultrasonographically as B-lines. B-lines can be seen in the presence of pulmonary edema but are also seen in multiple other pathologies (pulmonary contusions and hemorrhage, pneumonia, acute lung injury, acute respiratory distress syndrome). The presence of more than three or coalescing B-lines at a single site (especially if present at more than one site) is considered abnormal and increases the index of suspicion for underlying pathology.[2] However, a small number of B-lines can be found in healthy patients. One study demonstrated that 12% of cats with normal thoracic radiographs and no respiratory signs had B-lines.[4]
- Pleural effusion due to volume overload is common in cats. One study looking at volume overload in cats with urinary obstruction demonstrated that 73% of the cats with presumed volume overload had pleural effusion.[5]
- In cases of severe volume overload and/or right-sided congestive heart failure, gallbladder wall edema and ascites may be observed, although this finding is not well studied in cats.

Technique and tips to identify the appropriate ultrasound images/views:

- Start with the probe placed just caudal to the xiphoid cartilage in longitudinal axis, at roughly a 45° angle to the spine and with the marker pointed cranially.
- When the liver and diaphragm are identified, fan the probe gently to the right of midline until a longitudinal view of the CVC can be visualized as two horizontal parallel lines (walls of the CVC) crossing the diaphragm (**Figure 9.4.5a** and **b**).
- The ultrasound probe is then slowly fanned through all longitudinal planes across the CVC, to identify its luminal center, subjectively identified as the largest distance between the two walls of the CVC.
- The freeze and track functions to "slow" cineloop speed makes the subjective interpretation of CVC-CI easier.

Take-home messages

- Appropriate history and use of Cardiac POCUS, abdominal POCUS (namely CVC parameters), and PLUS together can provide valuable information and help detect signs of volume overload.
- The key sonographic findings in cats with volume overload are a non-collapsing CVC, enlarged LA:Ao ratio (>1.5), enlarged left ventricle lumen size (rather than hypertrophy as seen with hypertrophic cardiomyopathy), and bilateral B-lines on lung ultrasound. Ascites and gallbladder wall edema can be seen in severe cases.
- B-lines can be seen in multiple other disease processes and a small amount can be seen in healthy cats.

REFERENCES

1. Kendall A, Keenihan E, Kern ZT, Lindaberry C, Birkenheuer A, Moore GE, Vaden SL. Three-dimensional bladder ultrasound for estimation of urine volume in dogs compared with traditional 2-dimensional ultrasound methods. *J Vet Intern Med.* 2020 Nov;34(6):2460–2467. doi: 10.1111/jvim.15959. Epub 2020 Nov 6. PMID: 33156977; PMCID: PMC7694864.
2. Boysen SR, Gommeren K. Assessment of volume status and fluid responsiveness in small animals. *Front Vet Sci.* 2021;8:630643.
3. Hultman TM, Boysen SR, Owen R, Yozova ID. Ultrasonographically derived caudal vena cava parameters acquired in a standing position and lateral recumbency in healthy, lightly sedated cats: A pilot study. *J Feline Med Surg.* Dec 2021. doi: 10.1177/1098612X211064697.
4. Lisciandro GR, Fulton RM, Fosgate GT, Mann KA. Frequency and number of B-lines using a regionally based lung ultrasound examination in cats with radiographically normal lungs compared to cats with left-sided congestive heart failure. *J Vet Emerg Crit Care.* 2017 Sep;27(5):499–505. Epub 2017 Aug 1. PMID: 28763158.
5. Ostroski CJ, Drobatz KJ, Reineke EL. Retrospective evaluation of and risk factor analysis for presumed fluid overload in cats with urethral obstruction: 11 cases (2002–2012). *J Vet Emerg Crit Care.* 2017 Sep;27(5):561–568. doi: 10.1111/vec.12631. Epub 2017 Jul 28. PMID: 28752928.

SECTION 5

ABDOMINAL POINT-OF-CARE ULTRASOUND

Section Editor Erin Binagia

CHAPTER 10

INTRODUCTION TO ABDOMINAL POINT-OF-CARE ULTRASOUND

Erin Binagia

INTRODUCTION

The abdominal point-of-care ultrasound (POCUS) exam uses the views from the original abdominal-focused assessment with sonography for trauma (AFAST) exam and shares many common clinical applications.[1] While AFAST was designed only to detect the presence and quantity or absence of free fluid,[1,2] abdominal POCUS was designed to be a much broader point-of-care ultrasound examination. Abdominal POCUS has the capability of answering multiple questions per site. Veterinarians should use the abdominal POCUS exam to obtain quick answers and triage and initiate appropriate treatments in a timely manner. However, a detailed full abdominal ultrasound or computed tomography scan should be performed by a radiologist once the patient is stabilized.

This chapter briefly reviews what is entailed in an abdominal POCUS exam and how to obtain images. See additional references for more detailed information.[3]

PATIENT POSITION, PROBE, MACHINE SETTINGS

Abdominal POCUS can be performed in all positions including standing, so it is important to practice the exam in all positions. Ultimately, patient positioning is determined by the stability and temperament of the patient. Choose the position that is most comfortable for the patient. Dorsal recumbency should be avoided in critical patients, especially those with respiratory or cardiovascular distress. Compared to right lateral, in left lateral recumbency the right kidney is easier to find and the POCUS exam is completed faster, but there is no difference in the detection of free fluid.[4] An unstable patient should be placed in a position that will cause minimal distress. Consider the effects of gravity in your patient based on their position and visualize where fluid falls and gas rises.

A microconvex/curvilinear probe is used for abdominal POCUS exams, with a frequency of between 5 (large patients) and 7.5 MHz (small patients <15 kg).

Adjust the gain by using fluid in the urinary bladder or gallbladder as a reference – fluid should appear anechoic (black). Adjust the depth to fill at least three-quarters of the screen with the organ of interest only. Adjust the frequency depending on the depth and detail required. A higher frequency will improve the resolution or detail of the image, and a lower frequency will improve tissue penetration at the expense of detail. Decrease the frequency for larger dogs or deeper organs and choose a higher frequency when wanting to visualize better details (intestinal wall layers, renal parenchyma, etc.). See Chapter 1 for more information.

DOI: 10.1201/9781003436690-51

Patient preparation should be quick for this exam. Usually, parting the fur and applying alcohol as the coupling agent are sufficient. If proper contact is not possible with this method, gel can be applied and/or the fur can be shaved.

ABDOMINAL POCUS TECHNIQUE

The abdominal POCUS uses the four sites original described[1] with an additional modified view.[3] The sites used include:

1) The *subxiphoid* site is located caudal to the xiphoid process. Structures and organs to identify include the diaphragm, liver, gallbladder, ventral stomach wall, caudal vena cava, pleural space, caudal lung surface, heart, and pericardial space. This site is often evaluated first as it can assess and answer many of the binary questions from a single window.
2) The *left paralumbar* site is located caudal to the last rib and ventral to hypaxial muscles. Organs to visualize include the spleen, left kidney, and intestines.
3) The *urinary bladder* site is located between the pelvic limbs caudal to the umbilicus and includes the urinary bladder, uterine body, colon, and prostate.
4) The *right paralumbar* site is located just caudal to the last rib just below the hypaxial muscles. Organs to visualize include the right caudal liver lobe, right kidney, and intestines.
5) The *umbilical* site is a modified view by placing the probe at the umbilicus. This view is used to improve the detection of gravity-dependent fluid. Other organs identified include the intestines and spleen.

Place the probe on the five regions of the abdomen (**Figure 10.1**) in a systematic approach that is consistent for the individual examiner. Most choose to perform in a clockwise manner, but the order can also be directed based on clinical suspicion established by history and physical exam. At each site, rock and fan the probe through 45° angles in the short and long axis.[1,3] The patient position and probe placement can be modified depending on the pathology in question to increase the chance of detection (e.g., gas rises, fluid falls). To increase the likelihood of finding abdominal fluid, search the most gravity-dependent areas and fan, rock, and slide the probe.[1,3]

The abdominal fluid score (AFS) can be calculated by giving 1 point for each of the original four sites that fluid is found (for total a score of 0-4).[2] The AFS has been reported to predict need of blood transfusion and can assist in serial monitoring.[2]

Binary questions to ask via abdominal POCUS[3] (see **Figure 10.1**)

- Is free abdominal fluid present? (Cases 11.3, 11.6, 11.7, 11.11, 11.16)
  - All five sites, especially gravity-dependent areas.
  - Perform serial exams if deemed necessary.
- Is free abdominal air present? (Case 11.9)
  - All five sites, especially areas where gas may rise.
- Does the patient have ileus? (Case 11.15)
  - Right and left paralumbar view, subxiphoid, and umbilical view.
- Is the patient producing urine? (Case 11.13)
  - Urinary bladder view.
  - Perform serial exams after fluid administration.
- Does the patient have a gallbladder halo sign? (Case 11.10)
  - Subxiphoid view.

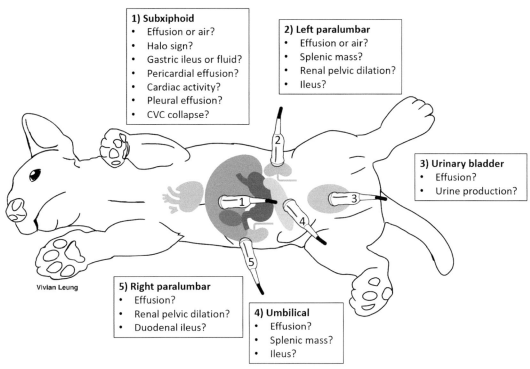

FIGURE 10.1 Answering specific binary questions using the five views of abdominal POCUS. Usually performed in a clockwise approach (starting with #1 and ending with #5). Baseline image courtesy of Vivian Leung and modified with permission.

- Is there renal pelvic dilation? (Cases 11.4, 11.12)
  - Right and left paralumbar view.
- Is there a pyometra? (Case 11.5)
  - Urinary bladder and umbilical view.
- Is there a splenic mass? (Case 11.7)
  - Left paralumbar view and umbilical view.
- Is there pleural or pericardial effusion? Is there cardiac activity? Is CVC collapsing? (Cases 5.21, 7.4, 9.2, 15.5)
  - Subxiphoid view.

Serial abdominal POCUS exams should be performed to monitor the appearance of fluid if none is found on initial exam of suspicious cases or to monitor the progression/resolution of abdominal fluid in fluid-positive cases.[2] How frequent the exam is repeated depends on the stability of the patient, with more frequent scans performed in an unstable patient or a repeated exam when patient status suddenly becomes unstable. Otherwise performing every 4 hours is sufficient in stable cases.[2]

The abdominal POCUS is meant to be a point-of-care exam performed on the triage table and requires minimal experience, but it has limitations. The initial exam could miss scant free fluid either due to hypovolemia, severe dehydration, or the poor quality of the ultrasound machine. Serial exams should be performed after rehydration, and if free fluid is still not found but suspected based on clinical presentation, a full abdominal ultrasound should be performed. A negative abdominal POCUS does not rule out pathology, and more advanced imaging should be performed if there is clinical suspicion for abdominal pathology. Ultimately, the abdominal POCUS should not replace a full abdominal ultrasound or computed tomography scan by a radiologist. See Table 10.1 for an abbreviated list of Abdominal POCUS normal and abnormal findings.

Table 10.1 Chart identifying the normal and abnormal findings and measurements discussed in this section

ABDOMINAL POCUS FINDINGS AND MEASUREMENTS

ORGAN	NORMAL	ABNORMAL	OTHER INFORMATION
Liver	Hypoechoic to spleen Usually located within the rib cage	Enlarged if extending distal to rib cage	Most liver pathology should be evaluated by an imaging specialist
Gallbladder Case 11.10	Wall <1 mm	Wall edema: >3–3.5 mm	"Double rim" or "halo" sign
Pancreas Case 15.4	Isoechoic or slightly hypoechoic to surrounding mesenteric fat Dogs: 1–3 cm wide, 1 cm thick Cats: 0.3–0.6 cm thick, body and left lobe 0.5–0.9 cm thick	Acute pancreatitis: enlargement, hypoechoic parenchyma, hyperechoic mesentery, peripancreatic free fluid	Normal appearance does not rule out pancreatitis
Spleen Cases 11.7, 15.6	Homogeneous, finely textured parenchyma, hyperechoic to liver	Torsion: hypoechoic parenchyma, starry night pattern, splenomegaly	No established guidelines for US measurements in the dog
Kidneys (cats)[5,6] Cases 11.4, 11.12	Cortex is isoechoic or hyperechoic to liver Kidney length: 3.1–5.1 cm RPW: 0.8–3.2 mm Ureter width: 0–0.4 mm	AKI: renomegaly, hyperechoic cortices, pyelectasia RPW >3–3.5 mm	AKI has multiple etiologies which cannot be distinguished based on US appearance
Kidneys (dogs)[5,6] Case 11.14	Medulla is hypoechoic to cortex. Echogenicity of cortex is isoechoic or hyperechoic to liver and hypoechoic to kidney RPW <3 mm Ureter width: <1.8 mm	RPW >3–4 mm AKI appears similar to cats	No established guidelines for US measurements in the dog
Urinary bladder[5] Case 11.13	Mean BWT (dogs): 2.3 mm (minimally distended), 1.4 mm (moderately distended) Mean BWT (cats): 1.3–1.7 mm	Empty: >3 mm (dogs) and >1.7 mm (cats) Distended: >2 mm	Urinary bladder volume equation*: length × width × height × 0.2π[7]
Gastrointestinal tract[8] Cases 11.15, 15.2, 15.3	Peristaltic contractions/min: Stomach: 4–5 Duodenum: 4–5 Small intestines: 1–3 Colon: not detectable Thickness+: Stomach: 3–5 mm (dogs), 2–4.4 mm (cats) Duodenum: 5 mm (dog), 2–4 mm (cat) Jejunum: 2–3 mm Ileum: 2.5–3.2 mm (cat) Colon: 1–2 mm (dog), 1.5–2 mm (cats)	Ileus: fewer contractions than normal, dilation, GRV >10 mL/kg	Fasting causes decreased frequency of peristaltic contractions

(Continued)

Table 10.1 **(Continued) Chart identifying the normal and abnormal findings and measurements discussed in this section**

ABDOMINAL POCUS FINDINGS AND MEASUREMENTS

ORGAN	NORMAL	ABNORMAL	OTHER INFORMATION
Uterus Case 11.5	Located between bladder and colon. Usually not visible in abdominal POCUS	Fluid-filled, distended	Fluid-filled uterus can be pyometra, mucometra, hydrometra, etc.
Fetal heart rate Case 11.8	FHR: 180–220 bpm	Fetal distress: <180 bpm	FHR may fluctuate during contractions, obtain multiple measurements

Note: See individual cases listed in the left column for more references.

Notes:

*Multiple formulas for calculating urinary bladder volume exist, additional equations include L × W × H × 0.52[9] (high accuracy in humans) and L × W × ([DL + DT]/2) × 0.625.[10] There is no consensus on which formula is most accurate in veterinary medicine; however, a recent study reported the human formula to be the most accurate of the three formulas, although this was a small sample size.[11]

†Not commonly part of abdominal POCUS evaluation but included for completeness.

US: ultrasound; RPW: renal pelvic width; AKI: acute kidney injury; BWT: bladder wall thickness; FHR: fetal heart rate; GRV: gastric residual volume.

REFERENCES

1. Boysen SR, Rozanski EA, Tidwell AS, Holm JL, Shaw SP, Rush JE. Evaluation of a focused assessment with sonography for trauma protocol to detect free abdominal fluid in dogs involved in motor vehicle accidents. *J Am Vet Med Assoc* Oct 15 2004;225(8):1198–1204. doi:10.2460/javma.2004.225.1198
2. Lisciandro GR, Lagutchik MS, Mann KA, et al. Evaluation of an abdominal fluid scoring system determined using abdominal focused assessment with sonography for trauma in 101 dogs with motor vehicle trauma. *J Vet Emerg Crit Care* Oct 2009;19(5):426–437. doi:10.1111/j.1476-4431.2009.00459.x
3. Gommeren K, Boysen SR. Chapter 189: Point-of-care ultrasound in the ICU. In: Silverstein DC, Hopper K, eds. *Small Animal Emergency and Critical Care*. 3rd ed. Elsevier; 2023, pp. 1076–1092.
4. McMurray J, Boysen S, Chalhoub S. Focused assessment with sonography in nontraumatized dogs and cats in the emergency and critical care setting. *J Vet Emerg Crit Care* 2016;26(1):64–73. doi:10.1111/vec.12376
5. Nyland TG, Widmer WR, Mattoon JS. Chapter 16: Urinary tract. In: Mattoon JS, Nyland TG, eds. *Small Animal Diagnostic Ultrasound*. Saunders; 2015, pp. 557–607.
6. Cole L, Humm K, Dirrig H. Focused ultrasound examination of canine and feline emergency urinary tract disorders. *Vet Clin North Am Small Anim Pract* Nov 2021;51(6):1233–1248. doi:10.1016/j.cvsm.2021.07.007
7. Lisciandro GR, Fosgate GT. Use of urinary bladder measurements from a point-of-care cystocolic ultrasonographic view to estimate urinary bladder volume in dogs and cats. *J Vet Emerg Crit Care (San Antonio)* Nov 2017;27(6):713–717. doi:10.1111/vec.12670
8. Nyland TG, Neelis DA, Matoon JS. Chapter 12: Gastrointestinal tract. In: Mattoon JS, Nyland TG, eds. *Small Animal Diagnostic Ultrasound*. Saunders; 2015, pp. 468–500.

9. Araklitis G, Paganotto M, Hunter J, Thiagamoorthy G, Robinson D, Cardozo L. Can we replace the catheter when evaluating urinary residuals? *Neurourol Urodyn* Apr 2019;38(4):1100–1105. doi:10.1002/nau.23963
10. Atalan G, Barr FJ, Holt PE. Assessment of urinary bladder volume in dogs by use of linear ultrasonographic measurements. *Am J Vet Res* Jan 1998;59(1):10–15.
11. Kendall A, Keenihan E, Kern ZT, et al. Three-dimensional bladder ultrasound for estimation of urine volume in dogs compared with traditional 2-dimensional ultrasound methods. *J Vet Intern Med* Nov 2020;34(6):2460–2467. doi:10.1111/jvim.15959

CHAPTER 11
ABDOMINAL POINT-OF-CARE CASES

CHAPTER 11 – CASE 1

PEKINESE WITH A 7-DAY HISTORY OF PROGRESSIVE OBTUNDATION, POLYURIA, POLYDIPSIA, DYSURIA, STRANGURIA, AND POLLAKIURIA

Georgina Hall and Erica Tinson

HISTORY, TRIAGE, AND STABILIZATION

A 2-year-old female spayed Pekinese presented with a 7-day history of progressive obtundation, polyuria, polydipsia, dysuria, stranguria, and pollakiuria. She had a reduced body condition score (BCS) of 3/9 and was smaller than her littermates.

Triage exam findings (vitals):
- Mentation: obtunded
- Respiratory rate: 24 breaths per minute
- Heart rate: 110 breaths per minute
- Mucous membranes: pink, tacky, 2 sec capilllary refill time
- Femoral and dorsal pedal arterial pulses: strong, synchronous, normodynamic
- Temperature: 38.4°C (101.1°F)
- Abdomen: painful on bladder palpation

POCUS exam(s) to perform:
- Urinary bladder assessment

Abnormal POCUS exam results:
See **Figure 11.1** and **Videos 11.1.1** and **11.1.2**

Additional point-of-care diagnostics and initial management:
Blood glucose was within normal limits, a point-of-care blood ammonia test was high. Urinalysis revealed isosthenuria, proteinuria, hematuria, and ammonium urate crystals.

QUESTIONS AND ANSWERS

1. What are the differentials to rule in or out with POCUS based on history and physical exam?
2. What are the sonographic findings?
3. What is the sonographic diagnosis?
4. If necessary, what additional sonographic examination or findings would help rule in or out the differential diagnoses?

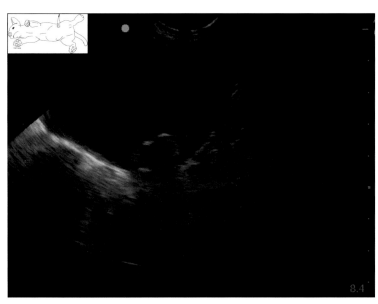

FIGURE 11.1.1, VIDEO 11.1.1, AND VIDEO 11.1.2 Abdominal POCUS urinary bladder view with the dog in right lateral recumbency, using a curvilinear probe placed longitudinally with marker placed cranially with the depth set at 8 cm and the frequency set at 7.5 MHz.

1. **Differential diagnosis to rule in or rule out with POCUS:**
   - Differentials for the dog's lower urinary tract signs and abdominal pain include urinary tract infection, cystoliths, cystitis, crystalluria, bladder neoplasia, and proximal urethral obstruction. POCUS can identify cystoliths, blood clots, bladder masses, and crystalluria. Marked bladder wall thickening and sediment on POCUS may be suggestive of bacterial or hemorrhagic cystitis. A urinary sample would further aid a final diagnosis.
   - The dog's obtunded mentation with normal cardiovascular status could be due to metabolic causes (hypoglycemia, urosepsis, hepatic encephalopathy, hypoadrenocorticism) or central neurologic disease. It is not possible to rule these in or out with POCUS.
2. **Describe your sonographic findings:**
   - A hyperechoic round structure with a distal acoustic shadow is seen in the dependent region of the bladder (**Figure 11.1.1** and **Figure 11.1.2**).
3. **What is your sonographic diagnosis?**
   - Cystolith.
4. **What additional sonographic examination or finding would help you rule in or rule out your differential diagnoses (if necessary)?**
   - To rule out a urinary tract infection and crystalluria, a urine sample may be taken via ultrasound-guided cystocentesis. A wet prep and stained sediment examination alongside culture and sensitivity would identify these differentials.
   - A full abdominal ultrasound with a boarded radiologist should be performed to assess the full urinary tract and to interrogate for portosystemic shunt (PSS) as a cause of obtundation and cystolith.

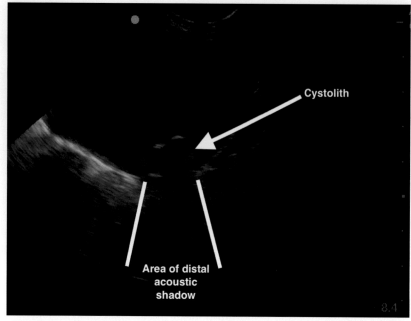

FIGURE 11.1.2 A hyperechoic round structure with a distal acoustic shadow is seen in the dependent region of the bladder.

CLINICAL INTEGRATION

Suspected diagnosis based on ultrasound findings:

Cystolith. With associated obtundation, low BCS, hyperammonemia, and ammonium urate crystals, a portosystemic shunt with urate cystolithiasis is most likely.[1]

Sonographic interventions, monitoring, and outcome:
- Abdominal ultrasound confirmed the presence of an extra-hepatic PSS.
- Urine culture was negative.
- The dog was stabilized with dietary and medical management prior to surgical ligation of the PSS and cystotomy.

Important concepts regarding the sonographic diagnosis of cystoliths:
- Sonography can identify smaller (<3 mm) cystoliths of any radio-opacity that may not be seen radiographically.
- Cystoliths appear as hyperechoic irregular or rounded structures that fall to the dependent region of the bladder. It is important to identify the distal acoustic shadow to differentiate the sonographic finding from a blood clot or neoplasm.
- Blood clots may also fall to the dependent region, whereas neoplasms will be attached to a fixed point of the bladder wall and will not move on repositioning the animal.[2]
- Unlike radiography, sonography cannot identify cystolith characteristics such as mineral composition, radiopacity, number, shape, and surface contour.[3] It can also overestimate the dimensions of cystoliths by 68% compared to radiographs.[4]

Techniques and tips to identify the appropriate ultrasound images/views:
- Assess the urinary bladder view with the patient in two positions to check that the structure seen is gravity dependent.

- Examination is optimized when the bladder is moderately full, reducing artifactual wall thickening.
- Acoustic shadowing may be absent with small cystoliths or with a low-frequency probe. To improve detection, increase the probe frequency, position your focal point at the level of the cystolith, and direct the beam perpendicularly. Tiny stones still may not shadow.
- Agitate the probe on the bladder to induce movement of calculi away from the wall.
- If color Doppler is available, assess for the "twinkle artifact." This artifact appears as a comet tail–shaped area of rapidly changing color (akin to turbulent flow) and is seen persistently behind the stone in the region of the acoustic shadow.[5]

Take-home messages

- Remember to assess for sonographic evidence (distal acoustic shadowing and gravity dependence) of a cystolith to differentiate from a blood clot or neoplasm.
- Radiography, urinalysis, blood work, a systemic work up, and cystolith analysis may all be needed to identify the type of cystolith and the underlying etiology causing its formation.

REFERENCES

1. Caporali EHG, Phillips H, Underwood L, Selmic LE. Risk factors for urolithiasis in dogs with congenital extrahepatic portosystemic shunts: 95 cases (1999–2013). *J Am Vet Med Assoc* 2015;246(5):530–536. doi:10.2460/JAVMA.246.5.530
2. Cole L, Humm K, Dirrig H. Focused ultrasound examination of canine and feline emergency urinary tract disorders. *Vet Clin North Am Small Anim Pract* 2021;51(6):1233–1248. doi:10.1016/J.CVSM.2021.07.007
3. Weichselbaum RC, Feeney DA, Jessen CR, Osborne CA, Dreytser V, Holte J. Relevance of sonographic artifacts observed during in vitro characterization of urocystolith mineral composition. *Vet Radiol Ultrasound* 2000;41(5):438–446. doi:10.1111/J.1740-8261.2000.TB01868.X
4. Byl KM, Kruger JM, Kinns J, Nelson NC, Hauptman JG, Johnson CA. In vitro comparison of plain radiography, double-contrast cystography, ultrasonography, and computed tomography for estimation of cystolith size. *Am J Vet Res* 2010;71(3):374–380. doi:10.2460/AJVR.71.3.374
5. Lee, JY, Kim, SH, Cho, JY, Han D. Color and power doppler twinkling artifacts from urinary stones: Clinical observations and phantom studies. *Am J Roentgenol* 2001 Nov 23;176(6):1441–1445.

CHAPTER 11 – CASE 2

MALTESE WITH A 2-DAY HISTORY OF LETHARGY, VOMITING, INAPPETENCE, AND ACUTE ABDOMINAL PAIN

Andrea Armenise

HISTORY, TRIAGE, AND STABILIZATION

An 11-year-old intact male Maltese with 2 days of lethargy, vomiting, inappetence, and acute abdominal pain.

Triage exam findings (vitals):
- Mentation: depressed
- Respiratory rate: 68 breaths per minute
- Heart rate: 140 beats per minute
- Mucous membranes: pink, 2–3 sec capillary refill time
- Dehydration estimate: 7%
- Femoral and dorsal pedal arterial pulses: normal to bounding
- Temperature: 37.6°C (99.6°F)

POCUS exam(s) to perform:
- Abdominal POCUS
- PLUS
- Cardiac POCUS

Abnormal POCUS exam results:
Day 1: See **Figure 11.2.1**.
Day 3: See **Figure 11.2.2** and **Videos 11.1.1** and **11.2.2**.

Additional point-of-care diagnostics and initial management:
Blood chemistry panel showed ALP 1254 UI/L (20–120), AST 130 UI/L (12–54), ALT 844 UI/L (15–64), GGT 70.3 UI/L (2.0–8.0), and TRIG 957 mmol/L (22–110). The patient was given a 15 mL/kg lactated Ringer's solution (LRS) bolus and a 7% replacement over 12 h with LRS was started. Day 1 POCUS exam revealed the image seen in **Figure 11.2.1**. Serial abdominal POCUS exams were performed to monitor the clinical evolution. After 2 days of treatment and serial abdominal POCUS monitoring, abdominal POCUS findings changed. (**Figure 11.2.2** and **Videos 11.2.1** and **11.2.2**)

QUESTIONS AND ANSWERS

1. What are the differentials to rule in or out with POCUS based on history and physical exam?
2. What are the sonographic findings?
3. What is the sonographic diagnosis?
4. If necessary, what additional sonographic examination or findings would help rule in or out the differential diagnoses?

DOI: 10.1201/9781003436690-54

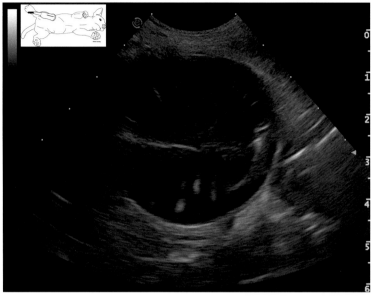

FIGURE 11.2.1 Day 1: Right paralumbar approach to the liver for the evaluation of the gallbladder. Patient is in left lateral recumbency. The probe is a microconvex curvilinear probe, the marker is directed cranially, the depth is set at 6 cm, and the frequency is set at 10 MHz. If the gallbladder cannot be seen, slide the probe toward the xiphoid process.

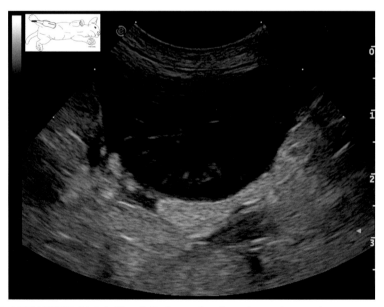

FIGURE 11.2.2 AND VIDEO 11.2.1 Day 3: Left paralumbar view using a microconvex curvilinear probe with the depth set at 4 cm and the frequency set at 10 MHz.

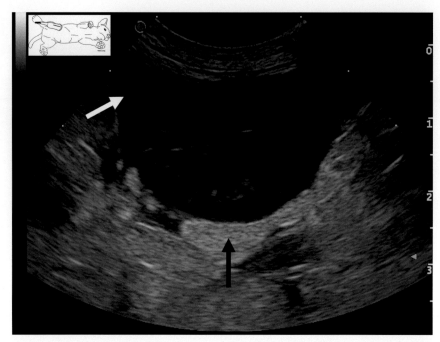

FIGURE 11.2.3 Labeled image from Figure 11.2.2. The black arrow indicates the mucoid fragment of the GBM. The white arrow indicates a small amount of anechoic fluid.

1. **Differential diagnosis to rule in or rule out with POCUS:**
   - Based on initial history and clinical findings, the most likely differentials included pancreatitis, peritonitis, gastrointestinal (GI) perforation, gastroenteritis, GI foreign body, and gallbladder obstruction. Gallbladder mucocele (GBM) rupture was much less likely on initial presentation.
2. **Describe your sonographic findings:**
   - Day 1: Transverse image of a distended gallbladder with hyperechoic striations radiating toward its center. The central hyperechoic area is slightly heterogeneous and immobile. This appearance is generally referred to as a kiwi-like pattern and stellate combination (**Figure 11.2.1**).
   - Day 3: A crescent-shaped anechoic structure with fine radiating striations is found lateral to the spleen and left kidney, which represents a fragment of a GBM that migrated after gallbladder rupture. A hyperechoic peritoneum with effusion was present (**Figure 11.2.3**, **Video 11.2.1**).
   - A small section of the right side of the gallbladder contains isoechoic liquid (**Figure 11.2.4** and **Video 11.2.2**).
3. **What is your sonographic diagnosis?**
   - Ruptured GBM
4. **What additional sonographic examination or finding would help you rule in or rule out your differential diagnoses (if necessary)?**
   - Serial abdominal POCUS exams.

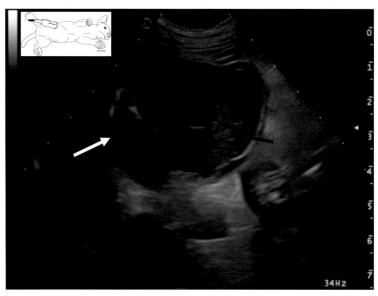

FIGURE 11.2.4 AND VIDEO 11.2.2 Day 3: Right paralumbar approach to the liver for the evaluation of the gallbladder. Same settings as Figure 11.2.1. The black arrow indicates the isoechoic fluid within the GB. The white arrow indicates the mucoid residual content of the GB.

CLINICAL INTEGRATION

Suspected diagnosis based on ultrasound findings:
Ruptured GBM with bile peritonitis.

Sonographic interventions, monitoring, and outcome:
- Serial abdominal POCUS was performed and free fluid was discovered.
- Ultrasound-guided fine needle aspiration of peritoneal effusion revealed the presence of degenerative neutrophils and basophilic amorphous mucinous materials without bile crystals with no bacteria found.
- A cholecystectomy was performed with peritoneal lavage to remove mucinous materials found in the abdomen. Patient was discharged 3 days post-operatively.

Important concepts regarding the sonographic diagnosis of mucocele:
- The gallbladder is susceptible to and influenced by venous return, so fluid therapy can potentially affect it causing physical distension and resultant ischemic necrosis of the gallbladder wall.[1] Early ultrasonographic detection of GBM and serial monitoring are essential to correctly manage these patients.
- Several GBM ultrasound patterns have been described, and there is no correlation between GBM appearance and GBM rupture.[2]
- There is significantly lower mortality (2%) in dogs undergoing elective cholecystectomy compared to those undergoing emergency cholecystectomy (22%–40%).[3] Therefore, early detection and surgical intervention is key to the successful management of these cases.
- When a gallbladder mucocele is found, serial abdominal POCUS exams to monitor the gallbladder evolution are essential. In those cases, fluid therapy can potentially cause gallbladder wall rupture increasing venous return, due to its typical fixed and

immobile wall and the cystic duct obstruction (author hypothesis).

Techniques and tips to identify the appropriate ultrasound images/views:
See **Figure 11.2.1** and **Case 11.10**.

Take-home messages
- Gallbladder evaluation during abdominal POCUS is essential in a patient with acute abdomen. Evaluate for increased wall thickness/edema (Case 11.10) and GBM presence.
- Serial abdominal POCUS exams are crucial to monitor clinical evolution and identify complications early.
- Patients with a GBM often present with similar history and physical laboratory findings when compared to several other diseases. A systemic POCUS exam together with serial monitoring and ultrasound-guided abdominocentesis can impact the clinical decision-making process.

REFERENCES

1. Pike FS, Berg J, King NW, Penninck DG, Webster CRL. Gallbladder mucocele in dogs: 30 cases (2000–2002). *J Am Vet Med Assoc* 2004; 224(10):1615–1622.
2. Choi J, Kim A, Keh S, Oh J, Kim H, Yoon J. Comparison between ultrasonographic and clinical findings in 43 dogs with gallbladder mucoceles. *Vet Radiol Ultrasound* 2014; 55(2):202–207.
3. Youn G, Waschak MJ, Kunkel KAR, Gerard PD. Outcome of elective cholecystectomy for the treatment of gallbladder disease in dogs. *J Am Vet Med Assoc* 2018; 252(8):970–975.

CHAPTER 11 – CASE 3

SHIH TZU WITH ACUTE COLLAPSE AND ABDOMINAL PAIN FOLLOWING 10-DAYS OF INAPPETENCE, VOMITING, POLYURIA, AND POLYDIPSIA

Esther Gomez Soto, Thom Watton, and Laura Cole

HISTORY, TRIAGE, AND STABILIZATION

A 2-year-old male intact Shih Tzu presented with acute collapse and abdominal pain with a 10-day history of inappetence, vomiting, polyuria, and polydipsia.

Triage exam findings (vitals):
- Mentation: obtunded, laterally recumbent
- Respiratory rate: 60 breaths per minute
- Heart rate: 232 beats per minute
- Mucous membranes: pale pink, 2 sec capillary refill time
- Femoral and dorsal pedal arterial pulses: weak and synchronous
- Temperature: 37.4°C (97.5°F)

POCUS exam(s) to perform:
- Abdominal POCUS
- Cardiac POCUS

Abnormal POCUS exam results:
See **Figures 11.3.1** and **11.3.2**.

Additional point-of-care diagnostics and initial management:
Point-of-care blood work showed mild anemia with a packed cell volume (PCV) of 31%, total solids (TS) of 60 g/L, and hyperlactatemia 4.2 mmol/L. The blood smear revealed a platelet count of eight to ten platelets per high power field. An electrocardiogram (ECG) showed ventricular tachycardia and the dog presented with hypertension with a systolic blood pressure (SBP) of 250 mmHg.

During initial stabilization, a bolus of 10 mL/kg of Hartmann's solution was given. The abdominal pain was treated with methadone 0.2 mg/kg IV. To convert the ventricular tachycardia into sinus rhythm, three boluses of lidocaine 2 mg/kg IV followed by an infusion of magnesium sulfate 0.2 mEq/kg IV over 30 minutes were given.

QUESTIONS AND ANSWERS

1. What are the differentials to rule in or out with POCUS based on history and physical exam?
2. What are the sonographic findings?
3. What is the sonographic diagnosis?
4. If necessary, what additional sonographic examination or findings would help rule in or out the differential diagnoses?

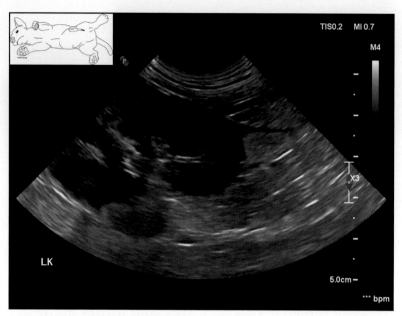

FIGURE 11.3.1 Abdominal POCUS left paralumbar view with the patient in right lateral recumbency. The fur was unshaven, the hair parted, and a microconvex ultrasound probe with coupling gel was placed directly on the skin. The probe was positioned caudal to the last rib in a dorsal location to identify the left kidney. Once the sagittal view was obtained, the probe was angled dorsally and ventrally to assess for peritoneal and/or retroperitoneal free fluid and to rule out obvious renal parenchymal abnormalities. The frequency and depth used were 8 MHz and 5 cm, respectively.

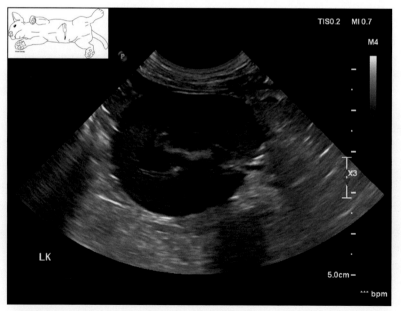

FIGURE 11.3.2 With the patient in the same position (right lateral recumbency) and using the same settings, the probe was rotated 90° to obtain a transverse view. The probe was swept cranially and caudally to further assess the renal parenchyma.

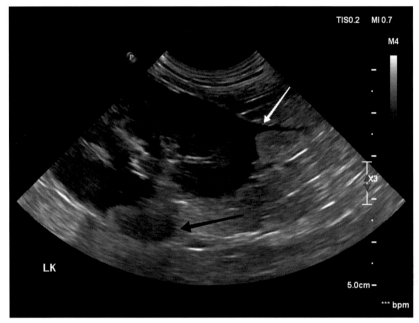

FIGURE 11.3.3 A small volume of anechoic free fluid is present at the caudal pole of the left kidney, extending caudo-dorsally within the retroperitoneum (white arrow). A rounded hypoechoic structure is identified medial to the kidney (black arrow).

1. **Differential diagnosis to rule in or rule out with POCUS:**
   - Hemoabdomen (e.g., splenic torsion/neoplasia), kidney disease (acute or acute-on chronic kidney disease), hepatopathy, septic peritonitis, pancreatitis, prostatitis, neoplasia, primary or secondary cardiac disease.
2. **Describe your sonographic findings:**
   - Retroperitoneal effusion (**Figures 11.3.3 and 11.3.4**).
   - Hypoechoic structure medial to the cranial pole of the left kidney (**Figures 11.3.3 and 11.3.4**).
3. **What is your sonographic diagnosis?**
   - Retroperitoneal effusion (**Figures 11.3.3 and 11.3.4**, white arrow).
   - Left peri-renal hypoechoic mass (**Figures 11.3.3 and 11.3.4**, black arrow).
4. **What additional sonographic examination or finding would help you rule in or rule out your differential diagnoses (if necessary)?**
   - Complete abdominal ultrasound to evaluate the adrenal glands as the presence of retroperitoneal effusion, ventricular tachycardia, hypertension, and a mass medial to the cranial pole of the left kidney could be consistent with adrenal neoplasia, specifically pheochromocytoma.
   - Examination of the caudal vena cava for thrombus or tumor invasion (color flow Doppler) if an adrenal gland mass is documented.
   - Echocardiography to evaluate for pheochromocytoma-induced HCM.[1]

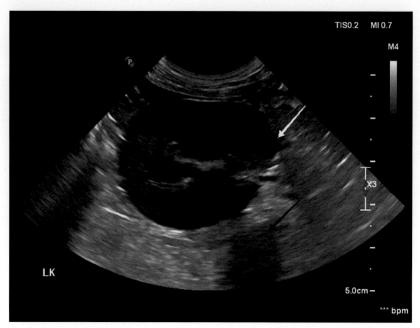

FIGURE 11.3.4 A 2 × 3 cm hypoechoic rounded structure is identified medial to the cranial pole of the left kidney (black arrow). A thin halo of anechoic peri-renal free fluid can be seen at the ventral aspect of the kidney (white arrow).

CLINICAL INTEGRATION

Suspected diagnosis based on ultrasound findings:
Pheochromocytoma.

Sonographic interventions, monitoring, and outcome:
- A complete abdominal ultrasound showed a well-defined mass (2.1 × 2.7 cm) within the cranial pole of the left adrenal gland and not associated with invasion of the vena cava.
- No worsening of the retroperitoneal fluid nor development of abdominal effusion was noted on serial abdominal POCUS exams during hospitalization.
- Urine and plasma metanephrine levels were compatible with a pheochromocytoma.
- The dog was started on phenoxybenzamine and adrenalectomy was performed 2 weeks later.

Important concepts regarding the sonographic diagnosis of retroperitoneal effusion:
- Differentials for retroperitoneal fluid include acute kidney injury, adrenal gland neoplasia, non-adrenal gland neoplasia, renal or adrenal hematoma, coagulopathy, and trauma.[2,3] These differentials can be narrowed based on clinical history, physical examination, and laboratory findings.
- Retroperitoneal fluid is not part of the abdominal fluid score, but it can be measured using the largest length, width, and height to describe it.[4]
- If a large enough volume of retroperitoneal fluid is present, it should be sampled and analyzed to help refine the differentials. Fluid associated with an adrenal mass is most often either hemorrhagic secondary

to rupture of the tumour[5] or a modified transudate secondary to tumor necrosis.[6]
- Peritoneal effusion can be present alongside retroperitoneal fluid. This should also be sampled and analyzed. Invasion or compression of the vena cava often leads to the development of a modified transudate and can help narrow the differentials further.

Techniques and tips to identify the appropriate ultrasound images/views:

- To distinguish between peritoneal and retroperitoneal fluid, the patient can be moved to a standing position and rescanned to determine whether the fluid remains in the same area (likely retroperitoneal) or flows toward the ventral abdomen (peritoneal).[4]
- The assessment of the right kidney (and the adjacent retroperitoneal space) can be difficult as it is located more cranially than the left kidney. An intercostal approach may be required.
- To assess for a small volume of retroperitoneal fluid, the patient may be placed in lateral recumbency so that the intestinal tract falls away from the uppermost kidney.[4]
- Fluid formed within the retroperitoneum tends to form a characteristic shape, typically linear or triangular to oval.[7]
- A "marble-like" pattern may be noticed as fluid dissects within the retroperitoneal fat.[7]

Take-home message

- The differential diagnosis for retroperitoneal effusion in dogs and cats is relatively narrow and its presence can be invaluable in the clinician's decision-making process.

REFERENCES

1. Edmondson EF, Bright JM, Halsey CH and Ehrhart EJ. 2008. Pathologic and cardiovascular characterization of pheochromocytoma-associated cardiomyopathy in dogs. *Vet Pathol*, 52(2), pp. 338–343.
2. Stoneham A, O'Toole T, de Laforcade A, et al. 2004. Retroperitoneal effusion in dogs and cats. *J Vet Emerg Crit Care*, 14(S1), p. 1.
3. Vendrell JR, Alcover J, Alcaraz A, et al. 1996. Unilateral spontaneous adrenal hematoma: An unusual cause of retroperitoneal hemorrhage. *Actas Urol Esp*, 20(1), pp. 59–62.
4. Lisciandro GR. 2021. POCUS: AFAST Introduction and image acquisition. In: Lisciandro GR, ed. *Point-of-Care Ultrasound Techniques for the Small Animal Practitioner*. Hoboken: John Wiley & Sons, Inc, p. 87.
5. Whittemore JC, Preston CA, Kyles AE, Hardie EM and Feldman EC. 2001. Nontraumatic rupture of an adrenal gland tumor causing intra-abdominal or retroperitoneal hemorrhage in four dogs. *J Am Vet Med Assoc*, 219(3), p. 329.
6. Valenciano AC and Rizzi TE. 2020. Abdominal, thoracic and pericardial effusions. In: Valenciano AC and Cowell RL, eds. *Diagnostic Cytology and Hematology of the Dog and Cat*. Missouri: Elsevier Inc, p. 236.
7. D'Anjou MA and Penninck D. 2015. Kidney and ureters. In: D'Anjou MA and Penninck D, eds. *Atlas of Small Animal Ultrasonography*. Ames, Iowa: John Wiley & Sons Inc, p. 357.

CHAPTER 11 – CASE 4

DOMESTIC SHORTHAIR WITH POLYURIA, POLYDIPSIA, VOMITING, AND HYPOREXIA

Xiu Ting Yiew

HISTORY, TRIAGE, AND STABILIZATION

A 12-year-old male neutered Domestic Shorthair presented for extreme lethargy with reported polyuria, polydipsia, vomiting, and hyporexia over the previous week. Referral blood work showed non-regenerative anemia (HCT 27%), moderate left shift neutrophilic leukocytosis (segmented neutrophils 48.9×10^9/L, band neutrophils 1.9×10^9/L), and severe azotemia (creatinine 664 umol/L [7.5 mg/dL], urea 42.7 mmol/L [256.5 mg/dL]).

Triage exam findings (vitals):
- Mentation: dull, responsive
- Respiratory rate: 36 breaths per minute
- Heart rate: 192 beats per minute
- Mucous membranes: pale pink gums, 1–2 sec capillary refill time
- Femoral pulses: strong, synchronous
- Temperature: 40.2°C (104.4°F)
- Abdominal palpation: painful enlarged right kidney, normal left kidney, non-turgid small bladder

POCUS exam(s) to perform:
- Abdominal POCUS with urinary tract assessment
- PLUS
- Cardiac POCUS

Abnormal POCUS exam results:
See **Figures 11.4.1** and **11.4.2**.

Additional point-of-care diagnostics and initial management:
Point-of-care blood work showed hemoconcentration, normokalemia, and high anion gap metabolic acidosis. Intravenous fluid rehydration (Plasma-Lyte A), analgesia (fentanyl CRI), and an antibiotic (ampicillin) were initiated following cystocentesis urine sample collection.

QUESTIONS AND ANSWERS

1. What are the differentials to rule in or out with POCUS based on history and physical exam?
2. What are the sonographic findings?
3. What is the sonographic diagnosis?
4. If necessary, what additional sonographic examination or findings would help rule in or out the differential diagnoses?

DOI: 10.1201/9781003436690-56

Domestic Shorthair with Polyuria, Polydipsia, Vomiting, and Hyporexia

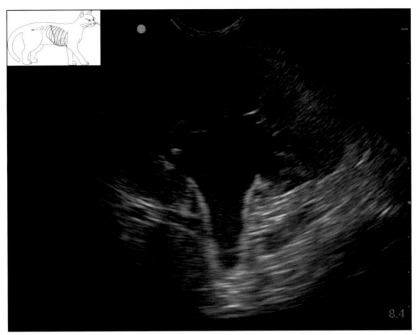

FIGURE 11.4.1 Abdominal POCUS – right paralumbar dorsal plane view. A 4.0–11.0 MHz microconvex array probe is placed parallel to the long axis of the kidney with the reference mark directed toward the head. High-frequency (7.5–10 MHz) linear array probes may be used in cats for better image resolution, given the superficial position of the kidneys.[1-3] In cats and thin dogs, both kidneys might be imaged together on the same side to avoid mistaking one kidney for the other – recall that the left kidney is located caudal to the right kidney[1] and deep to the great vessels[3] when imaged from the right. Unlike dogs, both feline kidneys are located entirely caudal to the ribs, so an intercostal approach is not necessary but easily displaced by mild transducer pressure.[1-3] The readily recognizable dorsal and sagittal plane views of the kidneys are useful reference points for novice sonographers. Once the desired image is identified, the depth is adjusted until the entire kidney is centered within the ultrasound field of view. The probe is repeatedly fanned in either direction from "capsule to capsule" through various angles of the kidney to fully assess the appearance and echogenicity of the renal cortex, medulla, and sinus.[2] This still image was obtained at the level of the renal sinus.

1. **Differential diagnosis to rule in or rule out with POCUS:**
   - Pyelonephritis ± pyonephrosis, dry form feline infectious peritonitis, partial or complete, unilateral or bilateral ureteral obstruction, renal lymphoma, perinephric pseudocyst.
2. **Describe your sonographic findings:**
   - Enlarged right kidney (measuring 6.9 cm in length) with smooth margination but loss of normal renal architecture and corticomedullary distinction (**Figures 11.4.3** and **11.4.4**).
   - Right renal pelvis dilation (up to 18.0 mm) and proximal ureteral distension (up to 11.5 mm) with dilated pelvic diverticula and intraluminal gravity-dependent hyperechoic material with a fluid debris level that do no cast a distal acoustic shadow (**Figures 11.4.3** and **11.4.4**).

- No renal cysts, nodules, infarcts, masses, perirenal (subcapsular) or retroperitoneal fluid.

3. **What is your sonographic diagnosis?**
   - Marked right renomegaly.
   - Severe right pyelectasia (hydronephrosis) with hydroureter – suspect pyonephrosis.
   - Partial or complete right ureteral obstruction.

4. **What additional sonographic examination or finding would help you rule in or rule out your differential diagnoses (if necessary)?**
   - Detailed ultrasonography with tracing of bilateral ureters from the renal pelvis to the ureterovesicular junctions (bladder) by an experienced sonographer using a higher-frequency (7.5–10 MHz) linear array probe for the identification and localization of obstructive lesions.
   - Ultrasound-guided pyelocentesis for cytology and bacterial culture and antimicrobial susceptibility testing.
   - Ultrasound-guided fluoroscopy-assisted antegrade pyelography for the localization and confirmation of partial or complete ureteral obstruction (if equivocal ultrasound findings).

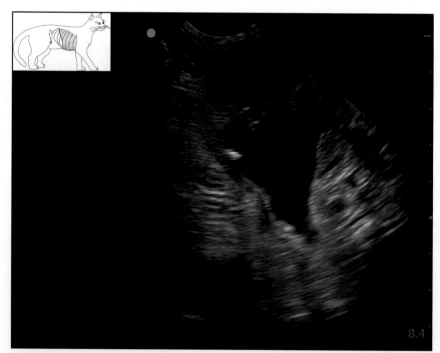

FIGURE 11.4.2 Abdominal POCUS – right paralumbar transverse plane view. The probe is rotated 90° counter-clockwise to the kidney's long axis with the reference mark directed toward the spine. The probe is repeatedly fanned from the cranial pole to the caudal pole for a full assessment of the renal cortex, medulla, and sinus.[2] This still image was obtained at the level of the renal sinus.

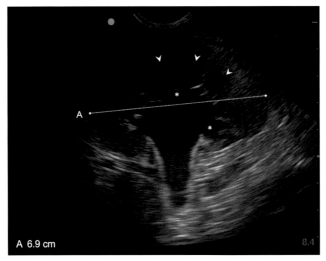

FIGURE 11.4.3 Marked renomegaly. Normal feline kidneys measure 3.0–4.4 cm in length bilaterally.[1-3] Calipers are placed across the maximum craniocaudal length of the kidney (dotted line). Note the loss of normal renal architecture and corticomedullary distinction (arrowheads) with dilated pelvic diverticula (asterisks).[3,4]

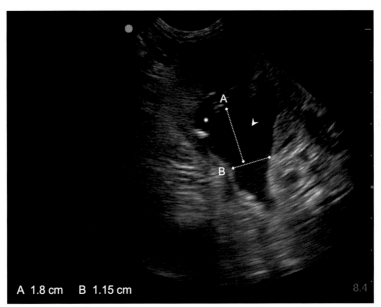

FIGURE 11.4.4 Severe pyelectasia (hydronephrosis) with hydroureter – suspect pyonephrosis. Normal renal pelvises (range 0.8–3.2 mm)[5] and ureters (range 0.0–0.4 mm)[6] are not readily visualized.[1-3] Pyelectasia and hydroureter are easier to identify in the transverse plane.[1,4] Maximal renal pelvic width is measured in the transverse plane to avoid an oblique measurement.[5] Calipers are placed at the inner surface of the renal pelvis margins without including the proximal ureter (dotted line A) and at the largest proximal ureteral diameter (dotted line B).[5,6] Note the dilated pelvic diverticula (asterisk) and intraluminal gravity-dependent echogenic content with a fluid debris level (arrowhead) within the renal collecting system.[7] Acoustic enhancement should be considered when assessing renal pelvis echogenicity if there is significant peritoneal or retroperitoneal fluid.[7]

CLINICAL INTEGRATION

Suspected diagnosis based on ultrasound findings:
Acute kidney injury secondary to pyelonephritis, pyonephrosis, and unilateral partial or complete ureteral obstruction from inflammatory debris, ureteritis, or concurrent ureterolithiasis.

Sonographic interventions, monitoring, and outcome:
- Ultrasound-guided pyelocentesis drained 3 mL of viscous pus from the right renal pelvis. The right ureter traced along its entire length showed multifocal amorphous hyperechoic material without distal acoustic shadowing.
- Parenteral third-generation cephalosporin (cefotaxime) was initiated for adequate renal tissue concentration and broader Gram-negative bacterial coverage. Cytology revealed highly degenerative neutrophils with abundant intracellular rod bacteria. Urine and renal pelvis sample grew beta-hemolytic *Escherichia coli* susceptible to third-generation cephalosporin, fluroquinolones, and beta-lactamases.
- Rechecked blood work showed significant improvements (WBC 13.7×10^9/L, segmented neutrophils 12.9×10^9/L, creatinine 176 umol/L [2 mg/dL], urea 18.7 mmol/L [112.3 mg/dL]). Serial abdominal POCUS showed gradual resolution of right hydronephrosis and hydroureter.
- The patient was discharged with 6 weeks of oral fluroquinolone (pradofloxacin). Blood work 1-month later revealed complete resolution of previous derangements.

Important concepts regarding the sonographic diagnosis of pyelonephritis/pyonephrosis:
- Mild-to-moderate renal pelvis dilation is termed pyelectasia. Hydronephrosis is associated with obstruction and is reserved for severe renal pelvis dilation with pelvic diverticula.[1,2] Hydroureter refers to abnormal ureteral dilation with fluid.[4]
- Pyelonephritis implies renal pelvis and parenchymal inflammation secondary to hematogenous or ascending urinary tract infection.[2,4] Acute pyelonephritis may not have detectable ultrasonographic abnormalities.[1] Ultrasonographic findings may include renomegaly with irregular margination and hyperechoic cortices, loss of normal renal architecture with blunted poorly defined diverticula, mild-to-moderate pyelectasia, proximal ureteral dilation, and gas within the kidney (with emphysematous pyelonephritis).[1,2,4,8]
- Pyonephrosis involves pus accumulation in the hydronephrotic kidney secondary to urinary stasis and subsequent infection.[7] Specific ultrasonographic findings include hydronephrosis, loss of normal renal architecture, and the presence of dispersed, completely filling, or dependent echogenic content with a fluid debris level within the renal collecting system.[7] Anechoic renal collecting contents cannot rule out pyonephrosis.[7]
- Ultrasonography alone cannot definitively diagnose pyelonephritis,[1,2] determine if renal pelvic fluid is purulent or infected,[7] or confirm the cause of urinary outflow obstruction.[9]
- Notable variations and overlap of renal pelvic widths and proximal ureteral diameters exist in cats and dogs with normal renal function, physiological or pathological diuresis, pyelonephritis, renal insufficiencies, and ureteral obstruction.[5,6]
- Cats with pyelonephritis (46.2%) and ureteral obstruction (81.8%) are more likely to have measurable proximal ureteral dilation than stable chronic kidney disease (6%) or normal (0%) cats.[6] Although a renal pelvic width ≥13 mm is 100% predictive

of obstruction,[5,6] minimal pyelectasia or ureteral dilation does not exclude obstruction.[10]
- Diagnostic utility of ultrasonography varies with obstructive lesions: ureteroliths (sensitivity 98%, specificity 96%, positive predictive value (PPV) 98%), strictures (sensitivity 44%, specificity 95%, PPV 88%), solidified blood clots (7/7 incorrect diagnosis).[9]

Techniques and tips to identify the appropriate ultrasound images/views:
See **Figure 11.4.1**.

Take-home messages
- Renal pelvis dilation ≥3.5 mm in cats and ≥4 mm in dogs with any degree of ureteral dilation is abnormal,[4–6] and a detailed evaluation by an experienced sonographer is warranted.
- Pyelonephritis is diagnosed through compatible history, clinical signs, laboratory results (blood work, urinalysis, bacterial culture), ultrasonographic findings, and response to therapy. Ultrasound-guided pyelocentesis is necessary to confirm pyonephrosis.

REFERENCES

1. Widmer WR, Biller DS, Adams LG. Ultrasonography of the urinary tract in small animals. *J Am Vet Med Assoc* 2004;225(1):46–54. doi:10.2460/javma.2004.225.46
2. Debruyn K, Haers H, Combes A, et al. Ultrasonography of the feline kidney: Technique, anatomy and changes associated with disease. *J Feline Med Surg* 2012;14(11):794–803. doi:10.1177/1098612X12464461
3. Griffin S. Feline abdominal ultrasonography: What's normal? What's abnormal? The kidneys and perinephric space. *J Feline Med Surg* 2020;22(5):409–427. doi:10.1177/1098612X20917598
4. Griffin S. Feline abdominal ultrasonography: What's normal? What's abnormal? Renal pelvis, ureters and urinary bladder. *J Feline Med Surg* 2020;22(9):847–865. doi:10.1177/1098612X20941786
5. D'Anjou MA, Bédard A, Dunn ME. Clinical significance of renal pelvic dilatation on ultrasound in dogs and cats. *Vet Radiol Ultrasound* 2011;52(1):88–94.
6. Quimby JM, Dowers K, Herndon AK, Randall EK. Renal pelvic and ureteral ultrasonographic characteristics of cats with chronic kidney disease in comparison with normal cats, and cats with pyelonephritis or ureteral obstruction. *J Feline Med Surg* 2017;19(8):784–790. doi:10.1177/1098612X16656910
7. Choi J, Jang J, Choi H, Kim H, Yoon J. Ultrasonographic features of pyonephrosis in dogs. *Vet Radiol Ultrasound* 2010;51(5):548–553. doi:10.1111/j.1740-8261.2010.01702.x
8. Gould EN, Cohen TA, Trivedi SR, Kim JY. Emphysematous pyelonephritis in a domestic shorthair cat. *J Feline Med Surg* 2016;18(4):357–363. doi:10.1177/1098612X15600481
9. Wormser C, Reetz JA, Drobatz KJ, Aronson LR. Diagnostic utility of ultrasonography for detection of the cause and location of ureteral obstruction in cats: 71 cases (2010–2016). *J Am Vet Med Assoc* 2019;254(6):710–715. doi:10.2460/javma.254.6.710
10. Lemieux C, Vachon C, Beauchamp G, Dunn ME. Minimal renal pelvis dilation in cats diagnosed with benign ureteral obstruction by antegrade pyelography: A retrospective study of 82 cases (2012–2018). *J Feline Med Surg* 2021;23(10):892–899. doi:10.1177/1098612X20983980

CHAPTER 11 – CASE 5

CHIHUAHUA WITH A 2-DAY HISTORY OF LETHARGY, ANOREXIA, AND VAGINAL DISCHARGE

LM Bacek and Kendon Wu Kuo

HISTORY, TRIAGE, AND STABILIZATION

A 7-year-old female intact Chihuahua presented with a 2-day history of lethargy, anorexia, and vaginal discharge. The last known estrus cycle was approximately 8 weeks prior.

Triage exam findings (vitals):
- Mentation: obtunded
- Respiratory rate: 50 breaths per minute with normal effort
- Heart rate: 170 beats per minute
- Mucous membranes: pink, tacky, >3 sec capillary refill time
- Femoral and dorsal pedal arterial pulses: weak
- Temperature: 103.8°F (39.9°C)

POCUS exam(s) to perform:
- Abdominal POCUS with urinary bladder assessment

Abnormal POCUS exam results:
See **Figures 11.5.1** and **11.5.2**.

Additional point-of-care diagnostics and initial management:
Additional initial point-of-care testing included packed cell volume/total solids (PCV/TS) of 55% and 5.2 g/dL, respectively, blood glucose of 175 mg/dL, and lactate of 4.9 mmol/L. Other diagnostics included an initial electrocardiogram (ECG) showing a sinus tachycardia and a Doppler blood pressure of 75 mmHg. A 20 mL/kg IV isotonic crystalloid bolus was given and a rechecked Doppler blood pressure was 90 mmHg.

QUESTIONS AND ANSWERS

1. What are the differentials to rule in or out with POCUS based on history and physical exam?
2. What are the sonographic findings?
3. What is the sonographic diagnosis?
4. If necessary, what additional sonographic examination or findings would help rule in or out the differential diagnoses?

Chihuahua with a 2-Day History of Lethargy, Anorexia, and Vaginal Discharge

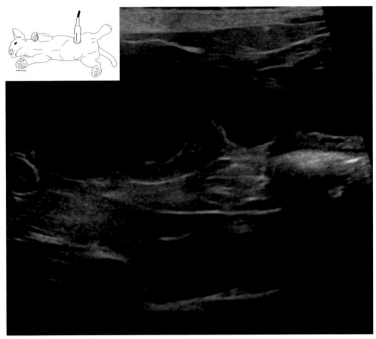

FIGURE 11.5.1 Abdominal POCUS umbilical view with the patient in right lateral recumbency. The microconvex probe is placed in long axis with the marker pointing toward the head. The depth is set at 5 cm.

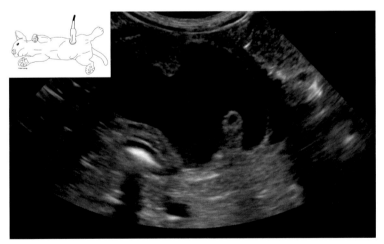

FIGURE 11.5.2 Abdominal POCUS umbilical view with the patient in right lateral recumbency. The microconvex probe is placed in short axis with the marker pointing dorsally. The probe is then fanned both cranially and caudally.

1. **Differential diagnosis to rule in or rule out with POCUS:**

   The most likely differential diagnoses based on signalment, history, and physical examination include pyometra, mucometra, hydrometra, or pregnancy. If peritoneal fluid is found, a ruptured pyometra may be diagnosed based on cytology.

2. **Describe your sonographic findings:**

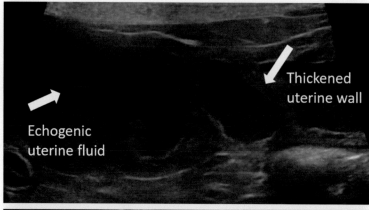
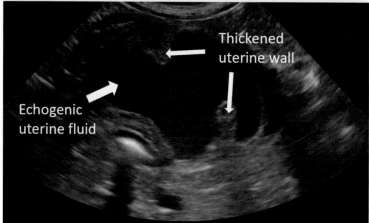

FIGURE 11.5.3 Labeled images of Figures 11.5.1 and 11.5.2. Note the fluid-filled uterus with thickened uterine wall.

- Fluid-filled, distended structure with thickened wall consistent with uterus (**Figure 11.5.3**).
- No free abdominal fluid.

3. What is your sonographic diagnosis?
   - Pyometra vs mucometra.

4. What additional sonographic examination or finding would help you rule in or rule out your differential diagnoses (if necessary)?
   None.

CLINICAL INTEGRATION

Suspected diagnosis based on ultrasound findings:
Pyometra is the most likely diagnosis due to the clinical signs and ultrasound findings.

Sonographic interventions, monitoring, and outcome:
- A repeat POCUS was performed to evaluate for the development of free abdominal fluid; this exam was also negative for free fluid.
- Broad spectrum antibiotics (ampicillin sulbactam 30 mg/kg IV q8hr) and pain medication (methadone 0.2 mg/kg IV q6hr) were started.
- An exploratory laparotomy and ovariohysterectomy were performed without complication.

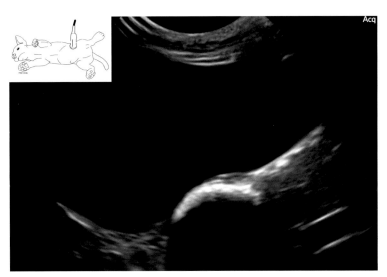

FIGURE 11.5.4 Normal urinary bladder view as part of the abdominal POCUS exam with the patient in right lateral recumbency. The image is taken with the probe placed transversally with the marker directed dorsally. In this case, the bladder and colon appear normal with no uterus visualized.

- The patient recovered well and was discharged on supportive care 24 h after surgery.

Important concepts regarding the sonographic diagnosis of pyometra:

- A fluid-filled uterus may be consistent with pyometra, mucometra, hydrometra, etc. The index of suspicion should increase with a consistent history and physical examination findings.[1,2]
- Ultimately, the final diagnosis of pyometra is based on histopathologic findings.[1]

Techniques and tips to identify the appropriate ultrasound images/views:

- Recall that the uterus is normally between the bladder and the colon. These landmarks can help locate the uterus (see **Figure 11.5.4**).
- Multiple cross sections of a fluid-filled uterus (compartmentalization) may be confused with free fluid. However, the fluid will be contained within the uterine wall (see **Figure 11.5.3**), which should be distinguished from the more obvious multilayered small intestinal wall. The echogenicity of the fluid may range from anechoic to echogenic with slowly swirling contents.

Take-home messages

- A fluid-filled uterus can be found in other disease states such as mucometra or hydrometra. Ideally, ultrasound imaging should be interpreted in conjunction with estrus history, physical examination findings, and supportive clinicopathologic data such as blood work and vaginal cytology. Vaginal discharge will not be present in a closed cervix pyometra.[2]
- In patients with a ruptured uterus, free abdominal fluid may be identified as well.
- Rarely, gas in the uterine wall may be diagnosed and consistent with emphysematous pyometra.[3]

REFERENCES

1. Hagman R. Pyometra in small animals 2.0. *Vet Clin North Am Small Anim Pract* 2022 May;52(3):631–657.
2. Jitpean S, Ambrosen A, Emanuelson U, Hagman R. Closed cervix is associated with more severe illness in dogs with pyometra. *BMC Vet Res* 2017 Jan 5;13(1):11.
3. Mattei C, Fabbi M, Hansson K. Radiographic and ultrasonographic findings in a dog with emphysematous pyometra. *Acta Vet Scand* 2018 Oct 29;60(1):67.

CHAPTER 11 – CASE 6

SAMOYED WITH A 4-DAY HISTORY OF VOMITING AFTER PASSING PARTS OF A RUBBER BALL

Anais Allen-Deal and Daria Starybrat

HISTORY, TRIAGE, AND STABILIZATION

A 4-year-old male neutered Samoyed presented with a 4-day history of vomiting, lethargy, and shaking. The owner reported that the dog passed part of a rubber ball 48 h prior to presentation.

Triage exam findings (vitals):
- Mentation: quiet, alert, and responsive with a hunched posture
- Respiratory rate: 80 breaths per minute
- Heart rate: 200 beats per minute
- Mucous membranes: pale pink with a 3 sec capillary refill time
- Femoral and dorsal pedal arterial pulses: femoral pulses weak but synchronous; absent dorsal pedal pulses
- Temperature: 36.0°C (96.8°F)
- Other: marked abdominal pain

POCUS exam(s) to perform:
- Abdominal POCUS
- PLUS
- Cardiac POCUS

Abnormal POCUS exam results:
See **Figure 11.6.1** and **Videos 11.6.1** and **11.6.2**.

Additional point-of-care diagnostics and initial management:
A large bore intravenous cannula was placed and a blood sample was obtained:

- Packed cell volume (PCV): 32%.
- Total solids: 70 g/L (7.0 g/dL).
- Blood glucose: 81 mg/dL (4.5 mmol/L).
- Lactate: 4 mmol/L.
- Blood smear: neutrophilia, five platelets per high powered field (HPF).
- Prothrombin time/activated partial thromboplastin time (PT/aPTT): within reference ranges.

Methadone (0.2 mg/kg) was given intravenously. Blood pressure was unreadable and therefore a 10 mL/kg bolus of lactated Ringer's solution was administered.

QUESTIONS AND ANSWERS

1. What are the differentials to rule in or out with POCUS based on history and physical exam?
2. What are the sonographic findings?
3. What is the sonographic diagnosis?
4. If necessary, what additional sonographic examination or findings would help rule in or out the differential diagnoses?

1. **Differential diagnosis to rule in or rule out with POCUS:**
   - Abdominal POCUS should be performed to assess for the presence of free fluid. The most likely causes of peritoneal effusion here include septic peritonitis and/or pancreatitis and any peritoneal effusion should be sampled for analysis. Other causes of peritoneal effusion may include hemoabdomen, bile peritonitis, uroabdomen, and neoplasia. A distended stomach and/or intestinal loops with no or minimal free fluid can be seen with obstructive gastrointestinal foreign bodies.
   - Aspiration pneumonia should be suspected in a vomiting patient with tachypnea and panting. This can be evaluated via lung ultrasound, and will present as ventral B-lines found commonly over the right middle lung lobe site.
   - Cardiac POCUS should be performed to evaluate for hypovolemia. A small left ventricle and atrium and a decreased

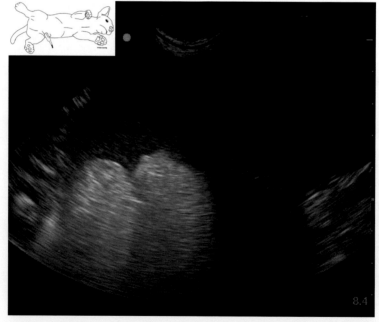

FIGURE 11.6.1 Urinary bladder view. With the dog in left lateral recumbency, the curvilinear probe was positioned longitudinally on the caudal abdomen with enough depth to view the entire bladder without compressing it, in this case 8.4 cm. The fur was not clipped and gel was used as the coupling agent.

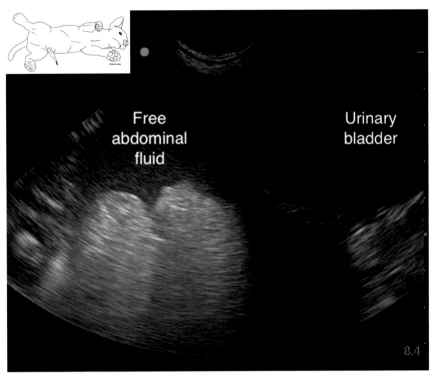

FIGURE 11.6.2 Labeled image of Figure 11.6.1. Note the free fluid cranial to the bladder and hyperechoic peritoneum.

left atrial to aortic ratio imply hypovolemia.[1] Increased contractility of the left ventricle can be seen in hypovolemia, while decreased movement indicates poor systolic function.[1] The caudal vena cava can be assessed at the suprailiac, right intercostal, and subxiphoid level.[1] A decreased diameter or a more distensible caudal vena cava is consistent with hypovolemia.

2. **Describe your sonographic findings:**
   - Echogenic free fluid cranial to the bladder with hyperechoic peritoneum. Spleen, small intestinal loops, and bladder appear normal (**Figures 11.6.1** and **11.6.2**).
   - Free fluid identified by the spleen and small intestinal loops. Largest pocket of free fluid identified cranial to the bladder (**Video 11.6.1** and **Figure 11.6.3**).
   - Shadowing obstructs the image of the spleen. As this originates from the edge of the peritoneum, it is consistent with free abdominal gas (**Video 11.6.2** and **Figure 11.6.4**).

3. **What is your sonographic diagnosis?**
   - Free peritoneal fluid with an abdominal fluid score of 2/4 and pneumoperitoneum.

4. **What additional sonographic examination or finding would help you rule in or rule out your differential diagnoses (if necessary)?**
   - None.

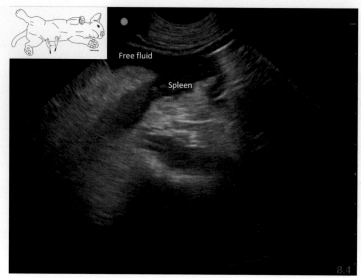

FIGURE 11.6.3 AND VIDEO 11.6.1 Left paralumbar and urinary bladder view. With the dog in left lateral recumbency, the curvilinear probe was held longitudinally and moved caudally along the gravity-dependent side of the abdomen from the umbilicus until the urinary bladder view was obtained. In this case, the abdomen was clipped and gel was used as the coupling agent. Note free fluid near spleen.

CLINICAL INTEGRATION

Suspected diagnosis based on ultrasound findings:

Septic peritonitis is highly likely in this patient. Echogenic free abdominal fluid is caused by small particulates in the fluid and is associated with septic, inflammatory, or hemorrhagic effusions. Free abdominal gas increases the likelihood of septic peritonitis as it is caused by the perforation of a hollow organ such as the gastrointestinal tract.

Sonographic interventions, monitoring, and outcome:

- Ultrasound-guided abdominocentesis of the peritoneal fluid cranial to the bladder was performed. A sample was prepared for cytological evaluation and intracellular bacteria were identified confirming the diagnosis of septic peritonitis.
- An exploratory laparotomy was subsequently performed and a proximal jejunal perforation was identified. A resection and anastomosis was performed. The volume of peritoneal effusion increased 24 h post-surgery and a repeat abdominocentesis confirmed the recurrence of septic peritonitis.
- The owners elected for euthanasia.

Important concepts regarding the sonographic diagnosis of septic peritonitis:

- Holding the probe longitudinally and moving it from the cranial abdomen caudally along the gravity-dependent part of the abdomen while the patient is in lateral recumbency assists with the identification of the largest pocket of free fluid for abdominocentesis.

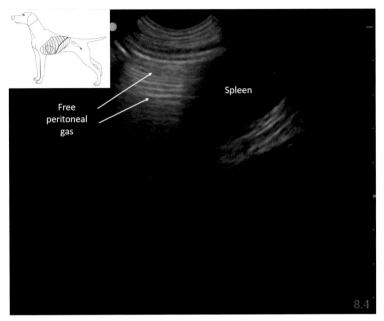

FIGURE 11.6.4 AND VIDEO 11.6.2 Left paralumbar view. With the dog standing, the curvilinear probe is held longitudinally, ventral to lumbar musculature on the left-hand side. The fur was not clipped and gel was used as the coupling agent. Note the free abdominal gas artifact obstructing the image of the spleen.

- Free abdominal gas appears as a reverberation coming from the edge of the peritoneum.[2] This, in combination with free fluid, supports a diagnosis of hollow organ rupture if there is no history of recent abdominal surgery. Care should be taken not to misdiagnose free abdominal gas with intestinal luminal gas.
- Serial abdominal fluid scoring should be performed in the post-operative patient as a sudden increase in volume raises concern for dehiscence.[3] Other causes of increasing fluid score include hemorrhage and third spacing secondary to hypoalbuminemia. To calculate the abdominal fluid score, each of four sites (subxiphoid view, left paralumbar view, right paralumbar view, and urinary bladder view) is graded as 0 (negative), half (weak positive), or 1 (strong positive) for free fluid.[3]

Techniques and tips to identify the appropriate ultrasound images/views:
- If free fluid is not immediately apparent, this can be due to hypovolemia and dehydration. Repeat examination should be performed after fluid resuscitation.
- Fanning the probe in each site can increase the chance of identifying small volume effusions.

Take-home message
- Abdominal POCUS can identify free abdominal fluid and free abdominal gas in critically ill patients and allows ultrasound-guided sampling, which is essential in confirming the diagnosis of septic peritonitis.

REFERENCES

1. S. R. Boysen and K. Gommeren, "Assessment of Volume Status and Fluid Responsiveness in Small Animals," *Frontiers in Veterinary Science*, 8:630643, 1–23, 2021.
2. S. Y. Kim, K. T. Park, S. C. Yeon and H. C. Lee, "Accuracy of Sonograph Diagnosis of Pneumoperitoneum Using the Enhanced Peritoneal Stripe Sign in Beagle Dogs," *Journal of Veterinary Science*, 15(2), pp. 195–198, 2014.
3. G. R. Lisciandro, "AFAST Target-Organ Approach and Fluid Scoring System in Dogs and Cats," *Veterinary Clinics of North America: Small Animal Practice*, 51(6), pp. 1217–1231, 2021.

CHAPTER 11 – CASE 7

LABRADOR RETRIEVER WITH ACUTE COLLAPSE AND A 24-HOUR HISTORY OF LETHARGY AND ANOREXIA

Charles T Talbot and Liz-Valérie S Guieu

HISTORY, TRIAGE, AND STABILIZATION

A 6-year-old female spayed Labrador Retriever presented after being found weak and acutely collapsed, with a 24-h history of lethargy and anorexia.

Triage exam findings (vitals):
- Mentation: quiet, alert, responsive
- Respiratory rate: 52 breaths per minute
- Heart rate: 140 beats per minute
- Mucous membranes: pale pink, tacky, 3 sec capillary refill time
- Femoral and dorsal pedal arterial pulses: weak, thready
- Temperature: 36.9°C (98.4°F)

POCUS exam(s) to perform:
- Abdominal POCUS with splenic assessment
- Cardiac POCUS

Abnormal POCUS exam results:
See **Figures 11.7.1–11.7.3**.

Additional point-of-care diagnostics and initial management:
A point-of-care electrocardiogram revealed a sinus tachycardia, packed cell volume/total protein (PCV/TP) of 38%/5.2 g/dL, systolic blood pressure of 105 mmHg, and venous blood gas showed a metabolic acidosis with an increased anion gap secondary to a hyperlactatemia (pH 7.25, HCO_3 12 mmHg, lactate 6.3 mmol/L, anion gap 26). C5 coagulation panel was within normal limits (prothrombin time, activated partial thromboplastin time, D-dimer, antithrombin, fibrinogen). A 10 mL/kg isotonic crystalloid bolus was given, with an improvement in cardiovascular parameters (heart rate 100 bpm, blood pressure 110 mmHg, capillary refill time <2 sec, and mucous membrane remained pale pink to pink).

QUESTIONS AND ANSWERS

1. What are the differentials to rule in or out with POCUS based on history and physical exam?
2. What are the sonographic findings?
3. What is the sonographic diagnosis?
4. If necessary, what additional sonographic examination or findings would help rule in or out the differential diagnoses?

DOI: 10.1201/9781003436690-59

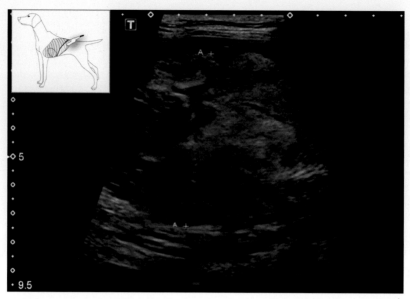

FIGURE 11.7.1 Splenic assessment. Patient in dorsal recumbency. Image taken using a linear probe with the depth set at 9.5 cm and the frequency set at 8 Hz, with the probe placed longitudinally and the marker directed cranially, caudal to the left thirteenth rib. In large breed dogs, it might be necessary to position the patient in right lateral recumbency to identify the head of the spleen underneath the rib cage through a left intercostal approach. The fur was not clipped and alcohol was used as the coupling agent.

1. **Differential diagnosis to rule in or rule out with POCUS:**
   - Pericardial effusion, hemoabdomen, anaphylaxis.
2. **Describe your sonographic findings:**
   - A cross-sectional 7.5 × 6.1 cm heterogeneous mass arising from the splenic head, having internal hyperechoic, isoechoic, hypoechoic, and anechoic regions (**Figure 11.7.1**).
   - Mild to moderate free fluid adjacent to the splenic head mass (**Figure 11.7.3**).
   - The body of the spleen appears normal (**Figure 11.7.2**).
3. **What is your sonographic diagnosis?**
   - Splenic mass.
   - Hemoabdomen.
   - Abdominal fluid score 1 out of 4.
4. **What additional sonographic examination or finding would help you rule in or rule out your differential diagnoses (if necessary)?**
   - Cardiac POCUS to examine the right auricle. Concurrent right atrial hemangiosarcoma occurs in approximately 25% of dogs with splenic hemangiosarcomas.[1,2]
   - Liver assessment. Metastasis may present as target lesions, which exhibit a hypoechoic margin with a hyperechoic center. When multiple lesions present, the positive predictive value of malignancy is 74%–81%; however, surgery should still be considered.[3] Leyva et al. (2018) have identified that approximately 30% of dogs undergoing surgery with both gross hepatic and splenic lesions have a favorable prognosis. While malignant neoplasia is most likely in cases with both splenic and hepatic malignancy, benign or treatable causes must be considered possible in each dog.[4]

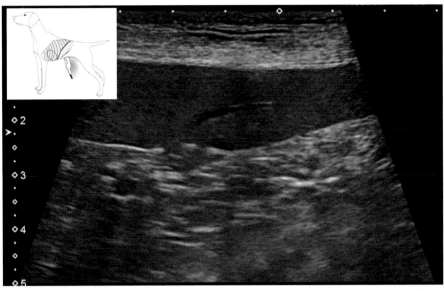

FIGURE 11.7.2 Splenic assessment. Patient in dorsal recumbency. Image taken using a linear probe with the depth set at 5 cm and the frequency set at 9 Hz, with the probe placed longitudinally and the marker directed cranially. While maintaining the probe in a linear position, with the marker directed cranially, move ventrally from the head of the spleen toward the umbilicus in order to visualize the entirety of the splenic body, as well as the splenic tail. This view is of the body of the spleen, land marked cranial and dorsal to the umbilical view, with the probe positioned horizontally. There is visualization of the greater vasculature using a linear probe with the depth set at 5 cm and the frequency set at 9 Hz. The fur was not clipped and alcohol was used as the coupling agent.

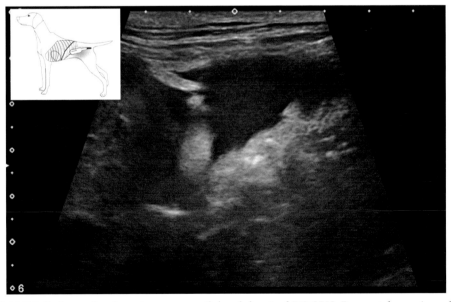

FIGURE 11.7.3 Left paralumbar view as part of the abdominal POCUS. Image taken using a linear probe with the depth set at 6 cm and the frequency set at 9 Hz, with the probe placed longitudinally and the marker directed cranially, caudally to the splenic head. The fur was not clipped and alcohol was used as the coupling agent.

CLINICAL INTEGRATION

Suspected diagnosis based on ultrasound findings:
Ruptured hemangiosarcoma vs hematoma are the most likely diagnoses. Confirmation is required via histopathology.

Sonographic interventions, monitoring, and outcome:
- Abdominocentesis was performed and the fluid PCV/TP was 37%/5.1 g/dL, consistent with an acute abdominal hemorrhagic bleed.
- Maropitant and a fentanyl CRI were administered. The patient had a run of ventricular tachycardia at >180 bpm, which responded to a 2 mg/kg IV lidocaine bolus. Blood type and cross-match diagnostics were performed.
- Cytology, blood glucose, and blood lactate were not performed on the abdominal effusion. Patient remained normotensive above 90 mmHg (systolic).
- The patient went to surgery for a splenectomy, with the histopathological diagnosis confirming a hemangiosarcoma.

Important concepts regarding the sonographic diagnosis of splenic masses:
- A normal spleen cannot be assessed in one ultrasonographic window. A complete trace from the splenic head to tail, or vice versa, is required for full evaluation.
- One study evaluating dogs with non-traumatic hemoabdomens found that the source of hemorrhage was splenic in 75/83 (90%).[5]
- Splenic masses that have not ruptured are more likely benign.[6-8]
- In patients with a non-traumatic hemoabdomen secondary to a ruptured splenic mass, 63% are hemangiosarcomas, 27% splenic hematomas, 5% splenic torsion, and 5% are neoplasias other than hemangiosarcoma.[9]
- There is no significant difference in erythrocyte morphology (acanthocyte identification) in analysis of the abdominal effusion or peripheral blood between dogs with a splenic hemangiosarcoma and those that do not have a hemangiosarcoma related to a non-traumatic hemoabdomen.[10]
- Nodular hyperplasia is difficult to differentiate on POCUS examination and requires further diagnostics such as histopathology.
- Myelolipomas are typically incidentally found. They appear as hyperechoic nodules and are most commonly in the region of the hilus vasculature.
- Hematomas may be associated with abdominal trauma and have variable ultrasonographic appearances; therefore, they cannot be differentiated from neoplasia via sonography.
- Splenic abscessations are rare and cannot be distinguished from benign or malignant masses. Air production secondary to gas-producing bacteria within the abscess (as seen ultrasonographically) will aid in this differentiation.
- Additional neoplastic lesions such as myeloproliferative diseases create a diffuse, coarsely mottled appearance to the splenic parenchyma, which is typically described as honeycomb, moth-eaten, or focal or multi-focal hypoechoic or complex lesions.

Techniques and tips to identify the appropriate ultrasound images/views:
- The normal spleen is a homogeneous organ, with the parenchyma finely textured, with a medium to high echogenic pattern.

- Comparisons of the spleen can be made with the adjacent liver parenchyma, in which the spleen will be hyperechoic to the liver.
- Additional comparisons can be made to the cortex of the kidney, in which the liver will be hyperechoic to the cortex.
- Subtle differences in echogenicity to the spleen, liver, and the cortex of the kidney parenchyma can be difficult to identify, as one or more of these organs can be concurrently abnormal.
- There are no established guidelines for splenic width, thickness, or length in the dog.
- The spleen is typically located in the left cranial abdomen, and typically follows the greater curvature of the stomach.
  - The actual position of the spleen itself will be dependent on the distension of the stomach, as well as size of the surrounding organs.
  - The head of the spleen is located under the border of the rib cage, in which it is best to place the probe in a horizontal plane with the marker facing cranially at the caudal margin of the 13th rib.
  - By moving the probe ventrally, adjacent to the 13th rib, the body (**Figure 11.7.2**) and tail of the spleen can be visualized running along the left body wall and across the ventral abdomen.
  - From an anatomical standpoint, the spleen will lie lateral and caudal to the stomach and ventral or lateral to the left kidney. When enlarged or displaced by a mass, it may cross the ventral midline and extend toward the bladder view.
  - In larger dogs, repositioning into right lateral recumbency will aid in the identification of the spleen, which is located under the rib cage on the left-hand side. It will enable an enhanced visualization of the cranial aspect of the head of the spleen within the eleventh and twelfth intercostal spaces.
  - Ensure that the entire spleen is systemically scanned through sagittal and transverse planes.
- The splenic vasculature consists of the splenic vein and the splenic artery.
  - The splenic vein can be visualized near the hilum, and should be assessed for thrombosis, as well as Doppler ultrasonography for both patency and location.
- The capsule surrounding the splenic parenchyma is typically smooth and regular.
  - It is important to identify the echogenic splenic capsule, as this will enable an appropriate origin determination of suspected masses, whether they arise from the spleen itself or an adjacent organ.
  - If the suspected mass originates from the spleen, typical ultrasonographic changes are consistent with an interruption of the splenic capsule which will deviate outward, with a continuation of the splenic parenchyma (**Figure 11.7.1**).
  - If the mass is from outside the spleen, the echogenic capsule will remain intact.
- Schick et al. (2019) evaluated a risk prediction model for hemangiosarcoma diagnosis in dogs presenting with non-traumatic hemoabdomen, utilizing body weight, total plasma protein, platelet count, and thoracic radiographic findings, establishing the HeLP score.[11]
  - The HeLP score has been shown to perform well in aiding in the establishment of a hemangiosarcoma diagnosis for dogs presenting with a non-traumatic hemoabdomen, in particular in those dogs that are at a lower risk for this diagnosis, thereby facilitating appropriate treatment.[11]

Take-home messages
- Ultrasonographic findings alone cannot rule in or out a benign or malignant splenic mass; however, multiple studies have identified a greater incidence of hemangiosarcomas with splenic masses that present with hemoperitoneum.[6,12–14]
- Dogs that underwent a splenectomy with an evaluation of the size to malignancy correlations, dogs with benign masses had a significantly greater mass to splenic volume ratio. Mass to splenic volume ratio and splenic weight as a percentage of body weight may be useful in differentiating between hemangiosarcoma and benign lesions.[12,15]
- Up to 67% of all dogs presenting with splenic masses have malignant disease, with hemangiosarcomas comprising the majority of these malignant diagnoses.[12]

REFERENCES

1. Waters DJ, Caywood DD, Hayden DW, Klausner JS. Metastatic pattern in dogs with splenic haemangiosarcoma: Clinical implications. *J Small Anim Pract* 1988;29(12):805–14. doi: 10.1111/j.1748-5827.1988.tb01907.x.
2. Clifford CA, Mackin AJ, Henry CJ. Treatment of canine hemangiosarcoma: 2000 and beyond. *J Vet Intern Med* Sep–Oct 2000;14(5):479–85. doi: 10.1892/0891-6640(2000)014<0479:tochab>2.3.co;2.
3. Cuccovillo A, Lamb CR. Cellular features of sonographic target lesions of the liver and spleen in 21 dogs and a cat. *Vet Radiol Ultrasound* May–Jun 2002;43(3):275–8. doi: 10.1111/j.1740-8261.2002.tb01003.x.
4. Leyva FJ, Loughin CA, Dewey CW, Marino DJ, Akerman M, Lesser ML. Histopathologic characteristics of biopsies from dogs undergoing surgery with concurrent gross splenic and hepatic masses: 125 cases (2012–2016). *BMC Res Notes* 2018 Feb 13;11(1):122. doi: 10.1186/s13104-018-3220-1.
5. Lux CN, Culp WT, Mayhew PD, Tong K, Rebhun RB, Kass PH. Perioperative outcome in dogs with hemoperitoneum: 83 cases (2005–2010). *J Am Vet Med Assoc* May 15 2013;242(10):1385–91. doi: 10.2460/javma.242.10.1385.
6. Hammond TN, Pesillo-Crosby SA. Prevalence of hemangiosarcoma in anemic dogs with a splenic mass and hemoperitoneum requiring a transfusion: 71 cases (2003–2005). *J Am Vet Med Assoc* Feb 15 2008;232(4):553–8. doi: 10.2460/javma.232.4.553.
7. Fife WD, Samii VF, Drost WT, Mattoon JS, Hoshaw-Woodard S. Comparison between malignant and nonmalignant splenic masses in dogs using contrast-enhanced computed tomography. *Vet Radiol Ultrasound* Jul–Aug 2004;45(4):289–97. doi: 10.1111/j.1740-8261.2004.04054.x.
8. Eberle N, von Babo V, Nolte I, Baumgärtner W, Betz D. Splenic masses in dogs. Part 1: Epidemiologic, clinical characteristics as well as histopathologic diagnosis in 249 cases (2000–2011). *Tierarztl Prax Ausg K Kleintiere Heimtiere* 2012;40(4):250–60. doi: 10.1055/s-0038-1623648.
9. Aronsohn MG, Dubiel B, Roberts B, Powers BE. Prognosis for acute nontraumatic hemoperitoneum in the dog: A retrospective analysis of 60 cases (2003–2006). *J Am Anim Hosp Assoc* Mar–Apr 2009;45(2):72–7. doi: 10.5326/0450072.
10. Wong RW, Gonsalves MN, Huber ML, Rich L, Strom A. Erythrocyte and biochemical abnormalities as diagnostic markers in dogs with hemangiosarcoma related hemoabdomen. *Vet Surg* Oct 2015;44(7):852–7. doi: 10.1111/vsu.12361.
11. Schick AR, Hayes GM, Singh A, Mathews KG, Higginbotham ML, Sherwood JM. Development and validation of a hemangiosarcoma likelihood prediction model in dogs presenting with spontaneous hemoabdomen: The HeLP score. *J Vet Emerg Crit Care* May 2019;29(3):239–45. doi: 10.1111/vec.12838.
12. Griffin MA, Culp WTN, Rebhun RB. Canine and feline haemangiosarcoma. *Vet Rec* Nov 2021;189(9):e585. doi: 10.1002/vetr.585.
13. Pintar J, Breitschwerdt EB, Hardie EM, Spaulding KA. Acute nontraumatic hemoabdomen in the dog: A retrospective analysis of 39 cases (1987–2001). *J Am Anim Hosp Assoc* Nov–Dec 2003;39(6):518–22. doi: 10.5326/0390518.

14. Aronsohn MG, Dubiel B, Roberts B, Powers BE. Prognosis for acute nontraumatic hemoperitoneum in the dog: A retrospective analysis of 60 cases (2003–2006). *J Am Anim Hosp Assoc* Mar–Apr 2009;45(2):72–7. doi: 10.5326/0450072.

15. Mallinckrodt MJ, Gottfried SD. Mass-to-splenic volume ratio and splenic weight as a percentage of body weight in dogs with malignant and benign splenic masses: 65 cases (2007–2008). *J Am Vet Med Assoc* 2011 Nov 15;239(10):1325–7. doi: 10.2460/javma.239.10.1325.

CHAPTER 11 – CASE 8

PIT BULL TERRIER WITH DYSTOCIA

Igor Yankin

HISTORY, TRIAGE, AND STABILIZATION

A 1.5-year-old female intact Pit Bull Terrier presented for dystocia. Abdominal contractions were first noted about 12 h prior to presentation, and the dog delivered two puppies in the following 3 h. She continued to have weak abdominal contractions for the next 8 h without any results.

Triage exam findings (vitals):
- Mentation: bright, alert, and responsive
- Respiratory rate: 30 breaths per minute
- Heart rate: 130 beats per minute
- Mucous membranes: pink, moist, 1 sec capillary refill time
- Femoral and dorsal pedal arterial pulses: strong, synchronous
- Temperature: 38.2°C (100.7°F)
- Vaginal examination: clear vaginal discharge, no fetuses could be palpated in the birth canal

POCUS exam(s) to perform:
- Abdominal POCUS
- Evaluation of the fetal cardiac activity and fetal heart rate (FHR) in the M-mode

Abnormal POCUS exam results:
See **Figures 11.8.1** and **11.8.2** and **Video 11.8.1**.

Additional point-of-care diagnostics and initial management:
Venous blood gas, blood glucose, and ionized calcium were all within normal limits. Abdominal radiographs showed a gravid uterus with two mineralized fetal skeletons without radiographic evidence of fetal demise or fetal/maternal mismatch.

QUESTIONS AND ANSWERS

1. What are the differentials to rule in or out with POCUS based on history and physical exam?
2. What are the sonographic findings?
3. What is the sonographic diagnosis?
4. If necessary, what additional sonographic examination or findings would help rule in or out the differential diagnoses?

DOI: 10.1201/9781003436690-60

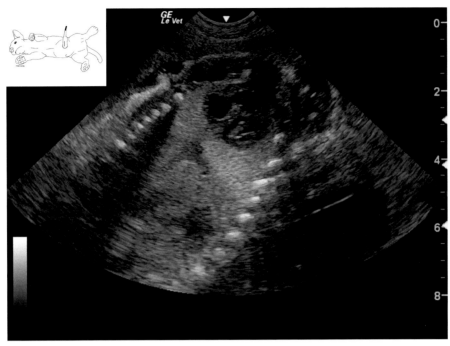

FIGURE 11.8.1 AND VIDEO 11.8.1 Ultrasound view of the fetal heart to evaluate fetal cardiac activity (Fetus #1). A microconvex transducer probe is positioned in the caudal abdomen (umbilical view) perpendicular to the ventral abdominal wall. The depth is set at 8 cm and the frequency is set at 8 MHz. The patient is positioned in right lateral recumbency.

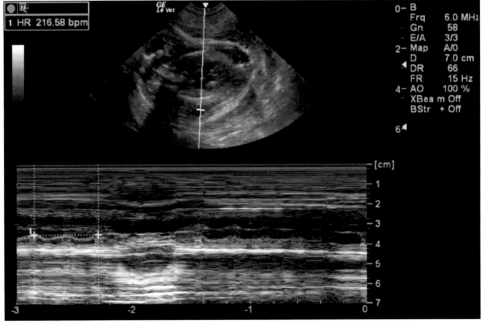

FIGURE 11.8.2 Ultrasound M-mode to calculate the FHR via the two beat peak-to-peak method (Fetus #2). Abdominal preset is used. A software calculation of the current FHR is depicted in the upper left corner of the image.

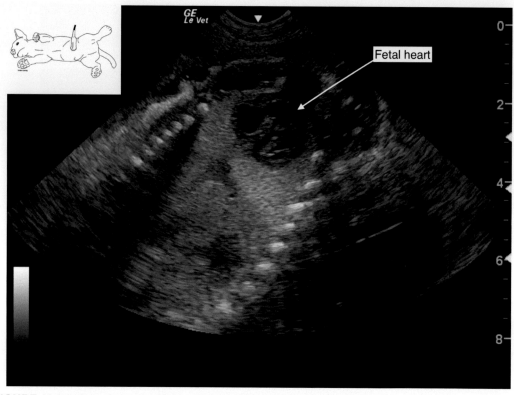

FIGURE 11.8.3 Same image as Figure 11.8.1 with fetal heart identified.

1. **Differential diagnosis to rule in or rule out with POCUS:**
   - Fetal distress and/or death, apparent fetal malformations, uterine rupture.
2. **Describe your sonographic findings:**
   - Fetus #1 has no cardiac activity (**Figures 11.8.1** and **11.8.3, Video 11.8.1**).
   - Fetus #2 has a heart rate of 216 bpm (**Figure 11.8.2**).
3. **What is your sonographic diagnosis?**
   - Deceased Fetus #1 (**Figure 11.8.1, Video 11.8.1**).
   - Alive Fetus #2 with a normal heart rate of 216 bpm (**Figure 11.8.2**).
4. **What additional sonographic examination or finding would help you rule in or rule out your differential diagnoses (if necessary)?**
   - Abdominal POCUS should be performed to rule out a significant amount of free peritoneal effusion that can be suggestive of a uterine rupture.

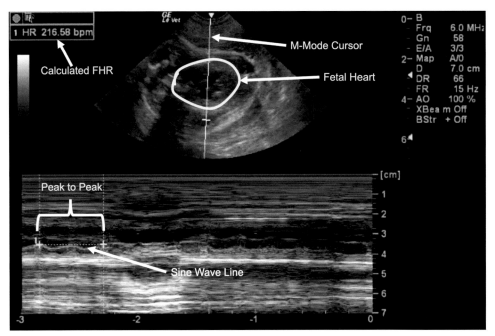

FIGURE 11.8.4 Still ultrasound image in the M-mode used to perform an automatic calculation of the fetal heart rate. The beating fetal heart was identified and optimized by putting it in the central position with the depth set at 6 cm. The abdominal preset was used. Alternatively, cardiac or OB preset can also be utilized. The M-mode was selected, and a single M-mode cursor beam was placed across the beating heart by using the trackpad. The freeze button was pressed after the sine wave line could be visualized. The caliper function to measure the FHR was used by measuring from peak to peak of two complete cardiac cycles. The software calculation turned this measurement into a calculated FHR depicted in the left upper corner of the image.

CLINICAL INTEGRATION

Suspected diagnosis based on ultrasound findings:

- Fetus #1: fetal death due to an unknown etiology (likely secondary to the prolonged labor).
- Fetus #2: live fetus with minimal to no evidence of fetal distress based on the FHR of 216 bpm.

Sonographic interventions, monitoring, and outcome:

- Due to the presence of non-productive abdominal contractions for 8 h prior to the presentation and the fetal death evident during POCUS assessment, an emergency Cesarean section was recommended; however, it was declined by the owner due to financial constraints.
- Since the alive fetus was not in overt distress, and there was no evidence of fetal/maternal mismatch on the abdominal radiographs, a medical management was attempted.
- After receiving fluid therapy, intravenous calcium gluconate, and two doses of intramuscular oxytocin (2 units), the dog delivered the deceased puppy with minimal assistance.
- The recheck FHR in the second fetus remained greater than 200 bpm. The

second puppy was delivered vaginally 2 h later.

Important concepts regarding determining fetal heart rates:
- The determination of fetal cardiac activity is an essential component of any basic ultrasound examination in pregnant animals presenting with dystocia or systemic illness.
- Zone et al. suggested that a normal FHR should be greater than 220 bpm. An FHR of between 180 and 220 bpm may indicate mild to moderate fetal distress, whereas values less than 180 bpm indicate severe fetal distress.[1]
- In a more recent study by Gil et al., it has been shown that the FHR transiently declined below 180–220 bpm as parturition approached even in healthy bitches without evidence of dystocia. Accelerations and decelerations and rapid and intermittent oscillations reaching a 119 bpm minimum were observed.[2] Similar to humans, the FHR in dogs may fluctuate during uterine contractions, and it is always recommended to obtain multiple measurements.

Techniques and tips to calculate fetal heart rate:
- Before a fetal heart rate can be calculated, a sonographer has to identify the fetuses and determine their viability by assessing their cardiac activity.
- To maximize the chances of identifying the majority of fetuses in the gravid uterus, the author prefers to locate the uterine body first in the caudal abdomen followed by tracing the uterine horns in lateral directions as they bifurcate off of the uterine body. Survey abdominal radiographs should be used for the most accurate determination of the fetal count.
- Once the fetus is identified, it is recommended to slowly scan its thoracic cavity in the longitudinal plane until the fetal heart is identified (**Figure 11.8.3**).
- Measurement of the FHR can be performed via B-mode (2D), pulsed-wave Doppler (PWD), or M-mode (author's preference).
- The M-mode ("motion" mode) provides a real-time representation of a moving object. When the M-mode is activated, a cursor line appears on the screen which detects motion through the tissue it intersects.[3–5]
- After the beam is placed over the best view of the beating heart, the resulting image represents the movement of the myocardium during cardiac cycles (a faint sinusoidal wave across the M-mode screen as depicted in **Figure 11.8.4**). By pressing the freeze button, the operator can use the caliper function to measure the FHR. This is performed by measuring the peak to peak (or valley to valley) of one or two subsequent cardiac cycles. It is important to note that different presets (abdominal vs cardiac) and different manufacturers may use either a one or two beat peak-to-peak calculation system. In the author's experience, the abdominal preset typically uses a two-beat calculation, whereas the cardiac preset uses a one-beat calculation.

Take-home messages
- The presence or absence of normal fetal cardiac activity in dogs and cats presenting with dystocia or systemic illness has significant management implications.
- The M-mode provides an easy and reliable way to determine a fetal heart rate in feline and canine patients within a short period of time.
- One should become familiar with a specific ultrasound machine's automatic FHR calculation system before interpreting the results.

REFERENCES

1. Zone MA, Wanke MM. Diagnosis of canine fetal health by ultrasonography. *J Reprod Fertil Suppl* 2001;57:215–9.
2. Gil EM, Garcia DA, Giannico AT, Froes TR. Canine fetal heart rate: Do accelerations or decelerations predict the parturition day in bitches? *Theriogenology* 2014 Oct 15;82(7):933–41.
3. Shah S, Adedipe A, Ruffatto B, Backlund BH, Sajed D, Rood K, Fernandez R. BE-SAFE: Bedside sonography for assessment of the fetus in emergencies: Educational intervention for late-pregnancy obstetric ultrasound. *West J Emerg Med* 2014 Sep;15(6):636–40.
4. Collins K, Collins C, Kothari A. Point-of-care ultrasound in obstetrics. *Australas J Ultrasound Med* 2019 Feb 19;22(1):32–9.
5. Jones R, Goldstein J. *Point-of-Care OB Ultrasound*. American College of Emergency Physicians, 2016 Sep.

CHAPTER 11 – CASE 9

GOLDEN RETRIEVER WITH A 2-DAY HISTORY OF LETHARGY AND HYPOREXIA FOLLOWING INGESTION OF AN OXTAIL TREAT

Nadine Jones and Erica Tinson

HISTORY, TRIAGE, AND STABILIZATION

A 10-month-old female entire Golden Retriever presented with a 2-day history of lethargy and hyporexia following ingestion of an oxtail treat.

Triage exam findings (vitals):
- Mentation: quiet, alert, responsive
- Respiratory rate: 28 breaths per minute
- Heart rate: 130 beats per minute
- Mucous membranes: injected pink, tacky, 1 sec capillary refill time
- Femoral and dorsal pedal arterial pulses: synchronous, bounding
- Temperature: 39.3°C (102.7°F)
- Abdominal palpation: abdominal pain, no palpable abnormalities

POCUS exam(s) to perform:
- Abdominal POCUS

Abnormal POCUS exam results:
See **Figures 11.9.1** and **11.9.2** and **Video 11.9.1**.

Additional point-of-care diagnostics and initial management:
- Ultrasound-guided abdominocentesis attempted and was unsuccessful due to small volume present.

QUESTIONS AND ANSWERS

1. What are the differentials to rule in or out with POCUS based on history and physical exam?
2. What are the sonographic findings?
3. What is the sonographic diagnosis?
4. If necessary, what additional sonographic examination or findings would help rule in or out the differential diagnoses?

DOI: 10.1201/9781003436690-61

Golden Retriever with a 2-Day History of Lethargy and Hyporexia

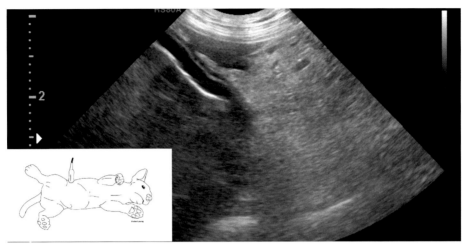

FIGURE 11.9.1 AND VIDEO 11.9.1 Abdominal POCUS with dog in left lateral position. Image obtained from the dog's right side using a microconvex curvilinear probe with a frequency of 8 MHz and the depth set to 5 cm. The fur was parted and alcohol was applied to allow contact between the probe and the skin. The dog was then position in left lateral recumbency, the fur clipped and a coupling gel applied to the skin. Video 11.9.1: Image obtained from the dog's right side using a microconvex curvilinear probe with a frequency of 8 MHz and the depth set to 5 cm.

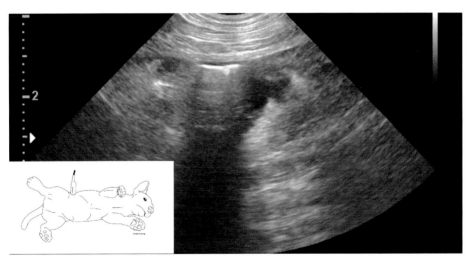

FIGURE 11.9.2 Abdominal POCUS in the caudal abdomen with dog in left lateral position. Same settings used as in Figure 11.9.1.

1. **Differential diagnosis to rule in or rule out with POCUS:**
   - Septic peritonitis, pyometra, gastrointestinal foreign body.

2. **Describe your sonographic findings:**
   - A pocket of peritoneal effusion was identified adjacent to a loop of intestine in the caudal abdomen on the dog's right side (**Figure 11.9.1**).

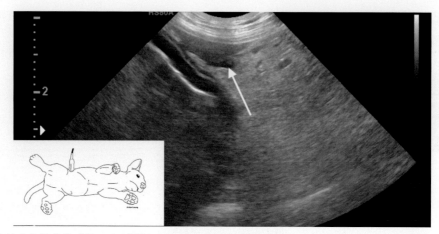

FIGURE 11.9.3 Labeled image of Figure 11.9.1. Pocket of peritoneal effusion (arrow) in the caudal abdomen.

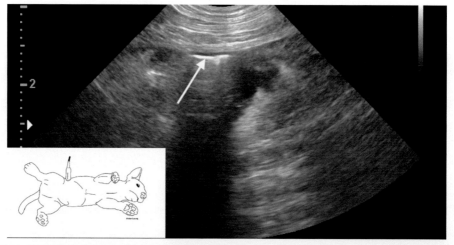

FIGURE 11.9.4 Labeled image of Figure 11.9.2. Enhanced peritoneal stripe sign (EPSS) indicating free peritoneal gas (arrow) on abdominal POCUS in the caudal abdomen.

- Free peritoneal air in the caudal abdomen (**Figures 11.9.2–11.9.5, Video 11.9.1**).
3. What is your sonographic diagnosis?
  - Scant peritoneal effusion with free peritoneal air.
4. What additional sonographic examination or finding would help you rule in or rule out your differential diagnoses (if necessary)?
  - None.

CLINICAL INTEGRATION

Suspected diagnosis based on ultrasound findings:

The free peritoneal air was confirmed by a board-certified radiologist. Based on the physical examination and abdominal POCUS, a ruptured viscus was considered most likely. With the recent history of ingestion of an oxtail, gastrointestinal rupture was considered most likely.

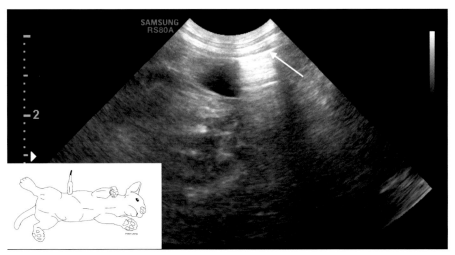

FIGURE 11.9.5 Still image from Video 11.9.1 at 14 sec showing an enhanced peritoneal stripe sign (EPSS) indicating free peritoneal gas (arrow). Image obtained from the dog's right side using a microconvex curvilinear probe with a frequency of 8 MHz and the depth set to 5 cm.

Sonographic interventions, monitoring, and outcome:

- The dog underwent an exploratory laparotomy which identified a colonic perforation by a penetrating foreign body (oxtail treat).

Important concepts regarding the sonographic diagnosis of free peritoneal air due to a ruptured viscus:

- Free peritoneal air within the peritoneal cavity, or pneumoperitoneum, can be diagnosed on abdominal POCUS. However, its diagnosis can be challenging and easily missed if the clinician is focusing on the presence or absence of free peritoneal fluid.
- Pneumoperitoneum is identified on abdominal ultrasound as a hyperechoic linear enhancement and thickening of the peritoneum (termed enhanced peritoneal stripe sign [EPSS]) with shadowing or multiple reverberation artifacts below.[1] These reverberation artifacts can be identified between the abdominal organs and obscure the clinician's field of view. Small volumes of gas may not interfere with abdominal ultrasonography, but large volumes of gas could hinder examination.[1]
- The two most common causes of pneumoperitoneum in dogs and cats are penetration of the abdominal wall (e.g., trauma or surgery) or disruption of a hollow abdominal viscus.[1] In dogs and cats without a history of trauma, pneumoperitoneum is most commonly associated with rupture of the gastrointestinal tract and therefore its identification is an indication for immediate surgical intervention.[2] Less common causes of pneumoperitoneum include urinary bladder rupture, splenic or hepatic abscesses, or the presence of gas-forming bacteria such as *Clostridium* sp.[2,3]
- The reported minimum volume of air that could be injected into the peritoneal cavity and subsequently be reliably detected by abdominal ultrasonography is 0.2 mL.[4] However, this was in the hands of an experienced ultrasonographer and it is likely that small volumes of air could easily be missed by both novices and non-specialists.
- Pneumoperitoneum may be present post-exploratory laparotomy for weeks following abdominal surgery.[1]

- Pneumoperitoneum may not always be detectable following viscus perforation; therefore, lack of identification on abdominal POCUS should not rule this out. Adjunctive imaging modalities such as abdominal computed tomography (CT) or radiography should be performed. Serial abdominal POCUS for the presence of peritoneal effusion or free peritoneal air could also be performed for monitoring.
- Identification of free peritoneal air by the emergency clinician is very useful. It places septic peritonitis at the top of the differential diagnoses list and gives urgency to initial stabilization steps while continuing further investigation to confirm the diagnosis.

Techniques and tips to identify the appropriate ultrasound images/views:

- To detect free peritoneal air on abdominal POCUS, the patient should be positioned in lateral recumbency and the ultrasound probe positioned on the most dorsal location on the abdominal wall. This patient and probe position gives the best chance of finding an air pocket because air will rise to the highest point.[1] Placing the dog in right lateral recumbency to assess the left paralumbar view has been cited as the best location for the detection of free peritoneal air.[5]
- Care should be taken to ensure a loop of intestine is not present as it can contain gas and be confused for free peritoneal air. Fanning of the ultrasound probe or patient repositioning may help differentiate between intestine and free peritoneal air.
- If there is doubt whether a patient has free peritoneal air on POCUS, an adjunctive imaging modality should be considered. For example, abdominal CT or radiography (including horizontal beam).
- Patients who have recently undergone exploratory laparotomy provide an excellent opportunity to practice the detection of free peritoneal air on abdominal POCUS.

Take-home messages
- Abdominal POCUS can be used to identify pneumoperitoneum and may be useful in cases where there is no peritoneal effusion or a scant effusion to aid clinical decision-making. This abdominal POCUS finding should be interpreted along with patient history and presenting clinical signs.
- Lack of identification of EPSS may not exclude pneumoperitoneum due to factors such as patient positioning, volume of air present, and operator experience. Care should be taken to avoid misinterpretation of potential EPSS from other potential artifacts or images, for example imaging of a rib and its associated reflection, "A lines" due to imaging a portion of lung, insufficient contact between the probe and the skin, or the presence of normal gas-filled viscus such as the intestine or colon. If the clinical index of suspicion is high for a ruptured gastrointestinal tract, additional imaging modalities should be utilized to further assess the patient.

REFERENCES

1. Steigner-Vanegas SM, Frank PM. Peritoneal space. In: Thrall DE, ed. *Textbook of Veterinary Diagnostic Radiology*. 7th ed. St Louis, MO: W.B. Elsevier; 2018, pp. 770–2.
2. Smelstoys JA, Davis GJ, Learn AE, Shofer FE, Brown DC. Outcome of and prognostic indicators for dogs and cats with pneumoperitoneum and no history of penetrating trauma: 54

cases (1988–2002). *J Am Vet Med Assoc* 2004 Jul 15;225(2):251–5.
3. Saunders WB, Tobias KM. Pneumoperitoneum in dogs and cats: 39 cases (1983–2002). *J Am Vet Med Assoc* 2003 Aug 15;223(4):462–8.
4. Kim SY, Park KT, Yeon SC, Lee HC. Accuracy of sonographic diagnosis of pneumoperitoneum using the enhanced peritoneal stripe sign in Beagle dogs. *J Vet Sci* 2014;15(2):195–8.
5. Lisciandro GR, Lisciandro SC. Global FAST for patient monitoring and staging in dogs and cats. *Vet Clin North Am Small Anim Pract* 2021 Nov;51(6):1315–33.

CHAPTER 11 – CASE 10
MIXED-BREED K9 WITH ACUTE VOMITING

Erin Binagia

HISTORY, TRIAGE, AND STABILIZATION

A 4-year-old 10 kg male neutered mixed-breed dog presented on referral for acute vomiting. The patient was extremely aggressive and required sedation with 0.3 mg/kg butorphanol and 6 mcg/kg dexmedetomidine IM before exam.

Triage exam findings (vitals):
- Mentation: bright, alert, responsive
- Respiratory rate: 20 breaths per minute
- Heart rate: 80 beats per minute (sedated)
- Mucous membranes: pink, tacky, <2 sec capillary refill time
- Femoral and dorsal pedal arterial pulses: strong, synchronous
- Temperature: 37.8°C (100°F)

POCUS exam(s) to perform:
- Abdominal POCUS with gallbladder assessment

Abnormal POCUS exam results:
See **Figure 11.10.1**.

Additional point-of-care diagnostics and initial management:
- Packed cell volume/total solids (PCV/TS) was 60%/8.0, and electrolytes, blood glucose, lactate, blood urea nitrogen, creatinine, and coagulation panel were normal.

QUESTIONS AND ANSWERS

1. What are the differentials to rule in or out with POCUS based on history and physical exam?
2. What are the sonographic findings?
3. What is the sonographic diagnosis?
4. If necessary, what additional sonographic examination or findings would help rule in or out the differential diagnoses?

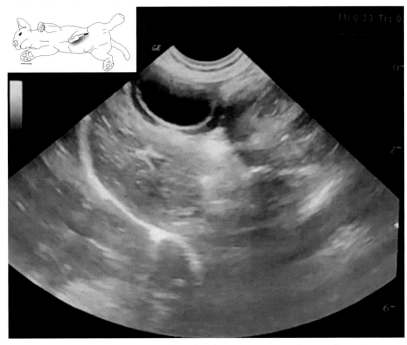

FIGURE 11.10.1 Gallbladder assessment, subxiphoid view. Image taken with a microconvex probe placed longitudinally and the marker directed cranially, immediately caudal to the xiphoid process. If the gallbladder is not in immediate view, the probe can either sweep or fan to the right. The depth was set to 6 cm for this image, but the user should have decreased the depth until the diaphragm was placed at the far side of the screen.

1. **Differential diagnosis to rule in or rule out with POCUS:**
   - The main differentials for acute vomiting that can be identified via an abdominal POCUS include intestinal foreign body obstruction, gastroenteritis, pancreatitis, and acute nephritis. Other differentials to consider that would be difficult to identify via abdominal POCUS include other extra-intestinal causes of vomiting (liver disease, Addison's disease, neoplasia, etc).
2. **Describe your sonographic findings:**
   - Gallbladder wall is diffusely hypoechoic with parallel hyperechoic lines on either side ("double rim" or "halo" sign) (**Figure 11.10.2**).
3. **What is your sonographic diagnosis?**
   - Gallbladder wall edema (also referred to as thickening).
4. **What additional sonographic examination or finding would help you rule in or rule out your differential diagnoses (if necessary)?**
   - Full abdominal ultrasound performed by a radiologist.

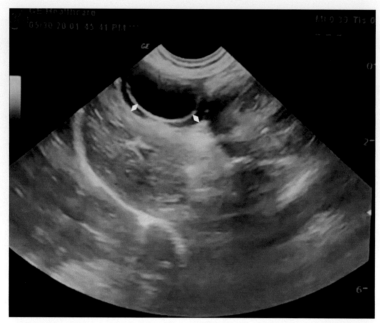

FIGURE 11.10.2 Labeled image of Figure 11.10.1. Gallbladder wall edema denoted by the two white diamonds.

CLINICAL INTEGRATION

Suspected diagnosis based on ultrasound findings:

Gallbladder wall edema in dogs is an important ultrasonographic finding but is not specific to any one disease. The edema is caused by alterations in Starling's forces and can be seen in primary gallbladder disease and systemic disease affecting the gallbladder.[1,2]

The four main causes (via alteration of Starling's Forces) of gallbladder wall edema include

- Inflammation of the gallbladder.
- Hepatic venous congestion.
- Obstruction of the venous or lymphatic drainage of the gallbladder.
- Hypoalbuminemia.

Inflammation of the gallbladder leads to increased vascular permeability and can be primary (acute and chronic cholecystitis and cholangiohepatitis) or secondary (pancreatitis).[1,2] Hepatic venous congestion causes increased hydrostatic pressure and therefore gallbladder wall edema. This can be caused by fluid overload, right heart failure,[3] pericardial effusion,[3] sedation with dexmedetomidine,[4] and portal hypertension secondary to cirrhosis, portal thrombosis, or anaphylaxis.[1,2,5] Obstruction of venous or lymphatic drainage of the gallbladder can be caused by pancreatitis, neoplasia, or other causes of biliary obstruction. Hypoalbuminemia leads to decreased oncotic pressure and edema.

Other reported diseases associated with gallbladder wall edema include chemotherapeutic drugs, renal failure, blood transfusion, immune-mediated hemolytic anemia, sepsis,[6] systemic inflammatory response syndrome, and disseminated intravascular coagulation.[4,5,7,8]

Sonographic interventions, monitoring, and outcome:

- Because gallbladder wall edema was identified on abdominal POCUS, a full

abdominal ultrasound by a radiologist was performed. The ultrasound confirmed true gallbladder wall edema and also identified mild gastrointestinal dilation and ileus.
- Suspected diagnosis was acute gastroenteritis of unknown cause and the patient was hospitalized with IV fluids and antiemetics. The patient did well overnight and a repeat gallbladder assessment the next morning showed resolution of the gallbladder wall edema; therefore, it was suspected that the underlying cause of this ultrasound finding was sedation with dexmedetomidine.
- Patient was discharged with no medications and clinical signs resolved.

Important concepts regarding the sonographic diagnosis of gallbladder wall edema:

- To most accurately measure the wall, orient the probe perpendicular to the nearest wall.[1,7]
- On ultrasound, the normal gallbladder wall is seen as a thin echogenic line and typically measures between 1 and 3 mm in thickness, but usually measures less than 1 mm.[1,7]
- Wall edema is diagnosed when the wall is thicker than 3–3.5 mm and typically looks like a hypoechoic central layer between two thin hyperechoic lines.[1,2,7] This is known as the "double rim" or "halo" sign.
- The presence of peritoneal effusion surrounding the gallbladder may mimic true gallbladder edema (also referred to as "pseudothickening" of the wall).[1,2,7] To differentiate this from true wall thickening, image the neck of the gallbladder or change the position of the animal to redistribute abdominal fluid.[7]
- If gallbladder wall edema is identified, it is extremely important to follow with a full abdominal POCUS to look for free abdominal fluid and pericardial effusion (via the subxiphoid view). It is also important to follow-up with a radiologist to confirm true edema and look for underlying causes.
- In a prospective observational study, dexmedetomidine (median dose 5 mcg/kg) was shown to lead to gallbladder wall thickening (2–5 mm) and peritoneal effusion in some dogs within 20–40 min of administration and resolved within 12–24 h.[4]

Techniques and tips to identify the gallbladder:

- If the gallbladder is not in immediate view, the probe can either sweep or fan to the right. In most animals, the gallbladder can be found in dorsal or lateral recumbency or standing with the probe placed caudal to the xiphoid process, using firm pressure, and then fanning or sweeping the probe to the right in the direction of the gallbladder.
- In deep-chested breeds, it is often difficult to find the gallbladder with this method. Instead, the gallbladder can be found on the caudal mid one-third of the right lateral thorax. Place the probe parallel to the ribs and then slide the probe laterally between rib spaces in the cranial and caudal directions until the gallbladder is found. This technique can also be used in patients that do not tolerate lateral recumbency.
- Once the gallbladder is found, fan the probe in either direction to visualize the wall from multiple angles to determine if edema is present.

Take-home messages

- Gallbladder wall edema is physically caused by changes in Starling's forces, including increased vascular permeability, increased hydrostatic pressure in the portal venous system, decreased oncotic pressure, or lymphatic obstruction.

- Gallbladder wall edema is almost always a pathologic finding. The edema should be confirmed via abdominal ultrasound and the cause should be determined.
- Many diseases can cause gallbladder wall edema. Complete diagnostic workup including a complete blood count (CBC), chemistry, prothrombin time/partial thromboplastin time (PT/PTT), and abdominal ultrasound should be performed to rule out these underlying causes.
- Cholecystitis should not be diagnosed solely on the presence of wall edema.[7]
- Dexmedetomidine may cause transient mild to moderate gallbladder wall edema and peritoneal effusion,[4] which can be misleading. In a patient sedated with dexmedetomidine, this ultrasound finding should be interpreted carefully in combination with history, physical exam, and other diagnostics. A repeat gallbladder assessment should be performed in 24 h to confirm resolution.[4]

REFERENCES

1. Mattoon JS, Larson MM, Nyland TG. Chapter 9: Liver. In: Mattoon JS, Nyland TG, eds. *Small Animal Diagnostic Ultrasound*, 3rd ed. Elsevier; 2015:332–399.
2. Larson MM. Ultrasound imaging of the hepatobiliary system and pancreas. *Vet Clin North Am Small Anim Pract* May 2016;46(3):453–480, v–vi. doi:10.1016/j.cvsm.2015.12.004
3. Lisciandro GR, Gambino JM, Lisciandro SC. Thirteen dogs and a cat with ultrasonographically detected gallbladder wall edema associated with cardiac disease. *J Vet Intern Med* May 2021;35(3):1342–1346. doi:10.1111/jvim.16117
4. Seitz MA, Lee AM, Woodruff KA, Thompson AC. Sedation with dexmedetomidine is associated with transient gallbladder wall thickening and peritoneal effusion in some dogs undergoing abdominal ultrasonography. *J Vet Intern Med* Nov 2021;35(6):2743–2751. doi:10.1111/jvim.16306
5. Quantz JE, Miles MS, Reed AL, White GA. Elevation of alanine transaminase and gallbladder wall abnormalities as biomarkers of anaphylaxis in canine hypersensitivity patients. *J Vet Emerg Crit Care* Dec 2009;19(6):536–544. doi:10.1111/j.1476-4431.2009.00474.x
6. Walters AM, O'Brien MA, Selmic LE, McMichael MA. Comparison of clinical findings between dogs with suspected anaphylaxis and dogs with confirmed sepsis. *J Am Vet Med Assoc* Sep 15 2017;251(6):681–688. doi:10.2460/javma.251.6.681
7. Spaulding K. Ultrasound corner: Gallbladder wall thickness. *Vet Radiol Ultrasound* 1993;34(4):270–272.
8. van Breda Vriesman AC, Engelbrecht MR, Smithuis RH, Puylaert JB. Diffuse gallbladder wall thickening: Differential diagnosis. *AJR Am J Roentgenol* Feb 2007;188(2):495–501. doi:10.2214/AJR.05.1712

CHAPTER 11 – CASE 11

HUSKY MIX WITH OBTUNDATION, LETHARGY, VOMITING, AND ANOREXIA 48 HOURS AFTER LANDING ON THE TAILGATE JUMPING INTO A TRUCK

Serge Chalhoub

HISTORY, TRIAGE, AND STABILIZATION

A 3-year-old spayed female Husky mix presented for crying after jumping out of a pickup truck 48 h prior. The owner did not see the dog jump out but thought the dog may not have landed on her feet based on the sound he heard. She seemed fine after a few minutes and ate well that night. However, 48 h later, she was not eating or ambulating and was vomiting.

Triage exam findings (vitals):
- Mentation: quiet, alert, responsive (QAR) to dull
- Respiratory rate: 24 breaths per miunte
- Heart rate: 140 beats per miunte
- Mucous membranes: pink, <2 sec capillary refill time
- Femoral and dorsal pedal arterial pulses: strong and synchronous
- Temperature: 38.7°C (101.6°F)
- Very painful abdomen

POCUS exam(s) to perform:
- Abdominal POCUS (performed first because of the painful abdomen)
- PLUS
- Cardiac POCUS

Abnormal POCUS exam results:
See **Figure 11.11.1** and **Video 11.11.1**.

Additional point-of-care diagnostics and initial management:
Packed cell volume (PCV) 49%, total solids (TS) 5.5 g/dL (55 g/L), and lactate 8.3 mmol/L. Azostix of peripheral blood was blood urea nitrogen (BUN) 15–26 mg/dL (5.3–9.3 mmol/L). Abdominocentesis revealed that a PCV of abdominal effusion was 8%. Azostix BUN of the uroabdomen fluid was 50–80 mg/dL (18–28.5 mmol/L) (highest reading).

DOI: 10.1201/9781003436690-63

QUESTIONS AND ANSWERS

1. What are the differentials to rule in or out with POCUS based on history and physical exam?
2. What are the sonographic findings?
3. What is the sonographic diagnosis?
4. If necessary, what additional sonographic examination or findings would help rule in or out the differential diagnoses?

1. **Differential diagnosis to rule in or rule out with POCUS:**
   - Hemoabdomen, uroabdomen, bile peritonitis, septic abdomen.
2. **Describe your sonographic findings:**
   - Free abdominal fluid near the urinary bladder (**Figure 11.11.2**).
3. **What is your sonographic diagnosis?**
   Abdominal effusion.

4. **What additional sonographic examination or finding would help you rule in or rule out your differential diagnoses (if necessary)?**
   PLUS and cardiovascular POCUS to look for pleural effusion and evaluate volume status. These were both performed and appeared normal.

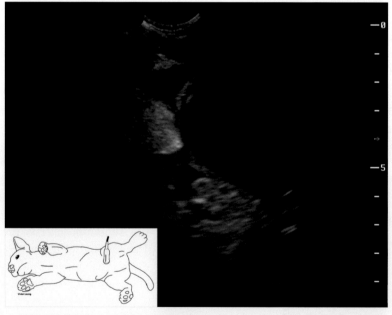

FIGURE 11.11.1 AND VIDEO 11.11.1 Urinary bladder view. Patient is in right lateral recumbency. The microconvex probe is placed on the caudal abdomen, oriented along the long axis and at the apex of the bladder, the marker directed cranially. Depth set at 5 cm. Image courtesy of Dr Søren Boysen. Video 11.11.1 Long-axis bladder cineloop with the same settings.

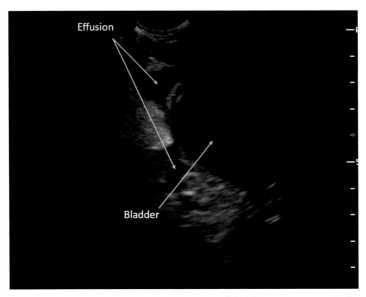

FIGURE 11.11.2 Labeled image of Figure 11.11.1. Urinary bladder view. Cranial, dorsal, and ventral to the bladder, free abdominal effusion is seen forming uncontained angles of hypoechoic fluid. This is especially visualized when fanning the probe wide in both the long and short axis of the bladder.

CLINICAL INTEGRATION

Suspected diagnosis based on ultrasound findings:
The most likely diagnosis is traumatic uroabdomen.

Sonographic interventions, monitoring, and outcome:
- The patient was placed on IV fluids and sent to surgery.
- A small apical bladder tear was identified and repaired.
- The patient recovered uneventfully and abdominal effusion resolved.

Important concepts regarding the sonographic diagnosis of uroabdomen:
- The findings of abdominal effusion in this case raised numerous differentials.
- Based on possible trauma, the top differentials included hemoabdomen, uroabdomen, and bile peritonitis[1] It is not possible to see an actual bladder tear using ultrasound as often these are too small, the bladder can still appear filled, and sometimes the tear seals itself with a clot or omentum. However, an irregular appearance of the urinary bladder wall or the inability to find it should raise suspicions of a uroabdomen.
- Ultimately, cystocentesis and fluid analysis are necessary for diagnosis. The fluid should be compared to peripheral blood. The abdominal fluid PCV was much lower than the peripheral PCV, making hemoabdomen unlikely. The abdominal fluid BUN (measured through Azostix here) was significantly greater than the peripheral BUN, making uroabdomen likely.
- Other tests that can help confirm uroabdomen include comparing abdominal to peripheral potassium ratio (>1.4:1 in dogs highly indicates uroabdomen), and abdominal to peripheral creatinine ratio (>2:1 in dogs highly suggests uroabdomen).[2]

Techniques and tips to identify the appropriate ultrasound images/views:

- On abdominal POCUS, there are five main regions to scan: subxiphoid, left and right paralumbar, umbilical, and bladder views when looking for free abdominal effusion.[1]
- Fanning the probe widely at each site is important to increase sensitivity in finding free fluid.
- At each site, changing the depth to make sure the organ/region of interest takes up at least 50%–75% of the screen is important to not miss pathology (author experience).
- At each site, consider how gravity affects where free fluid would be located.
- At the urinary bladder view, make sure to fan widely until the bladder disappears from left to right in long axis and from cranial to caudal in short axis.
- Free abdominal effusion is uncontained and will form sharp angles of demarcation when up against abdominal structures and organs.
- Abdominocentesis using POCUS and fluid analysis is important to identify the source of effusion.

Take-home messages

- Uroabdomen is possible with abdominal trauma.
- Free abdominal effusion can be detected with POCUS, but cystocentesis and confirmatory diagnostics are needed to identify the source.
- Bladder ruptures are not diagnosed on POCUS.

REFERENCES

1. Boysen SR, Rozanski EA, Tidwell AS, et al. Evaluation of a focused assessment with sonography for trauma protocol to detect free abdominal fluid in dogs involved in motor vehicle accidents. *J Am Vet Med Assoc.* 2004; 225(8): 1198–1204.

2. Schmiedt C, Tobias KM, Otto CM. Evaluation of abdominal fluid: Peripheral blood creatinine and potassium ratios for diagnosis of uroperitoneum in dogs. *J Vet Emerg Crit Care.* 2001; 11(4): 275–280.

CHAPTER 11 – CASE 12

DOMESTIC SHORTHAIR WITH A 2-DAY HISTORY OF LETHARGY AND VOMITING

Annelies Valcke and Laura Cole

HISTORY AND PRESENTING COMPLAINT:

A 5-year-old female spayed Domestic Shorthair presented with a 2-day history of lethargy and vomiting.

Triage exam findings (vitals):
- Mentation: stuporous, in lateral recumbency
- Heart rate: 125 breaths per minute
- Temperature: 35.9°C (96.6°F)
- Abdominal palpation: bilaterally enlarged, painful kidneys

POCUS exam(s) to perform:
- Abdominal POCUS with assessment of the kidneys (size, structure, renal pelvic dilation) and bladder (size)

Abnormal POCUS exam results:
See **Figures 11.12.1** and **11.12.2** and **Video 11.12.1**.

Additional point-of-care diagnostics and initial management:
- Electrocardiogram: bradycardia with absent P waves, tented T waves.
- Blood gas: hyperkalemia (8.5 mmol/L), azotemia (creatinine 14.97 mg/dl or 1342 umol/L)
- Emergency treatment: Gave 10% calcium gluconate 1 ml/kg IV

QUESTIONS AND ANSWERS

1. What are the differentials to rule in or out with POCUS based on history and physical exam?
2. What are the sonographic findings?
3. What is the sonographic diagnosis?
4. If necessary, what additional sonographic examination or findings would help rule in or out the differential diagnoses?

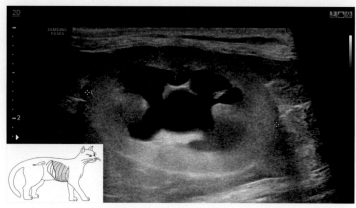

FIGURE 11.12.1 Abdominal POCUS right paralumbar view, sagittal image of kidney. Performed with the cat in sternal recumbency. In this case, a linear probe was placed longitudinally with the marker in cranial direction; however, a microconvex probe can also be used. The fur was not shaved. An ultrasound coupling gel was used and the probe was placed caudal to the last rib, ventral to the sublumbar muscles, and oriented dorsally to locate the kidneys. The left kidney lies caudomedial to the head of the spleen. The right kidney is situated more cranially. Once the kidney was found, the probe was fanned dorsally to ventrally to find the longest axis of the kidney and this location was used for measurement of renal length. The frequency and depth used were 10 Hz and 5 cm, respectively.

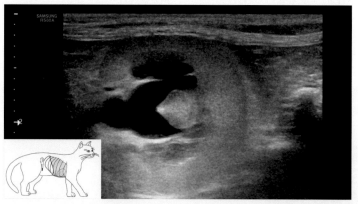

FIGURE 11.12.2 AND VIDEO 11.12.1 Abdominal POCUS right paralumbar view, transverse image of kidney. The scanning techniques was as described in Figure 11.12.1. To convert to transverse image the probe was rotated 90° from sagittal view, with the marker pointing ventrally. The renal crest was used as a landmark to identify the V-shaped renal pelvis.

1. Differential diagnosis to rule in or rule out with POCUS:
   - Acute kidney injury (AKI) (with major differential diagnosis: renal lymphoma, pyelonephritis, toxicity), bilateral ureteral obstruction, urethral obstruction.

2. Describe your sonographic findings:
   - Bilateral renomegaly (**Figure 11.12.3**).
   - Bilateral renal pelvic dilation (**Figure 11.12.4**).
   - Proximal ureteral dilation (**Figure 11.12.4**).
   - Bladder: small.

Domestic Shorthair with a 2-Day History of Lethargy and Vomiting

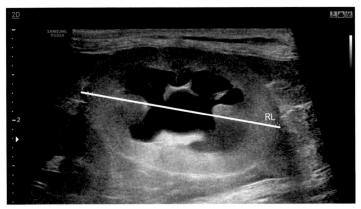

FIGURE 11.12.3 Labeled image of Figure 11.12.1. Sagittal image of right kidney. RL = renal length. Renal length was 5.2 cm.

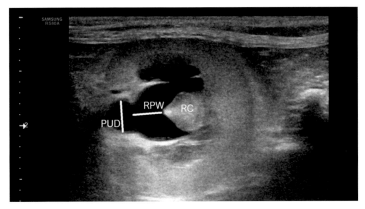

FIGURE 11.12.4 Labeled image of Figure 11.12.2. Transverse image of kidney. RPW = renal pelvic width; PUD = proximal ureteral diameter; RC = renal crest.

3. What is your sonographic diagnosis?
   - Bilateral renal pelvic dilation.
   - Bilateral proximal ureteral dilation.
   - Small bladder.
4. What additional sonographic examination or finding would help you rule in or rule out your differential diagnoses (if necessary)?

- Abdominal ultrasound: bilateral changes compatible with chronic kidney disease and bilateral moderate pyelectasia (renal pelvic width [RPW] 6.2 mm).
- Ultrasound-guided pyelocentesis and fluoroscopic-assisted antegrade pyelography (Figure 11.12.5).

CLINICAL INTEGRATION

Suspected diagnosis based on ultrasound findings:

Bilateral ureteral obstruction with post-renal azotemia and hyperkalemia.

Sonographic interventions, monitoring, and outcome:

- Emergency management of hyperkalemia consisted of 1 mL/kg 10% calcium gluconate IV, 0.2 IU/kg soluble insulin, and 1

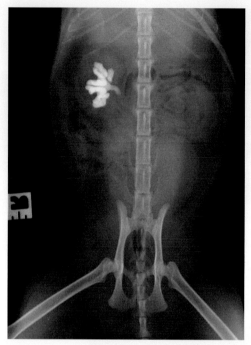

FIGURE 11.12.5 Fluoroscopic image of the right kidney and ureter pyelogram 10 min after a contrast agent (iohexol) was directly administered into the renal pelvis demonstrating renal pelvic dilation and proximal ureteral obstruction.

mL/kg 50% glucose IV followed by a 5% dextrose continuous rate infusion (CRI) at 1 mL/kg/h. Electrolytes and blood glucose levels were then monitored every 2–4 h.
- Cystocentesis and urine analysis were performed. The urine sediment and cytology were unremarkable. The patient went for surgery for bilateral subcutaneous ureteral bypass placement 8 h post-admit and was polyuric post-operatively. Fluid balance was assessed by monitoring urine output and body weight and the fluid rate adjusted accordingly.
- The patient's azotemia resolved prior to discharge.

Important concepts regarding the sonographic diagnosis of ureteral obstruction:

- Differentiate between ureteral and urethral obstruction, based on bladder size. When the bladder is small, a bilateral ureteral obstruction or anuric AKI is the most likely diagnosis. Urethral obstruction is usually associated with a large firm bladder. However, the bladder may be normal with ureteral obstruction, particularly unilateral obstruction, and the bladder may be small but firm with a urethral obstruction.
- For the detection of renomegaly use the sagittal view for the longest axis measurement. Normal feline renal lengths range from 3.1 to 5.1 cm.[1] Therefore, it is important to assess the shape, architecture, and symmetry of both kidneys when assessing for abnormalities.[2,3]
- Ureteral obstruction can occur with chronic kidney disease. Therefore, the kidneys may be small with reduced corticomedullary distinction.
- Use a transverse view to measure the renal pelvis and proximal ureter. Measurement of the renal pelvis is from the renal crest to the beginning of the ureter (**Figure 11.12.4**).
- Ureteral obstruction becomes increasingly more likely the greater the width of the renal pelvis.[4–7] However, cats with pyelonephritis and ureteral obstruction cannot be differentiated by renal pelvic width alone.[8] Proximal ureteral dilation does not occur in healthy animals.[3]

Techniques and tips to identify the appropriate ultrasound images/views:

- Standing or sternal recumbency facilitates visualization of both kidneys.
- Placement of the probe intercostally may be required in patients with small kidneys.
- Familiarize with normal renal anatomy (**Figures 11.12.6** and **11.12.7**).

Pearls and pitfalls:

- Ureteral obstruction can be unilateral or bilateral. Azotemia and electrolyte derangements can occur with unilateral

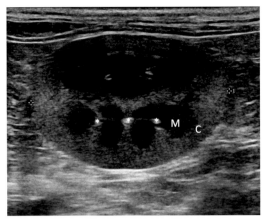 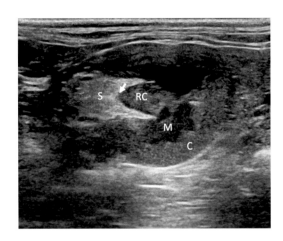

FIGURES 11.12.6 (LEFT) AND 11.12.7 (RIGHT)

Sagittal (11.12.6) and transverse (11.12.7) view of normal feline kidney, respectively. M = renal medulla; C = renal cortex; S = renal sinus; RC = renal crest; arrow = renal pelvis (no dilation present). Asterisks (*) denote the ventral sets of diverticula and interlobar vessels. Original source: Cole L, Humm K, Dirrig H. Focused Ultrasound Examination of Canine and Feline Emergency Urinary Tract Disorders (published correction appears in *Vet Clin North Am Small Anim Pract* 2022 May;52(3):xiii). *Vet Clin North Am Small Anim Pract* 2021;51(6):1233–1248.

obstruction if there is pre-existing renal disease.
- Radiopaque uroliths are the most common cause of ureteral obstruction, but an obstruction can also occur secondary to blood clots, stricture, or purulent material not identifiable on routine imaging.
- The renal pelvis should not be confused with normal renal medulla. The renal crest is the landmark for the identification of V-shaped renal pelvis (**Figures 11.12.4 and 11.12.6**).

Take-home message
- Renomegaly associated with renal pelvic and proximal ureteral dilation is strongly suggestive of ureteral obstruction.

REFERENCES

1. Debruyn K, Paepe D, Daminet S, Combes A, Duchateau L, Peremans K, Saunders JH. Renal dimensions at ultrasonography in healthy Ragdoll cats with normal kidney morphology: Correlation with age, gender and bodyweight. *J Feline Med Surg* 2013;15(12):1046–1051.
2. Beeston, D., Dirrig, H., & Cole, L. (2023). The utility of clinicopathological findings and point-of-care ultrasound in increasing the index of suspicion of ureteral obstruction in azotaemic cats presenting to the emergency room. *Journal of Small Animal Practice*.
3. Cole L, Humm K, Dirrig H. Focused ultrasound examination of canine and feline emergency urinary tract disorders. *Vet Clin North Am Small Anim Pract* 2021;51(6):1233–1248.
4. Lulich JP, Berent AC, Adams LG, Westropp JL, Bartges JW, Osborne CA. ACVIM small animal consensus recommendations on the treatment and prevention of uroliths in dogs and cats. *J Vet Intern Med* 2016;30(5):1564–1574.
5. Lamb CR, Cortellini S, Halfacree Z. Ultrasonography in the diagnosis and management of cats with ureteral obstruction. *J Feline Med Surg* 2017:1–8.

6. D'Anjou MA, Bédard A, Dunn ME. Clinical significance of renal pelvic dilatation on ultrasound in dogs and cats. *Vet Radiol Ultrasound* 2011;52(1):88–94.
7. Testault I, Gatel L, Vanel M. Comparison of nonenhanced computed tomography and ultrasonography for detection of ureteral calculi in cats: A prospective study. *J Vet Intern Med* 2021;35(5):2241–2248.
8. Quimby JM, Dowers K, Herndon AK, Randall EK. Renal pelvic and ureteral ultrasonographic characteristics of cats with chronic kidney disease in comparison with normal cats, and cats with pyelonephritis or ureteral obstruction. *J Feline Med Surg* 2017;19(8):784–790.

CHAPTER 11 – CASE 13

SPANIEL WITH A 3-DAY HISTORY OF LETHARGY AND A SINGLE EPISODE OF VOMITING

Olivia X Walesby and Daria Starybrat

HISTORY, TRIAGE, AND STABILIZATION

A 15 kg (33 lb) 4-year-old female spayed working Spaniel presented with a 3-day history of lethargy and a single episode of vomiting. No urination witnessed in 24 h. No previous medical history and no toxin access.

Triage exam findings (vitals):
- Mentation: quiet, responsive
- Respiratory rate: 28 breaths per minute
- Heart rate: 100 beats per minute
- Mucous membranes: icteric, 1.5 sec capillary refill time
- Femoral and dorsal pedal arterial pulses: strong and synchronous
- Temperature: 38.3°C (100.9°F)

POCUS exam(s) to perform:
- Abdominal POCUS with urinary bladder assessment
- PLUS

Abnormal POCUS exam results:
- Abdominal POCUS showed increased kidney echogenecity (image not available).
- Full urinary bladder (**Figure 11.13.1**).

Additional point-of-care diagnostics and initial management:

Further diagnostics showed packed cell volume (PCV) 47%, total solids (TS) 7.2 g/dL (72 g/L), and glucose, lactate, and electrolytes within normal ranges. Additionally, a thrombocytopenia was identified (manual estimate 22,500/μL). Biochemistry revealed urea 25 mmol/L (70 mg/dL) (RR 1.70–7.40 mmol/L or 4.76–20.72 mg/dL) and creatinine 195 μmol/L (2.2 mg/dL) (RR 22–115 μmol/L or 0.25–1.3 mg/dL). Blood pressure was 120 mmHg.

Intravenous fluid therapy with Hartmann's solution was initiated. An indwelling urinary catheter placement was considered but not performed due to thrombocytopenia in this patient.

QUESTIONS AND ANSWERS

1. What are the differentials to rule in or out with POCUS based on history and physical exam?
2. What are the sonographic findings?
3. What is the sonographic diagnosis?
4. If necessary, what additional sonographic examination or findings would help rule in or out the differential diagnoses?

DOI: 10.1201/9781003436690-65

1. **Differential diagnosis to rule in or rule out with POCUS:**
   From the history and physical exam, main differentials for azotemia with possible oliguria include acute kidney injury (AKI), e.g., due to pyelonephritis or leptospirosis and post-renal causes, e.g., ureteral obstruction.
   Abdominal POCUS can help to rule out uroabdomen. Additionally, kidneys should be assessed for evidence of hyperechogenicity, which may be suggestive of AKI, or marked pyelectasia with/without the presence of calculi, potentially indicative of ureteral obstruction. Serial urinary bladder assessments can be used to monitor the bladder size and therefore estimate urine production.
2. **Describe your sonographic findings:**
   - Urinary bladder volume (mL) can be estimated using the formula: length × width × height × 0.2π.[1] The maximal length and maximal height are obtained with the probe oriented longitudinally (**Figure 11.13.2**) and the maximal width with the probe oriented in transverse (**Figure 11.13.3**). Baseline urinary volume in this case was estimated as 53 ml (length 4.5 × width 3.2 × height 5.9 cm × 0.2π).
   - Using measurements from **Figures 11.13.4** and **11.13.5**, the urinary bladder volume can be estimated as 6 × 6 × 6.5 × 0.2π = 147 mL.
3. **What is your sonographic diagnosis?**
   Acute kidney injury with adequate urine production.

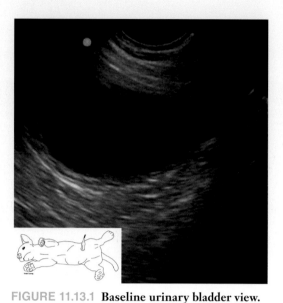

FIGURE 11.13.1 Baseline urinary bladder view. The patient was in right lateral recumbency with a curved probe oriented longitudinally on the caudal third of the abdomen. The marker is directed cranially and the probe has a frequency of 8 MHz. The image is at a depth of 5 cm and was achieved without fur clipping and by using ultrasound gel as the coupling agent.

- Renal hyperechogenicity suggested AKI.
- Serial ultrasound estimations of the urinary volume showed an increasing bladder size and therefore urinary volume ruling out anuria.

4. **What additional sonographic examination or finding would help you rule in or rule out your differential diagnoses (if necessary)?**
   None.

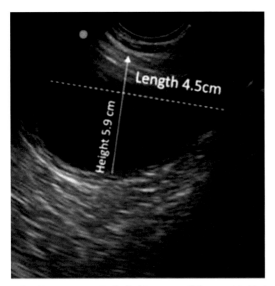

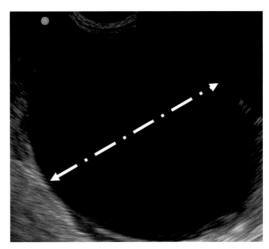

FIGURE 11.13.2 Labeled image of Figure 11.13.1 The bladder wall is smooth with no irregularities and no intra-luminal masses or uroliths noted. The bladder wall thickness will alter when the volume of urine within the bladder changes. The bladder measures 4.5 cm in length shown with the dotted line, and a height of 5.9 cm shown with the solid white line.

FIGURE 11.13.4 The patient is in right lateral recumbency and the probe is oriented in transverse. This image was taken 8 h after initiating intravenous fluid therapy at 5 mL/kg/h. The dot and dash line shows a width of 6 cm.

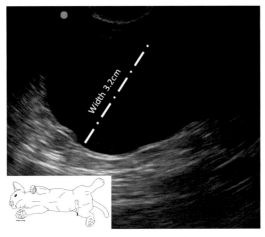

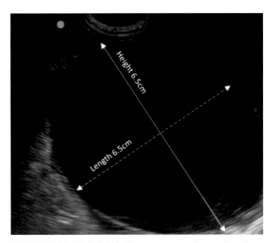

FIGURE 11.13.3 After obtaining the view from Figure 11.13.1 (and Figure 11.13.2), the probe was oriented in transverse to obtain the width. The width was measured as 3.2 cm.

FIGURE 11.13.5 This image was taken after 8 h on intravenous fluid therapy. The patient is in right recumbency and the probe is oriented longitudinally. The solid line represents the height of 6.5 cm and the dotted line shows a length of 6 cm.

CLINICAL INTEGRATION

Suspected diagnosis based on ultrasound findings:
A non-anuric AKI.

Sonographic interventions, monitoring, and outcome:
- Serial measurements of the bladder determined that the dog's bladder was increasing in size thus ruling out anuria. Anuria is associated with higher severity of AKI.[2]
- The patient was subsequently diagnosed with *Leptospira interrogans* serovar *canicola*, received appropriate antibiotic therapy and supportive care, and eventually recovered to discharge.

Important concepts regarding the sonographic diagnosis of urinary estimation in AKI:
- When placement of indwelling catheters or repeated catheterization is not possible, serial POCUS measurements of the bladder can aid in monitoring urine production and as an indirect measurement of urinary output.
- The formula used in this case for urinary bladder volume estimation [length (cm) × height (cm) × width (cm) × 0.2π] has been shown to underestimate the volume of urine by <10%.[1,3] Measurement bias can occur with very small and large bladder volumes.[1]
- Other formulas for urinary volume estimation exist, and the accuracy also varies with bladder shape and size.[4]

Techniques and tips to identify the appropriate ultrasound images/views:
- The urinary bladder view is achieved by holding the probe with the marker pointing cranially and gently applying to the caudal third of the abdomen. Rocking the probe cranially and caudally and fanning it dorsally and ventrally will allow assessment for free fluid in this view.
- Length and height can be achieved in the longitudinal plane. Turning the probe 90° will give the view needed to measure the width.
- Manual pressure exerted on the probe can affect bladder size measurements.

Take-home message
Serial measurements of the urinary bladder can aid monitoring of urinary production.

REFERENCES

1. Lisciandro, G.R., Fosgate, G.T. (2017). Use of urinary bladder measurements from a point-of-care cysto-colic ultrasonographic view to estimate urinary bladder volume in dogs and cats. *Journal of Veterinary Emergency and Critical Care*, 27(6), 713–717.
2. Rimer, D., Chen, H., Bar-Nathan, M., Segev, G. (2022). Acute kidney injury in dogs: Etiology, clinical and clinicopathologic findings, prognostic markers, and outcome. *Journal of Veterinary Internal Medicine*, 36(2), 609–618.
3. Atalan, G., Barr, F.J., Holt, P.E. (1998). Estimation of bladder volume using ultrasonographic determination of cross-sectional areas and measurements. *Veterinary Radiology and Ultrasound*, 39(5), 446–450.
4. Kendall A, Keenihan E, Kern ZT, et al. Three dimensional bladder ultrasound for estimation of urine volume in dogs compared with traditional 2-dimensional ultrasound methods. *J Vet Intern Med* Nov 2020;34(6):2460–2467. doi:10.1111/jvim.15959

CHAPTER 11 – CASE 14

STAFFORDSHIRE BULL TERRIER WITH 2 DAYS OF VOMITING, ANOREXIA, AND NO OBSERVED URINATION FOR 48 HOURS DESPITE WATER INTAKE

Alexandra Nectoux and Mark Kim

HISTORY, TRIAGE, AND STABILIZATION

A 3-year-old female spayed unvaccinated Staffordshire Bull Terrier presented with 2 days of vomiting and anorexia. The owner did not observe any urination for 2 days despite water intake.

Triage exam findings (vitals):
- Mentation: obtunded
- Respiratory rate: 40 breaths per minute
- Heart rate: 120 beats per minute
- Mucous membranes: pink, <2 sec capillary refill time
- Femoral and dorsal pedal arterial pulses: strong and regular
- Temperature: 36.8°C (96.8°F)
- Other findings: abdominal pain, halitosis

POCUS exam(s) to perform:
- Abdominal POCUS

Abnormal POCUS exam results:
See **Figures 11.14.1** and **11.14.2**.

Additional point-of-care diagnostics and initial management:
Biochemistry showed severe azotemia (urea 179.6 mg/dL, creatinine 10.47 mg/dL) and mild elevation of liver enzymes (ALT 250 U/L, ALKP 364 U/L). A Parvo SNAP (IDEXX) test was negative. A WITNESS® LEPTO (Zoetis) test was positive.

QUESTIONS AND ANSWERS

1. What are the differentials to rule in or out with POCUS based on history and physical exam?
2. What are the sonographic findings?
3. What is the sonographic diagnosis?
4. If necessary, what additional sonographic examination or findings would help rule in or out the differential diagnoses?

DOI: 10.1201/9781003436690-66

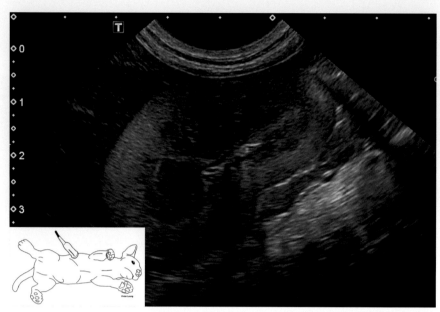

FIGURE 11.14.1 Abdominal POCUS, right paralumbar view over the right kidney. Image taken with a 7.5 MHz microconvex probe placed longitudinally and the marker directed cranially. The probe is placed in the middle of the triangle obtained by the last right rib and lumbar muscles. The probe can be oriented cranially and some pressure can be applied on the probe because the right kidney lies under the ribs. The depth depends on the size of the animal but usually a 5–6 cm depth is appropriate. To obtain this image, the fur is not clipped and alcohol is applied to the area.

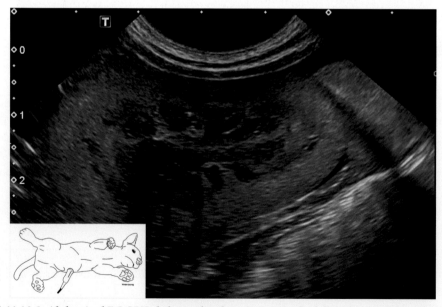

FIGURE 11.14.2 Abdominal POCUS, left paralumbar view over the left kidney. Image taken with a 7.5 MHz microconvex probe placed longitudinally and the marker directed cranially. The probe is placed in the middle of the triangle obtained by the last left rib and lumbar muscles. The depth depends on the size of the animal but usually a 5–6 cm depth is appropriate. To obtain this image, the fur is not clipped and alcohol is applied to the area.

1. **Differential diagnosis to rule in or rule out with POCUS:**
   - Gastrointestinal differentials include septic peritonitis, obstruction (foreign body, intussusception, volvulus), and gastrointestinal ulceration. Extra-intestinal differentials include pancreatitis, liver disease, and acute kidney injury (AKI) from prerenal, intrinsic, or postrenal causes.
   - Extra-intestinal causes include acute kidney injury (AKI) from prerenal, intrinsic, or postrenal causes, pancreatitis, or liver disease
2. **Describe your sonographic findings:**
   - Normal size and silhouette of the kidneys. No dilation of renal pelvises.
   - Bilateral reduced corticomedullary junction definition.
   - Perirenal effusion and mild peritoneal effusion (**Figures 11.14.3** and **11.14.4**).
3. **What is your sonographic diagnosis?**
   Acute bilateral nephropathy. Tubulo-interstitial acute nephritis of infectious (leptospirosis) or toxic origin (ethylene glycol, grapes).
4. **What additional sonographic examination or finding would help you rule in or rule out your differential diagnoses (if necessary)?**
   None.

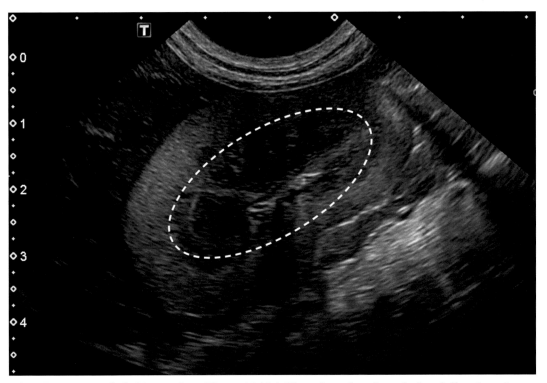

FIGURE 11.14.3 Labeled image from Figure 11.14.1. Note the reduced corticolmedullary junction definition (dotted line).

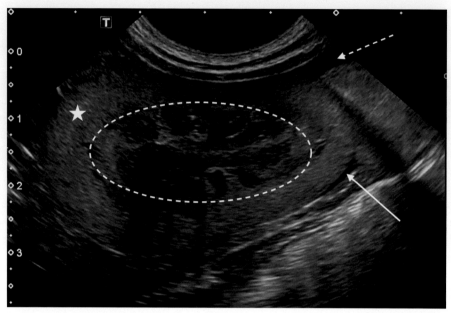

FIGURE 11.14.4 Labeled image from Figure 11.14.2. Note the focal hyperechoic foci in the caudal aspect of the left kidney (star). Bilateral reduced corticomedullary junction definition (dotted line). Bilateral increased medullary echogenicity. Perirenal effusion (solid arrow). Mild peritoneal effusion (dotted arrow).

CLINICAL INTEGRATION

Suspected diagnosis based on ultrasound findings:

Leptospirosis is the most likely diagnosis based on history, ultrasound findings, and bedside test results.

Sonographic interventions, monitoring, and outcome:

- The bladder was empty. An abdominocentesis was performed and the fluid total nucleated cell count and total solids were 1000/μL and 1.0 g/dL respectively, supportive of anuric AKI and fluid overload. Penicillin and maropitant were administered intravenously.
- Due to fluid overload, anuria, and severe acute kidney injury, renal replacement therapy was initiated and a total of three sessions resulted in the stabilization of the renal parameters and the resolution of oliguria. Daily serial 4-h abdominal fluid scores were performed to evaluate the evolution of peritoneal effusion, and the score decreased to 0 within 24 h.
- Patient was discharged after 6 days of monitoring and supportive care.

Important concepts regarding the sonographic diagnosis of leptospirosis:

- Sonographic abnormalities of the kidneys are seen in most dogs affected by leptospirosis.[1]
- Most common findings for AKI include increased cortical echogenicity, increased medullary echogenicity, reduced corticomedullary border distinction, thickening of the renal cortex, enlarged kidneys, and pelvic dilation. These abnormalities are not specific for leptospirosis.[1–3]
- Perirenal effusion is also frequently reported in leptospirosis.[4] Some dogs may

have both retroperitoneal and peritoneal effusions.[1-3]
- Leptospirosis is a multisystemic disease that most commonly affects the kidneys, the liver, and the lungs. The gastrointestinal tract and coagulation system can also be affected. Complete abdominal ultrasound, thoracic radiographs, and complete blood work are recommended to detect all possible lesions.[2,5]

Techniques and tips to identify the appropriate ultrasound images/views:
- Do not apply too much pressure on the probe, as this can push the effusion away.
- If tolerated, keep the dog in lateral recumbency to perform the POCUS examination, particularly both kidney sites. Scant effusions will be more easily detected on the dependent side due to gravity.
- Adjust the focus depth settings on the machine to focus on the kidney and to improve visualization of possible abnormalities.

Take-home messages
- All dogs with vomiting or abdominal discomfort should have a five-point abdominal POCUS performed.
- Perirenal effusion and increased cortical echogenicity are some ultrasonographic abnormalities that help the clinician suspect acute kidney injury.[1-4]
- Renal abnormalities on POCUS are not specific for leptospirosis.[1-3]
- History, physical exam, and bedside tests allow confirmation of a leptospirosis diagnosis.[6]

REFERENCES

1. Sonet J, Barthélemy A, Goy-Thollot I, Pouzot-Nevoret C. Prospective evaluation of abdominal ultrasonographic findings in 35 dogs with leptospirosis. *Vet Radiol Ultrasound*. Jan 2018;59(1):98–106. doi: 10.1111/vru.12571
2. Knöpfler S, Mayer-Scholl A, Luge E, et al. Evaluation of clinical, laboratory, imaging findings and outcome in 99 dogs with leptospirosis. *J Small Anim Pract*. Oct 2017;58(10):582–588. doi: 10.1111/jsap.12718
3. Forrest L, O'Brien R, Tremelling M, et al. Sonographic renal findings in 20 dogs with leptospirosis. *Vet Radiol Ultrasound*. Jul 1998;39(4):337–340. doi: 10.1111/j.1740-8261.1998.tb01617.x
4. Holloway A, O'Brien R. Perirenal effusion in dogs and cats with acute renal failure. *Vet Radiol Ultrasound*. Nov 2007;48(6):574–579. doi: 10.1111/j.1740-8261.2007.00300.x
5. Adin CA, Cowgill LD. Treatment and outcome of dogs with leptospirosis: 36 cases (1990–1998). *J Am Vet Med Assoc*. Feb 2000;216(3):371–375. doi: 10.2460/javma.2000.216.371
6. Lizer J, Velineni S, Weber A, Krecic M, Meeus P. Evaluation of 3 serological tests for early detection of leptospira-specific antibodies in experimentally infected dogs. *J Vet Intern Med*. 2018 Jan;32(1):201–207. doi: 10.1111/jvim.14865

CHAPTER 11 – CASE 15

ROTTWEILER WITH FREQUENT REGURGITATION AND ANOREXIA FOLLOWING GASTROINTESTINAL FOREIGN BODY REMOVAL

Nadine Jones and Erica Tinson

HISTORY, TRIAGE, AND STABILIZATION

A 5-month-old male intact Rottweiler was hospitalized following an exploratory laparotomy and enterotomy for a foreign body removal (sock). The dog was subsequently anorexic and regurgitating frequently. He is currently receiving methadone and maropitant.

Triage exam findings (vitals):
- Mentation: quiet, alert, responsive
- Respiratory rate: 36 breaths per minute
- Heart rate: 70 beats per minute
- Mucous membranes: pink, moist, 1.5 sec capillary refill time
- Femoral and dorsal pedal arterial pulses: strong, synchronous
- Temperature: 38.7°C (101.7°F)
- Abdominal auscultation: absent borborygmus

POCUS exam(s) to perform:
- Abdominal POCUS
- Gastrointestinal assessment (distension and motility)

Abnormal POCUS exam results:
See **Figure 11.15.1** and **Videos 11.15.1–11.15.3**.

Additional point-of-care diagnostics and initial management:
Thoracic radiographs obtained pre-exploratory laparotomy ruled out esophageal foreign body and were unremarkable. Electrolyte analysis identified hypokalemia (3 mmol/L, RI 3.6–4.6) and was managed with intravenous potassium chloride supplementation.

QUESTIONS AND ANSWERS

1. What are the differentials to rule in or out with POCUS based on history and physical exam?
2. What are the sonographic findings?
3. What is the sonographic diagnosis?
4. If necessary, what additional sonographic examination or findings would help rule in or out the differential diagnoses?

DOI: 10.1201/9781003436690-67

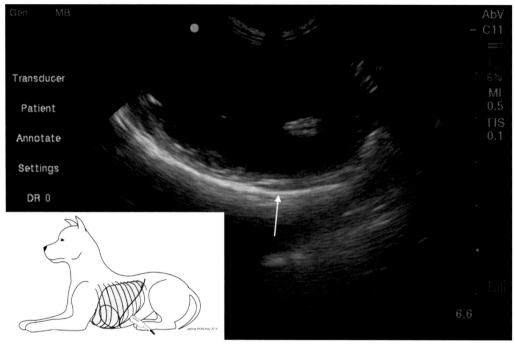

FIGURE 11.15.1 AND VIDEO 11.15.1 Assessment of the stomach with the patient in sternal recumbency. Image obtained with probe placed on the left abdominal wall with the marker directed cranially, above the xiphoid process and caudal to the liver. Standing position is acceptable. The white arrow demarcates the gastric wall. Images obtained using a microconvex curvilinear probe with a frequency of 7.5 MHz and a depth of 6.6 cm. The fur was parted and alcohol applied to allow contact between the probe and skin. Video 11.15.1: Assessment of the stomach. Images obtained using a microconvex curvilinear probe with a frequency of 7.5 MHz and a depth of 13 cm.

1. **Differential diagnosis to rule in or rule out with POCUS:**
   - Gastrointestinal ileus, intussusception, septic peritonitis, residual foreign body.
2. **Describe your sonographic findings:**
   - Abdominal POCUS did not identify peritoneal effusion.
   - Assessment of the stomach identified gross gastric distension with anechoic fluid with hyperechoic material and minimal to no peristalsis (**Figure 11.15.1, Video 11.15.1**). Compare these images to normal stomach and rugal folds (**Figures 11.15.2 and 11.15.3, Video 11.15.5**).
   - The small intestines were diffusely dilated with anechoic fluid and hyperechoic material and no peristalsis was observed (**Videos 11.15.2 and 11.15.3**).
   - Single loops of small intestine were observed for 2–3 min at a time and there were no visible contractions or peristalsis waves.
   - There was no evidence of intussusception.
   - Refer to the healthy dog still images (**Figures 11.15.2 and 11.15.3**) and videos (**Video 11.15.4 and 11.15.5**) provided for a normal comparison.
   - **Video 11.15.2** is an assessment of the small intestines. It is a long-axis view of small intestinal loop. Images were obtained using a microconvex

curvilinear probe with a frequency of 7.5 MHz and a depth of 5.2 cm. Single loops of small intestine were observed for 2–3 min at a time and the number of peristalsis waves or contractions counted. The fur was parted and alcohol applied to allow contact between the probe and skin.
- **Video 11.15.3** is an assessment of the small intestines, and is a transverse view of a small intestinal loop. Images were obtained using a microconvex curvilinear probe with a frequency of 7.5 MHz and a depth of 5.2 cm. Compare these videos to normal small intestinal peristalsis (**Video 11.15.4**).
- **Video 11.15.5** is an assessment of the small intestines in a healthy dog. It is a long-axis view of a small intestinal loop showing peristalsis and movement of ingesta. Images obtained using a microconvex curvilinear probe with a frequency of 7.5 MHz and a depth of 5.2 cm.

3. **What is your sonographic diagnosis?**
   - Delayed gastric emptying with generalized intestinal ileus.
4. **What additional sonographic examination or finding would help you rule in or rule out your differential diagnoses (if necessary)?**
   - None.

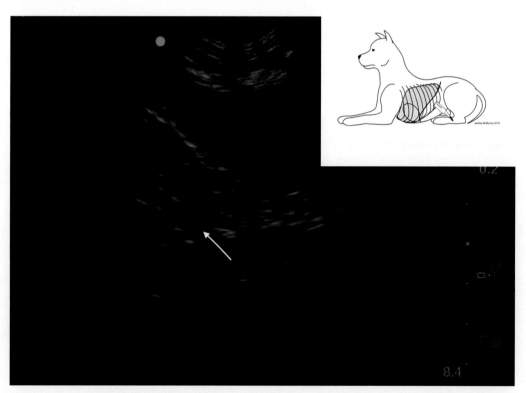

FIGURE 11.15.2 Assessment of the stomach in a healthy dog showing normal rugal folds (white arrow) and ingesta within the stomach lumen. Images obtained using a microconvex curvilinear probe with a frequency of 7.5 MHz and a depth of 8.4 cm.

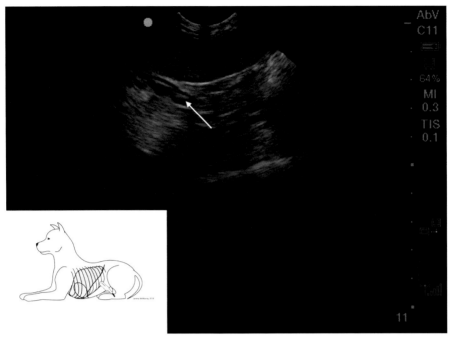

FIGURE 11.15.3 AND VIDEO 11.15.4 Assessment of the stomach in a healthy dog showing flattening of the rugal folds (white arrow in image) with the presence of ingesta. Movement of ingesta is observed in the video. Image and video obtained using a microconvex curvilinear probe with a frequency of 7.5 MHz and a depth of 11 cm.

CLINICAL INTEGRATION

Suspected diagnosis based on ultrasound findings:

Gastrointestinal ileus most likely secondary to a combination of factors such as opioids, neurogenic inhibition (e.g., stress, recent abdominal surgery), post-operative ileus, inflammation (e.g., pancreatitis, gastroenteritis, peritonitis), or electrolyte derangements (e.g., hypokalemia, hypomagnesemia).

Sonographic interventions, monitoring, and outcome:

- The dog had a nasogastric enteral feeding tube placed to allow gastric emptying followed by the introduction of enteral nutrition with daily increases in volume administered. A metoclopramide constant rate infusion (CRI) was started and the dog's hypokalemia was identified and corrected.
- Pain scores were persistently low; therefore, the methadone dose was decreased and administration discontinued after 24 h. The dog was also encouraged to walk frequently.
- Serial assessments of the gastrointestinal tract was performed and documented a decrease in gastric distension and the return of peristalsis.
- The dog was discharged after 3 days of hospital care once eating voluntarily.

Important concepts regarding the sonographic diagnosis of gastrointestinal ileus:

- Gastrointestinal dysmotility describes either hypermotility or hypomotility anywhere along the length of the gastrointestinal tract, with hypomotility common in critically ill small animals.[1] Hypomotility causes delayed gastric emptying and/or small intestinal ileus, and can be due to a physical obstruction (mechanical ileus) or vascular or neuromuscular abnormalities (functional ileus).[2]

- A presumptive diagnosis of gastrointestinal ileus can be made based on physical examination findings such as abdominal distension or pain, decreased borborygmus on auscultation of the abdomen, clinical signs such as anorexia or vomiting, regurgitation, or increased gastric residual volumes as measured by gastric emptying via a nasogastric enteral feeding tube.
- Abdominal ultrasound is a non-invasive imaging modality that can be used in the conscious patient to subjectively assess gastrointestinal motility and allow for serial monitoring.
- In healthy dogs, the mean number of peristaltic contractions of the stomach, duodenum, and the rest of the small intestinal tract are four to five, four to five, and one to three per minute, respectively.[3] Fasting results in a decrease in the frequency of peristalsis contractions.[4] Refer to **Video 11.15.5** for an example of normal small intestinal motility.
- Delayed gastric emptying and ileus can cause discomfort and nausea, increase the risk of aspiration of the gastric contents, and interfere with the delivery of enteral nutrition.[5] Ileus and gastrointestinal fluid sequestration can result in significant fluid losses and hypovolemia.[1] Some texts advocate the removal and measuring of gastric residual volume before an enteral feed[6]; however, there is currently no veterinary data regarding threshold gastric residual volume necessitating gastric decompression.[7] It has been proposed that animals with gastric residual volumes greater than 10 mL/kg should be treated with prokinetics and their nutrition plan adjusted so that the volume of feed is reduced temporarily.[7]

Techniques and tips to identify the appropriate ultrasound images/views:

- The appearance of the stomach and small intestines depends on the amount of distension and the amount of gas and ingesta present.[2]
- To visualize the stomach, the ultrasound probe is positioned caudal to the xiphoid and moved along the caudal aspect of the ribs along the right or left abdominal wall.[8] The normal stomach has rugal folds (**Figure 11.15.2**) which appear to flatten as the stomach fills with ingesta (**Figure 11.15.3**, **Video 11.15.4**). With increasing distension, the rugal folds are no longer visible (**Figure 11.15.1**, **Videos 11.15.1** and **11.15.4**). It can be difficult to appreciate gastric peristalsis given the varying degrees of gastric distension with ingesta and the movement of the stomach with breathing (**Video 11.15.4**).
- Identification of the different segments of the small intestinal tract (i.e., duodenum, jejunum, and ileum) may be more challenging. However, identification of the stomach and a loop of small intestine for the assessment of distension and motility is simple and can be practiced as part of POCUS.
  - For the small intestine, the probe is placed on the mid-abdomen and a loop of small intestine identified and followed, with the ultrasound probe oriented in a transverse or longitudinal plane. The loop is assessed subjectively for the degree of fluid or material distension.
  - A contraction is observed as an indent in the intestinal wall with the associated movement of the luminal contents. The contraction frequency is noted over 3 min.[4] If ileus is identified, it is important to note whether all loops versus a single loop of intestines are affected, as a focal small intestinal ileus might be

more consistent with a mechanical ileus secondary to obstruction.[2] Always defer to an imaging specialist if concerned in this instance.

Take-home messages
- Assessment of the gastrointestinal tract as part of abdominal POCUS is feasible and can be readily practiced. Assessment is subjective and patient factors affecting gastrointestinal distension and motility should be taken into account, such as a recent meal, fasting, or aerophagia.[2,4]
- Acoustic shadowing artifact from gas within the lumen of the gastrointestinal tract may hinder the assessment of the gastrointestinal tract.
- If there is clinical concern for a mechanical obstruction, then a full abdominal ultrasound or abdominal radiography is recommended.

REFERENCES

1. Whitehead K, Cortes Y, Eirmann L. Gastrointestinal dysmotility disorders in critically ill dogs and cats. *J Vet Emerg Crit Care*. 2016 Mar–Apr;26(2):234–253.
2. Riedesel EA. Small bowel. In: Thrall DE, ed. *Textbook of Veterinary Diagnostic Radiology*. 7th ed. St Louis, MO: W.B. Elsevier; 2018: 926–954.
3. Penninck DG, Nyland TG, Fisher PE, Kerr LY. Ultrasonography of the normal canine gastrointestinal tract. *Vet Radiol*. 1989;30(6):272–276.
4. Sanderson JJ, Boysen SR, McMurray JM, Lee A, Stillion JR. The effect of fasting on gastrointestinal motility in healthy dogs as assessed by sonography. *J Vet Emerg Crit Care*. 2017 Nov;27(6):645–650.
5. Husnik R, Gaschen F. Gastric motility disorders in dogs and cats. *Vet Clin North Am Small Anim Pract*. 2021 Jan;51(1):43–59.
6. Haskins SC, King LG. Positive pressure ventilation. In: King LG, ed. *Textbook of Respiratory Diseases in Dogs and Cats*. St Louis, MO: Elsevier; 2004, pp. 217–229.
7. Chan DL. Nutritional support in the mechanically ventilated small animal patient. In: *Nutritional Management of Hospitalized Small Animals*. 1st ed. West Sussex: Wiley-Blackwell Ltd; 2015, pp. 228–233.
8. Stieger-Vanegas SM, Frank PM. Stomach. In: Thrall DE, ed. *Textbook of Veterinary Diagnostic Radiology*. 7th ed. St Louis, MO: W.B. Elsevier; 2018, pp. 894–925.

CHAPTER 11 – CASE 16

BELGIAN SHEPHERD WITH 1 WEEK OF SEVERE EXERCISE INTOLERANCE FOLLOWING TRAUMA THAT OCCURRED WHILE JUMPING ONTO A BOAT

Pauline Jaillon and Kris Gommeren

HISTORY, TRIAGE, AND STABILIZATION

A 6-year-old female spayed Belgian Shepherd presented with 1 week of severe exercise intolerance. Symptoms started after a trauma that occurred while jumping onto a boat.

Triage exam findings (vitals):
- Mentation: quiet, alert, responsive
- Respiratory rate: 40 breaths per minute
- Heart rate: 140 beats per minute
- Mucous membranes: pale pink, tacky, capillary refill time 2.5 secs
- Femoral and dorsal pedal arterial pulses: bounding pulses
- Abdominal palpation: distended and tense abdomen
- Temperature: normal

POCUS exam(s) to perform:
- Abdominal POCUS
- PLUS
- Cardiac POCUS

Abnormal POCUS exam results:
See **Figures 11.16.1–11.16.3** and **Videos 11.16.1–11.16.4**.

Additional point-of-care diagnostics and initial management:
The systolic blood pressure was 70 mmHg at presentation. Hematology showed a regenerative anemia (hematocrit 28%) with a severe thrombocytopenia of 17 K/µL (148–484). Blood smear confirmed thrombocytopenia. The patient was given a 10 mL/kg lactated Ringer's bolus over 15 min, increasing the blood pressure to 80 mmHg.

QUESTIONS AND ANSWERS

1. What are the differentials to rule in or out with POCUS based on history and physical exam?
2. What are the sonographic findings?
3. What is the sonographic diagnosis?
4. If necessary, what additional sonographic examination or findings would help rule in or out the differential diagnoses?

Belgian Shepherd with 1 Week of Severe Exercise Intolerance

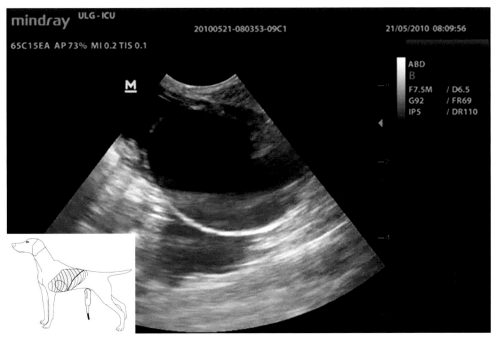

FIGURE 11.16.1 AND VIDEO 11.16.1 Urinary bladder view of an abdominal POCUS exam on a standing dog. A microconvex probe is placed longitudinally on the caudal part of the abdomen, with the marker directed cranially. The frequency used here is 5 MHz and the depth is 6 cm. For this exam the coupling agent is alcohol.

1. **Differential diagnosis to rule in or rule out with POCUS:**
   - Based on the history, differentials include uroabdomen due to ruptured urinary bladder and hemoabdomen due to trauma. After identifying peripheral anemia and hypotension, acute internal bleeding with secondary hypovolemia was more strongly suspected. Point-of-care ultrasound is indicated to look for free fluid, as well as to confirm the hypovolemic state.
2. **Describe your sonographic findings:**
   - Abdominal POCUS: small amounts of free fluid at all five key points. Bladder appears intact (**Figures 11.16.4–11.16.6**).
   - Caudal vena cava: flat and collapsing caudal vena cava (**Figure 11.16.7**, **Video 11.16.4**)
3. **What is your sonographic diagnosis?**
   - Peritoneal effusion with abdominal fluid score 4 out of 4.
   - Hypovolemia.
4. **What additional sonographic examination or finding would help you rule in or rule out your differential diagnoses (if necessary)?**
   - Abdominocentesis should be performed to confirm a hemoabdomen or to identify other types of free fluid.

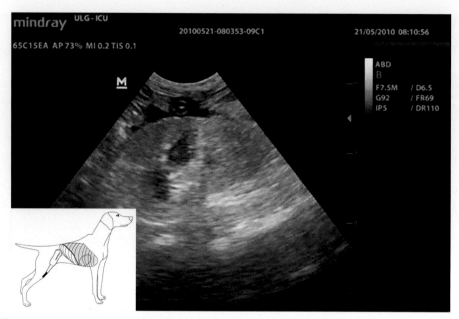

FIGURE 11.16.2 AND VIDEO 11.16.2 Right paralumbar view of an abdominal POCUS exam performed with the dog in the same standing position. The microconvex probe is placed longitudinally behind the last rib, on the most dorsal part of the abdomen. If the kidney is not visualized correctly, the probe can be fanned cranioventrally, as the right kidney is anatomically situated behind the last rib. The frequency used here is 5 MHz and the depth is 6 cm. Clipping the fur can sometimes help this view, especially in long-haired patients.

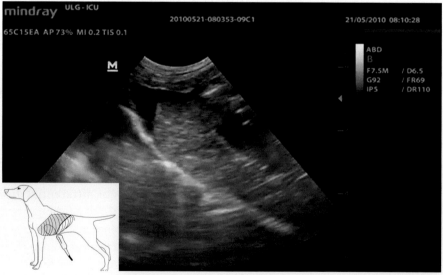

FIGURE 11.16.3 AND VIDEOS 11.16.3 and 11.16.4 Subxiphoid view of the abdominal POCUS exam obtained with the patient in the same standing position. The probe is placed longitudinal to the abdomen, with the marker directed cranially, immediately caudal to the xiphoid process. The probe should be fanned left and right to increase the likelihood of identifying free fluid. The frequency used here is 5 MHz and the depth is 6 cm.

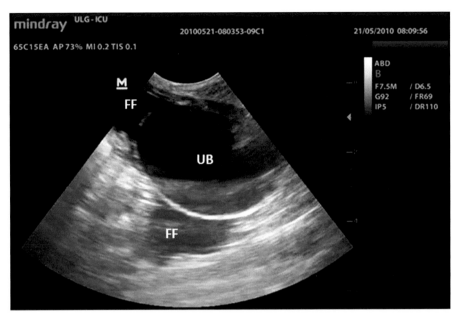

FIGURE 11.16.4 Free abdominal fluid is detected cranioventrally to the bladder, as it often tends to accumulate in the gravity-dependent regions. With the patient in standing position, free fluid is typically found at the midline and cranial to the bladder. FF: free fluid; UB: urinary bladder.

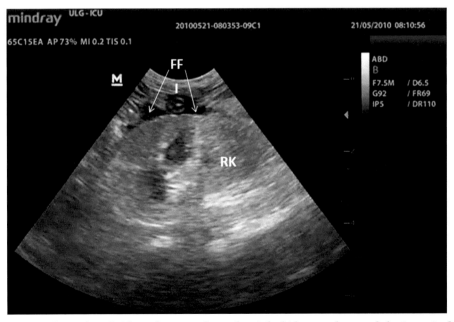

FIGURE 11.16.5 Scant free fluid can be seen above the right kidney and around the intestinal loop. FF: free fluid; I: intestine; RK: right kidney.

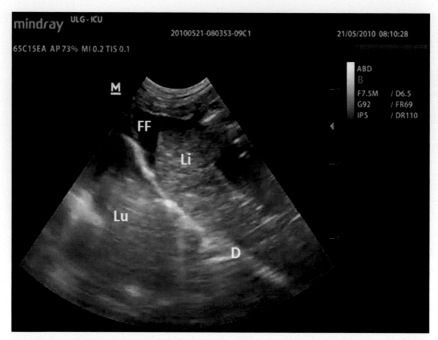

FIGURE 11.16.6 Free abdominal fluid is detected cranially to the liver. It can also sometimes be seen between the lobes of the liver. FF: free fluid; Li: liver; Lu: lung; D: diaphragm.

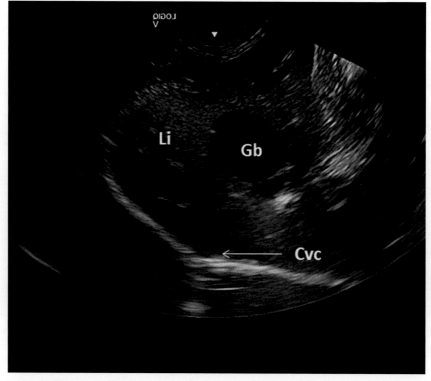

FIGURE 11.16.7 Still image of Video 11.16.4 showing the collapsed caudal vena cava. Li: liver; Gb: gallbladder; Cvc: caudal vena cava.

CLINICAL INTEGRATION

Suspected diagnosis based on ultrasound findings:
Traumatic hemoabdomen leading to hypovolemic shock.

Sonographic interventions, monitoring, and outcome:
- Abdominocentesis was performed, revealing macroscopically hemorrhagic fluid. The packed cell volume (PCV) of the fluid was 60% compared to the peripheral hematocrit of 28%, confirming a hemoabdomen.
- To confirm a hemoabdomen with active bleeding, the PCV of the patient's peripheral blood and the PCV of the free fluid can be compared, and their evolution over time can be tracked.[1] An actively bleeding patient will have a higher peripheral PCV compared to abdominal PCV, but the abdominal PCV will still be considerably high. With ongoing active bleeding, the PCV in the abdominal cavity tends to decrease, as blood with a decreasing PCV continues to accumulate. During the resorptive phase (no further active bleeding), the abdominal PCV increases as resorption of the abdominal fluid occurs faster than that of red blood cells.[2] Ongoing active bleeding is an indication for emergency surgery.
- Abdominal ultrasound by a boarded radiologist identified a large splenic hematoma (8 cm), which could have been caused by the previous trauma. The owner elected for a splenectomy, and histopathology confirmed the suspected diagnosis. During surgery, the dog received a packed red blood cell transfusion. The dog had an uneventful recovery and was discharged after 72 h of monitoring for ventricular arrythmias.

Important concepts regarding the sonographic diagnosis of hemoabdomen:
- According to literature[3] unstable patients presented in the emergency room often have free fluid detected. POCUS of the abdomen and thoracic cavity may help to rapidly detect internal blood loss.[4]
- Determining the abdominal fluid score can indicate the likelihood that a transfusion may be required.[5] Indeed, a patient with traumatic hemoabdomen with a fluid score of 3 or 4 is more likely to need a transfusion. For a score of 0–2, close monitoring and serial abdominal fluid scoring is recommended. This patient had a 4/4 score and required a transfusion in addition to surgical intervention.

Techniques and tips to identify the appropriate ultrasound images/views:
- Fluid tends to accumulate in gravity-dependent sites, especially in the pleural space. Therefore, consider patient position when looking for free fluid.
- The urinary bladder view is usually the easiest spot to detect free abdominal fluid in standing patients, even for a novice.
- Cardiac POCUS allows confirmation of hypovolemia (see Case 9.2), but is currently not considered gold standard to look for right auricular masses.

Take-home messages
- Evaluate all five views of the abdominal POCUS when suspicious of a hemoabdomen to avoid missing scant free fluid.
- Use abdominal fluid scoring, as it can be indicative of transfusion requirements. Moreover, by tracking the evolution of the abdominal fluid score, changes can be detected before clinical symptoms appear.
- POCUS allows assessment of the volume status.

REFERENCES

1. Sigrist N, Spreng D. Monitoring of traumatic hemoperitoneum by serial measurements of abdominal and venous hematocrit in a dog. *Tierarztl Prax Ausg K Kleintiere Heimtiere* 2007;35(5):371–374. doi: 10.1055/s-0038-1622644.
2. Sigrist N, Spreng D. Traumatic haemoabdomen. *Eur J Companion Anim Pract (EJCAP)* 2010;20(1):45–52. doi: 10.5167/uzh-123588.
3. McMurray J, Boysen S, Chalhoub S. Focused assessment with sonography in nontraumatized dogs and cats in the emergency and critical care setting. *J Vet Emerg Crit Care* Jan–Feb 2016;26(1):64–73. doi: 10.1111/vec.12376.
4. Boysen SR, Lisciandro GR. The use of ultrasound for dogs and cats in the emergency room: AFAST and TFAST. *Vet Clin North Am Small Anim Pract* 2013 Jul;43(4):773–797. doi: 10.1016/j.cvsm.2013.03.011.
5. Lisciandro GR, Lagutchik MS, Mann KA, Fosgate GT, Tiller EG, Cabano NR, Bauer LD, Book BP, Howard PK. Evaluation of an abdominal fluid scoring system determined using abdominal focused assessment with sonography for trauma in 101 dogs with motor vehicle trauma. *J Vet Emerg Crit Care* 2009 Oct;19(5):426–437. doi: 10.1111/j.1476-4431.2009.00459.x. PMID: 19821883.

SECTION 6

MISCELLANEOUS POINT-OF-CARE ULTRASOUND CASES

Section Editor Erin Binagia

CHAPTER 12

INTRODUCTION TO MISCELLANEOUS POINT-OF-CARE ULTRASOUND CASES

Erin Binagia

The point-of-care ultrasound (POCUS) exam is not limited to the standard lung, heart, and abdominal imaging as discussed in the previous sections. Ultrasound has been used to estimate intracranial pressure (Case 13.1),[1] diagnose subcutaneous emphysema and cellulitis (Case 13.2),[2,3] and identify diaphragmatic hernias.[4] POCUS can assist in challenging procedures such as venous access in the face of peripheral edema,[5,6] and can assist in nasogastric tube and endotracheal tube placement and confirmation.[7,8] As practitioners become more comfortable and advanced with ultrasound, more uses for POCUS will arise to facilitate rapid cageside assessment and performance of procedures. It is important to note that diligent training and practice will be necessary before relying solely on POCUS results, as the findings and conclusions are determined by the experience of the individual sonographer. Other imaging techniques evaluated by a boarded radiologist (radiology, ultrasound, computed tomography, magnetic resonance imaging) should remain the reference standard as final confirmation. This section presents case examples of how POCUS can be used outside of the standard imaging.

REFERENCES

1. Ilie LA, Thomovsky EJ, Johnson PA, et al. Relationship between intracranial pressure as measured by an epidural intracranial pressure monitoring system and optic nerve sheath diameter in healthy dogs. *Am J Vet Res* Aug 2015;76(8):724–731. doi:10.2460/ajvr.76.8.724
2. Stieger-Vanegas S. Focused ultrasound of superficial-soft tissue swellings, masses, and fluid collections in dogs and cats. *Vet Clin North Am Small Anim Pract* Nov 2021;51(6):1283–1293. doi:10.1016/j.cvsm.2021.07.009
3. Boysen SR, Gommeren K, Chalhoub S. PLUS image interpretation: Normal findings. In: Boysen SR, Gommeren K, Chalhoub S, eds. *The Essentials of Veterinary Point-of-Care Ultrasound: Pleural Space and Lung*. Groupo Asis Biomedia; 2022, pp. 35–62.
4. Spattini G, Rossi F, Vignoli M, Lamb CR. Use of ultrasound to diagnose diaphragmatic rupture in dogs and cats. *Vet Radiol Ultrasound* 2003;44(2):226–230. doi:10.1111/j.1740-8261.2003.tb01276.x
5. Chamberlin SC, Sullivan LA, Morley PS, Boscan P. Evaluation of ultrasound-guided vascular access in dogs. *J Vet Emerg Crit Care* 2013;23(5):498–503. doi:10.1111/vec.12102
6. Lee JA, Guieu LS, Bussières G, Smith CK. Advanced vascular access in small animal emergency and critical care. *Front Vet Sci* 2021;8:703595. doi:10.3389/fvets.2021.703595
7. Furthner E, Kowalewski MP, Torgerson P, Reichler IM. Verifying the placement and length of feeding tubes in canine and feline neonates. *BMC Vet Res* Jun 07 2021;17(1):208. doi:10.1186/s12917-021-02909-7
8. Herreria-Bustillo VJ, Kuo KW, Burke PJ, Cole R, Bacek LM. A pilot study evaluating the use of cervical ultrasound to confirm endotracheal intubation in dogs. *J Vet Emerg Crit Care* Sep 2016;26(5):654–658. doi:10.1111/vec.12507

CHAPTER 13

MISCELLANEOUS POINT-OF-CARE ULTRASOUND CASES

CHAPTER 13 – CASE 1

CAVALIER KING CHARLES WITH A GRAND MAL SEIZURE 2 HOURS PRIOR TO PRESENTATION AND A 1-MONTH HISTORY OF ABNORMAL BEHAVIOR

William Glenn Lane

HISTORY, TRIAGE, AND STABILIZATION

A 10-year-old male neutered Cavalier King Charles Spaniel weighing 10 kg presented with a 1-month history of abnormal behavior including vocalizing at night, standing in corners, and walking through objects. Grand mal seizure occurred 2 h before presentation.

Triage exam findings (vitals):
- Mentation: semi-comatose, sternal recumbency, absent menace, bilateral miosis with minimal pupillary light reflex (PLR)
- Respiratory rate: 32 breaths per minute
- Heart rate: 60 beats per minute
- Mucous membranes: pink, moist, <2 sec capillary refill time
- Femoral and dorsal pedal arterial pulses: strong and synchronous
- Temperature: 37.7°C (100°F)

POCUS exam(s) to perform:
- Optic nerve sheath ultrasound with dog in sternal position

Abnormal POCUS exam results:
See **Figures 13.1.1** and **13.1.2**.

Additional point-of-care diagnostics and initial management:
Blood pressure: 180 mmHg.
Blood glucose: 6.2 mmol/L.

QUESTIONS AND ANSWERS

1. What are the differentials to rule in or out with POCUS based on history and physical exam?
2. What are the sonographic findings?
3. What is the sonographic diagnosis?
4. If necessary, what additional sonographic examination or findings would help rule in or out the differential diagnoses?

DOI: 10.1201/9781003436690-72

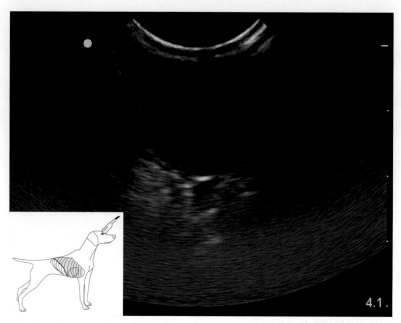

FIGURE 13.1.1 Left optic nerve sheath ultrasound. Image taken with a curvilinear probe through the upper eyelid. Depth set to 4 in. Probe held in transverse orientation, marker lateral, and tip pointing ventral toward the mandibular ramus. The fur is not shaved and a standard eye wash solution is used as the coupling agent.

1. **Differential diagnosis to rule in or rule out with POCUS:**
   - Optic nerve sheath dilation, which if absent would provide evidence against intracranial hypertension (ICH) and if present would support a clinical suspicion for intracranial hypertension.
2. **Describe your sonographic findings:**
   - Optic nerve sheath diameter (ONSD) measured 3 mm distal to the optic disk, which is above the reference interval of 1.15–1.99 mm (**Figure 13.1.3**).[1]
3. **What is your sonographic diagnosis?**
   - Optic nerve dilation consistent with increased intracranial pressure in the patient.
4. **What additional sonographic examination or finding would help you rule in or rule out your differential diagnoses (if necessary)?**
   - The finding should be confirmed on the contralateral side as there could be variation in image acquisition, patient variability, or unilateral disease.

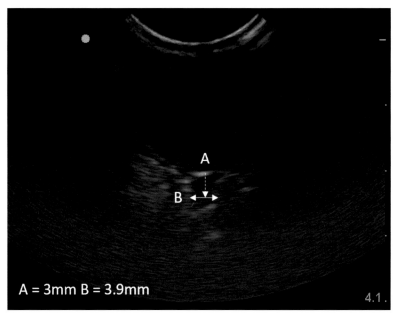

FIGURE 13.1.2 Same image as Figure 13.1.1 with measurements. Measurement A is a vertical distance, while B is horizonal across.

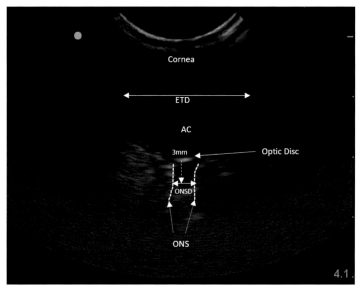

FIGURE 13.1.3 The anterior chamber (AC) appears anechoic with a hyperechoic optic disc and hypoechoic optic nerve distal to the optic disc. The optic nerve does not appear linear. The optic nerve sheath (ONS) is the hyperechoic border around the hypoechoic optic nerve. ONSD is measured 3 mm distal to the optic disk. Eyeball transverse diameter (ETD) is measured across the widest part of the eye.[2]

CLINICAL INTEGRATION

Suspected diagnosis based on ultrasound findings:
Increased intracranial pressure. Neoplasia considered most likely.

Sonographic interventions, monitoring, and outcome:
- Increased ONSD was used to support the presumptive diagnosis of ICH
- Mannitol was administered with improvement in clinical signs and ONSD improved to 2.9 mm (image not available).

Important concepts regarding the sonographic diagnosis of intracranial hypertension:
- Traditionally, mentation changes, brainstem reflexes (including pupil size and reactivity), and Cushing reflex have been used to support a clinical suspicion for ICH. These parameters are not sensitive or specific for ICH and may not readily improve with intervention even if ICH does improve.[2]
- The optic nerve is part of the central nervous system. Within the optic nerve sheath is a subarachnoid compartment that is continuous with the intracranial compartment.[3] In intracranial hypertension this compartment is expanded by cerebrospinal fluid (CSF) which subsequently dilates the optic nerve sheath.
- In dogs, ONSD has been shown to dilate with presumed ICH using MRI measurements.[4] In another canine study using direct intracranial pressure and experimentally induced ICH, the ONSD was shown to positively associate with increasing intracranial pressure.[5]
- POCUS-acquired reference ranges for ONSD in normal dogs have been published[1] and correlate primarily with body weight.

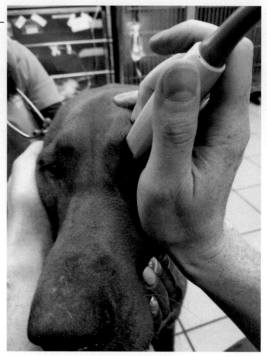

FIGURE 13.1.4 Image of proper probe position to obtain the optic nerve sheath diameter. The upper eyelid is held closed and standard eye wash solution is applied to soak the fur as the coupling agent. Probe is placed on the upper eyelid, held in transverse orientation, and pointed ventrally and medially toward the globe.

- To minimize variability from weight, the ONSD can be divided by the eyeball transverse diameter (ETD). The ONSD/ETD ratio has been shown to be effective in diverse populations of humans for identifying ICH.[2]

Techniques and tips to identify the appropriate ultrasound images/views:
- Hold probe in transverse orientation and apply the tip to the upper eyelid. Point ventrally toward the globe and medially[1] (Figure 13.1.4).
- Fan dorsoventral until a longitudinal image of the optic nerve with the largest diameter is obtained.

- Measure the ONSD 3 mm from the optic disk.[1]
- Globe diameter can be measured across the widest part to obtain the ETD. ONSD/ETD can then be calculated. ONSD/ETD >0.2 is abnormal.[2]
- Alcohol should be avoided as a coupling agent as it will cause ulcers.

Take-home messages
- Optic nerve sheath ultrasound can be used as a tool to screen for intracranial hypertension.
- ONSD correlates with body weight. Dogs weighing 5–25 kg have an ONSD of around 2 mm and the diameter increases to around 3 mm at 40 kg.[1]
- An ONSD/ETD ratio >0.2 may be indicative of ICH.[2]
- Further studies in dogs are needed to validate the use of ONSD and OSND/ETD ratios in patients with naturally occurring confirmed intracranial hypertension.

REFERENCES

1. Smith, J.J., Fletcher, D.J., et al. Transpalpebral ultrasonographic measurement of the optic nerve sheath diameter in healthy dogs. *J Vet Emerg Crit Care* 2018; 28(1):31–38.
2. Dupanloup, A., Osinchuk, S. Relationship between the ratio of optic nerve sheath diameter to eyeball transverse diameter and morphological characteristics of dogs. *Am J Vet Res* 2021; 82(8):667–675.
3. Soni, N.J. *Point-of-Care Ultrasound*, 2nd ed., Elsevier, Inc, 2020.
4. Scrivani, P.V., Fletcher, D.J., et al. T2-weighted magnetic resonance imaging measurements of optic nerve sheath diameter in dogs with an without presumed intracranial hypertension. *Vet Rad US* 2013; 54(3):263–270.
5. Ilie, L.A., Thomovsky, E.J., et al. Relationship between intracranial pressure as measured by an epidural intracranial pressure monitoring system and optic nerve sheath diameter in healthy dogs. *Am J Vet Res* 2015; 76(8):724–731.

CHAPTER 13 – CASE 2

DOMESTIC SHORT HAIR THAT PRESENTS WITH DYSPNEA AN HOUR AFTER RECOVERING FROM DENTAL TREATMENT PERFORMED UNDER GENERAL ANESTHESIA

Serge Chalhoub

HISTORY, TRIAGE, AND STABILIZATION

A 7-year-old female spayed Domestic Short Hair presented on emergency about an hour after recovering from general anesthesia for a comprehensive periodontal treatment.

Triage exam findings (vitals):
- Mentation: anxious, vocalizing loudly, panting
- Respiratory rate: 40 breaths per minute, dyspnea
- Heart rate: 240 beats per minute
- Mucous membranes: pink, <2 sec capillary refill time
- Femoral and dorsal pedal arterial pulses: strong and synchronous
- Temperature: not taken
- Cat's generalized subcutaneous tissue (especially dorsally) felt distended, seemed to have crepitus, and was painful

POCUS exam(s) to perform:
- PLUS

Abnormal POCUS exam results:
See **Figure 13.2.1** and **Video 13.2.1**.

Additional point-of-care diagnostics and initial management:
No other POCUS exams were completed. An IV catheter was placed and the patient was given butorphanol 0.4 mg/kg IV. The patient was then placed in an oxygen cage.

QUESTIONS AND ANSWERS

1. What are the differentials to rule in or out with POCUS based on history and physical exam?
2. What are the sonographic findings?
3. What is the sonographic diagnosis?
4. If necessary, what additional sonographic examination or findings would help rule in or out the differential diagnoses?

Domestic Short Hair that Presents with Dyspnea

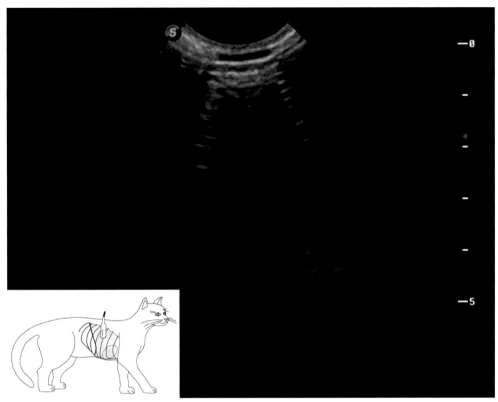

FIGURE 13.2.1 AND VIDEO 13.2.1 PLUS image and cineloop taken from the patient's mid-left hemithorax. The probe was quickly placed on the left hemithorax and oriented perpendicular to the ribs, then PLUS was aborted due to tachypnea and dyspnea.

1. **Differential diagnosis to rule in or rule out with POCUS:**
   - Subcutaneous emphysema.
2. **Describe your sonographic findings:**
   - At mid-thorax, right at the skin surface, B-lines were suspected. However, these do not originate from the lung surface and instead originate in the subcutaneous region. No structures are seen below except for reverberation artifact (**Figure 13.2.2**).
3. **What is your sonographic diagnosis?**
   - Subcutaneous emphysema.
4. **What additional sonographic examination or finding would help you rule in or rule out your differential diagnoses (if necessary)?**
   - Chest radiographs (**Figure 13.2.3**).

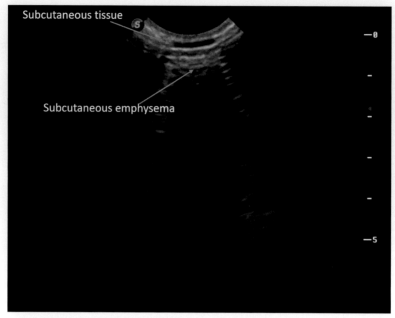

FIGURE 13.2.2 Labeled image of Figure 13.1. Suspected B-lines seen originating in the subcutaneous region consistent with subcutaneous emphysema.

CLINICAL INTEGRATION

Suspected diagnosis based on ultrasound findings:

The most likely diagnosis was traumatic tracheal tear leading to subcutaneous emphysema.[1]

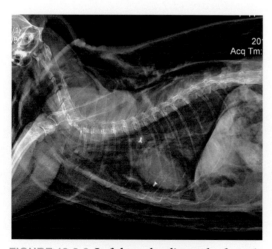

FIGURE 13.2.3 Left lateral radiograph of a cat's thorax with subcutaneous emphysema.

Sonographic interventions, monitoring, and outcome:

- A single lateral radiograph was taken after 20 min when the respiratory rate had reduced to 32 breaths/min (see **Figure 13.2.3**), which confirmed the suspected diagnosis of subcutaneous emphysema likely due to traumatic tracheal tear.
- The patient was placed on IV fluids and oxygen, with anxiolytics given frequently.v
- Tracheal tears secondary to ET tube trauma will often heal on their own within days and without further intervention.
- Patients need to be supported nutritionally and for pain.

Important concepts regarding the sonographic diagnosis of subcutaneous emphysema:

- B-lines are usually an artifact created at the lung surface (pleural line).

- In this case, the pleural line could not be visualized and instead we see B-lines at the skin surface. Air is present below the skin, leading to the inability to visualize anything below the skin and the presence of reverberation artifact.
- It is possible to press harder and therefore evaluate the air at the probe surface to then see the lung surface (as seen at the end of the video clip), but this could lead to discomfort for the patient.

Techniques and tips to identify the appropriate ultrasound images/views:
- On PLUS or abdominal POCUS, B-lines and reverberation artifacts seen below the skin surface may indicate subcutaneous emphysema.

Take-home messages
- Subcutaneous emphysema can occur secondary to tracheal trauma if an intubated patient is rotated without first disconnecting the endotracheal tube.[1]
- POCUS is unable to detect normal structures in the pleural space, lung, and abdomen because of the presence of air below the skin surface, which is an enemy of ultrasound.[2]

REFERENCES

1. Mitchell SL, McCarthy R, Rudloff R, Pernell RT. Tracheal rupture associated with intubation in cats: 20 cases (1996–1998). *J Am Vet Med Assoc* 2000 May 15;216(10):1592–1595.
2. Boysen S, Gommeren K, Chalhoub S. PLUS image interpretation: Normal findings. In: Boysen S, Gommeren K, Chalhoub S, editors. *The Essentials of Veterinary Point-of-Care Ultrasound: Pleural Space and Lung.* Zaragoza Spain: Groupo Asis Biomedia; 2022, pp. 35–62.

SECTION 7

SYSTEMIC AND ADVANCED POINT-OF-CARE ULTRASOUND CASES

Section Editor Erin Binagia

CHAPTER 14

INTRODUCTION TO SYSTEMIC AND ADVANCED POINT-OF-CARE ULTRASOUND CASES

Erin Binagia

In the previous sections, most cases have been presented in a way that only one point-of-care ultrasound (POCUS) exam is utilized for diagnosis. However, in critical case scenarios, multiple body systems are usually affected. To understand the status of the whole patient, multiple POCUS exams are performed on the triage table to obtain a diagnosis and initiate appropriate treatment.[1,2] In addition, there are many other clinical scenarios in which ultrasound can be used that require more advanced techniques, such as identifying thromboses, foreign bodies (Case 15.3), and pancreatitis (Case 15.4). These pathologies ideally require an imaging specialist (i.e. radiologist, cardiologist, internist, etc.) for a definitive diagnosis; however, in emergency situations in which that is not possible (such as overnight or an unstable patient that cannot be moved or requires immediate treatment), these techniques can be utilized to guide patient treatment rapidly and appropriately. It is important to note that further research to determine the appropriate method and accuracy is required before considering it a standard part of the POCUS exam.

This section presents clinical cases in which more advanced ultrasound techniques are utilized, as well as cases that require multiple POCUS exams to evaluate and treat the whole patient.

REFERENCES

1. Lau YH, See KC Point-of-care ultrasound for critically-ill patients: A mini-review of key diagnostic features and protocols. *World J Crit Care Med* Mar 9 2022;11(2):70–84. doi:10.5492/wjccm.v11.i2.70

2. Choi WJ, Ha YR, Oh JH, et al. Clinical guidance for point-of-care ultrasound in the emergency and critical care areas after implementing insurance coverage in Korea. *J Korean Med Sci* Feb 24 2020;35(7):e54. doi:10.3346/jkms.2020.35.e54

CHAPTER 15
SYSTEMIC AND ADVANCED POINT-OF-CARE ULTRASOUND CASES

CHAPTER 15 – CASE 1

GOLDEN RETRIEVER WITH ACUTE VOMITING, DIARRHEA, AND COLLAPSE 10 MINUTES AFTER GOING OUTSIDE

Erin Binagia

HISTORY, TRIAGE, AND STABILIZATION

A 5-year-old female spayed Golden Retriever presented with acute vomiting, diarrhea, and collapse 10 min after going outside. Before this episode the patient was normal.

Triage exam findings (vitals):
- Mentation: stuporous, lateral recumbency
- Respiratory rate: 50 breaths per minute
- Heart rate: 190 beats per minute
- Mucous membranes: pale, tacky, 3 sec capillary refill time
- Femoral and dorsal pedal arterial pulses: weak, thready
- Temperature: 36.4°C (97.5°F)

POCUS exam(s) to perform:
- Abdominal POCUS with gallbladder assessment
- Cardiac POCUS

Abnormal POCUS exam results:
See **Figures 15.1.1** and **15.1.2** and **Videos 15.1.1** and **15.1.2**.

Additional point-of-care diagnostics and initial management:
Initial point-of-care testing included an echocardiogram that showed sinus tachycardia, Doppler systolic blood pressure of 50 mmHg, packed cell volume/total solids (PCV/TS) of 60% and 4.0 g/dL, and venous blood gas showed a lactic acidosis (pH 7.15, lactate 8.0 mmol/L) and hyperglycemia (180 mg/dL). Coagulation profile was normal. Additional blood was drawn and saved for later clinicopathologic evaluation. The patient was given a 20 mL/kg lactated Ringer's solution (LRS) bolus and recheck blood pressure was 40 mmHg.

QUESTIONS AND ANSWERS

1. What are the differentials to rule in or out with POCUS based on history and physical exam?
2. What are the sonographic findings?
3. What is the sonographic diagnosis?
4. If necessary, what additional sonographic examination or findings would help rule in or out the differential diagnoses?

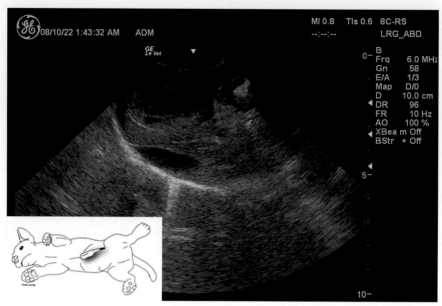

FIGURE 15.1.1 AND VIDEO 15.1.1 Gallbladder assessment in the subxiphoid view with the patient in left lateral recumbency. Image taken with a microconvex probe placed longitudinally and the marker directed cranially, immediately caudal to the xiphoid process. If the gallbladder is not in immediate view, the probe can either sweep or fan toward the right. The frequency was set at 6 MHz and the depth was set at 10 cm in this case; however, the depth could have been set to 6 cm for a better image.

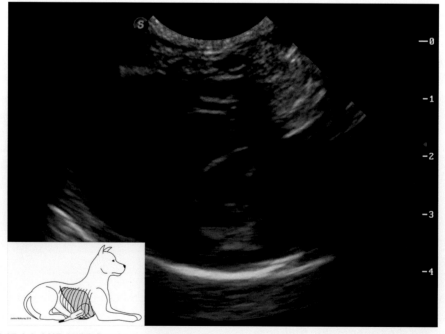

FIGURE 15.1.2 AND VIDEO 15.1.2 Cardiac POCUS performed to evaluate for pericardial effusion and hypovolemia. Right parasternal short-axis papillary view, image, and video. See Chapter 6, Chapter 8b, Case 7.2, and Case 9.2 for more information on how to obtain these images.

1. **Differential diagnosis to rule in or rule out with POCUS:**
   - The most likely differential diagnoses based on history and physical exam include anaphylaxis, spontaneous hemoabdomen, and pericardial effusion. Other differentials to consider include septic abdomen and portal vein thrombus. A gallbladder assessment is an important first diagnostic step. If gallbladder wall edema is found, the heart should be evaluated for pericardial effusion via the same subxiphoid view. Lastly, the abdomen should be evaluated for free fluid. If no pericardial or abdominal effusion is seen, anaphylaxis should be highly suspected unless the patient has a high-risk factor for a portal thrombus (such as an immune-mediated hemolytic anemia patient). If peritoneal fluid is found, septic abdomen can be ruled out by analyzing a fluid sample.
2. **Describe your sonographic findings:**
   - Gallbladder wall is diffusely hypoechoic with parallel hyperechoic lines on either side ("double rim" or "halo" sign) (**Figures 15.1.3** and **15.1.4**).
   - Scant to mild free fluid seen at the subxiphoid view (**Figures 15.1.3** and **15.1.4**).
   - No pericardial effusion.
   - Hyperdynamic cardiac contractions, subjectively underfilled ventricle in which the lumen disappears during systole (**Video 15.1.2**).
   - Left ventricle (LV) wall thickened compared to the LV lumen ("pseudohypertrophy") (**Figure 15.1.5**).
3. **What is your sonographic diagnosis?**
   - Gallbladder wall edema.
   - Abdominal fluid score 1 out of 4.
   - Hypovolemic shock.
4. **What additional sonographic examination or finding would help you rule in or rule out your differential diagnoses (if necessary)?**
   - If free fluid is present, abdominal fluid scoring can be used for serial monitoring.[1]
   - Hypovolemia can objectively be evaluated via the collapsibility of the caudal vena cava[2] (see Chapter 9 Case 9.2).
   - Lastly, evaluation of the portal vein by a radiologist would be ideal to look for other causes of acute portal hypertension and gallbladder wall edema (such as a portal thrombus).

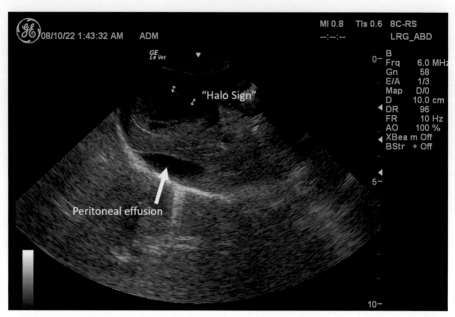

FIGURE 15.1.3 Same but labeled image of Figure 15.1.1 identifying peritoneal effusion (large white arrow) and outlining the gallbladder "halo sign" (small double arrows).

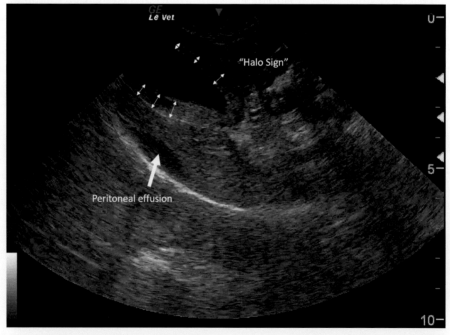

FIGURE 15.1.4 Still image of Video 15.1.1 taken at 4 sec (very end) showing the same peritoneal effusion (large white arrow) and outlining a better view of the gallbladder "halo sign" (small double arrows).

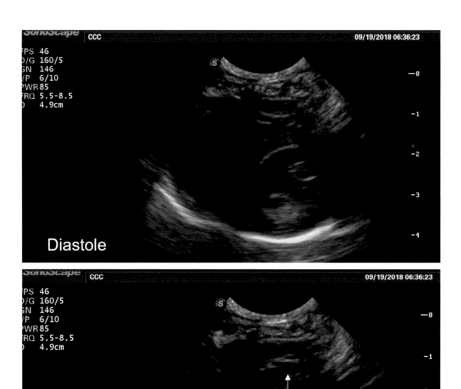

FIGURE 15.1.5 Still images of Video 15.1.2 taken during diastole and systole. Note that the left ventricular lumen (highlighted in yellow) is almost completely obliterated during systole. The right ventricular lumen (highlighted in green) is almost completely collapsed in both systole and diastole. The left ventricle wall appears thickened compared to its lumen during systole. This "pseudohypertrophy" is outlined by the white double arrows.

CLINICAL INTEGRATION

Suspected diagnosis based on ultrasound findings:
Anaphylaxis is the most likely diagnosis.

Sonographic interventions, monitoring, and outcome:
An abdominocentesis was performed and the fluid PCV/TS was 8%/1.0 g/dL, supporting anaphylaxis. Epinephrine (0.01 mg/kg IV bolus) followed by a continuous rate infusion (CRI) was initiated to normalize blood pressure. Diphenhydramine, famotidine, and maropitant were also administered. Blood pressure normalized within 1 h of epinephrine CRI, which was weaned and then discontinued soon after.

Serial 4-h abdominal fluid scores were performed to evaluate the progression or resolution of peritoneal effusion. The fluid score decreased to 0/4 within 8 h. The patient was on telemetry overnight and no significant arrhythmias were noted. The patient was discharged after 48 h of monitoring and supportive care.

Important concepts regarding the sonographic diagnosis of anaphylaxis:

- Diagnosis of anaphylaxis is often based on history, physical exam, and point-of-care testing and should be treated with epinephrine in a timely manner.
- Gallbladder wall edema is a good biomarker for the diagnosis of anaphylaxis (sensitivity 93%, specificity 98%) in dogs with hypersensitivity reactions,[3] but there are many other differentials that must also be considered (see Case 11.10).
- Peritoneal effusion can be a common finding in anaphylaxis, which can mimic right heart failure secondary to pericardial effusion, septic peritonitis, hemoabdomen, or portal thrombus.[3–8]

Techniques and tips to identify gallbladder wall edema:
See Case 11.10.

Techniques and tips to find small volume peritoneal effusion:

- If the abdominal POCUS technique is performed and no free fluid is found, but clinical suspicion is high for the presence of fluid, it is helpful to place the probe to the most gravity-dependent area of the mid-abdomen and apply firm pressure while the dog is in lateral recumbency. This is often where a small volume effusion is found.
- If still no free fluid is found, serial abdominal POCUS exams can be performed while the patient is on fluid therapy, and free fluid will be found within a few hours if it is indeed present.[1]

Take-home messages

- The common history for an anaphylaxis patient is often acute vomiting, diarrhea, and collapse after going outside. Common physical exam findings include pale gums, recumbent, depressed mentation, tachycardia, hypothermia, and hypotension.[3,5–7]
- This history and physical exam findings can appear similarly to a dog with septic abdomen,[4] portal thrombus, spontaneous hemoabdomen, and acute pericardial effusion. A common clinical complaint for both pericardial effusion and anaphylaxis cases is acute vomiting and collapse.[9]
- Gallbladder wall edema can also be found in these abovementioned disease processes. Therefore, after discovering gallbladder wall edema, always perform a cardiac POCUS (or extended subxiphoid view) to evaluate for pericardial effusion. Then perform a complete abdominal POCUS exam to evaluate for and obtain a sample of free abdominal fluid if present for cytological evaluation.
- The primary treatment indicated for true anaphylaxis is epinephrine, all other treatments are supplemental. The second most important therapy is fluid support.[7]

REFERENCES

1. Lisciandro GR, Lagutchik MS, Mann KA, et al. Evaluation of an abdominal fluid scoring system determined using abdominal focused assessment with sonography for trauma in 101 dogs with motor vehicle trauma. *J Vet Emerg Crit Care (San Antonio)* Oct 2009;19(5):426–437. doi:10.1111/j.1476-4431.2009.00459.x

2. DeFrancesco TC, Ward JL. Focused canine cardiac ultrasound. *Vet Clin North Am Small Anim Pract* Nov 2021;51(6):1203–1216. doi:10.1016/j.cvsm.2021.07.005
3. Quantz JE, Miles MS, Reed AL, White GA. Elevation of alanine transaminase and gallbladder wall abnormalities as biomarkers of anaphylaxis in canine hypersensitivity patients. *J Vet Emerg Crit Care (San Antonio)* Dec 2009;19(6):536–544. doi:10.1111/j.1476-4431.2009.00474.x
4. Walters AM, O'Brien MA, Selmic LE, McMichael MA. Comparison of clinical findings between dogs with suspected anaphylaxis and dogs with confirmed sepsis. *J Am Vet Med Assoc* Sep 15 2017;251(6):681–688. doi:10.2460/javma.251.6.681
5. Smith MR, Wurlod VA, Ralph AG, Daniels ER, Mitchell M. Mortality rate and prognostic factors for dogs with severe anaphylaxis: 67 cases (2016–2018). *J Am Vet Med Assoc* May 15 2020;256(10):1137–1144. doi:10.2460/javma.256.10.1137
6. Lisciandro GR, Gambino JM, Lisciandro SC. Thirteen dogs and a cat with ultrasonographically detected gallbladder wall edema associated with cardiac disease. *J Vet Intern Med* May 2021;35(3):1342–1346. doi:10.1111/jvim.16117
7. Shmuel DL, Cortes Y. Anaphylaxis in dogs and cats. *J Vet Emerg Crit Care (San Antonio)* 2013 Jul–Aug 2013;23(4):377–394. doi:10.1111/vec.12066
8. Turner K, Boyd C, Stander N, Smart L. Clinical characteristics of two-hundred thirty-two dogs (2006–2018) treated for suspected anaphylaxis in Perth, Western Australia. *Aust Vet J* Dec 2021;99(12):505–512. doi:10.1111/avj.13114
9. Fahey R, Rozanski E, Paul A, Rush JE. Prevalence of vomiting in dogs with pericardial effusion. *J Vet Emerg Crit Care (San Antonio)* Mar 2017;27(2):250–252. doi:10.1111/vec.12570

CHAPTER 15 – CASE 2

LABRADOODLE WITH LETHARGY AND INAPPETENCE PROGRESSING TO ACUTE HEMATOCHEZIA, TENESMUS, AND PRAYER POSTURING

Armi M Pigott

HISTORY, TRIAGE, AND STABILIZATION

A 7-month-old intact male labradoodle presented with diarrhea for 3 days, progressive lethargy, and inappetence. Acutely worsening on the afternoon of presentation with hematochezia, tenesmus, and prayer posturing. No veterinary care prior to presentation.

Triage exam findings (vitals):
- Mentation: quiet, alert, and responsive (QAR)
- Respiratory rate: 36 breaths per minute
- Heart rate: 150 beats per minute
- Mucous membranes: pink, tacky, 2 sec capillary refill time
- Femoral and dorsal pedal arterial pulses: strong and synchronous
- Temperature: 39.2°C (102.5°F)
- Abdominal palpation: note pertinent findings

POCUS exam(s) to perform:
- Abdominal POCUS
- Gastrointestinal (GI) assessment

Abnormal POCUS exam results:
See **Figures 15.2.1** and **15.2.2**.

Additional point-of-care diagnostics and initial management:
Methadone (0.25 mg/kg IV) and ketamine (1 mg/kg IV) were administered for pain control. Notable additional diagnostics included a negative point-of-care parvo test, and a complete blood count (CBC) with a normal white blood cell count. Fecal exam was positive for hookworms and roundworms.

QUESTIONS AND ANSWERS

1. What are the differentials to rule in or out with POCUS based on history and physical exam?
2. What are the sonographic findings?
3. What is the sonographic diagnosis?
4. If necessary, what additional sonographic examination or findings would help rule in or out the differential diagnoses?

Labradoodle With Lethargy, And Inappetence Progressing

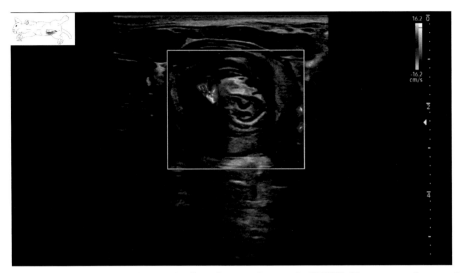

FIGURE 15.2.1 GI assessment. Patient in dorsal recumbency. A 15 MHz linear transducer oriented parallel to the spine with the marker cranial, 2 cm cranial to umbilicus on midline. Depth set at 5 cm. Color Doppler function is activated and evaluating the area inside the white square.

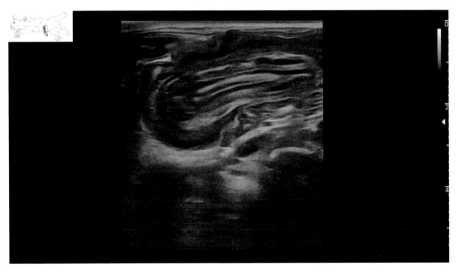

FIGURE 15.2.2 GI assessment. Same settings as Figure 15.2.1 with the transducer oriented perpendicular to the spine with the marker to the right side of the body.

1. **Differential diagnosis to rule in or rule out with POCUS:**
   - Intussusception, GI perforation/free abdominal fluid, gastroenteritis, gastrointestinal foreign body and/or obstruction.
2. **Describe your sonographic findings:**
   - The intussuscipiens (outer segment of bowel) contains invaginated small bowel, mesenteric fat, and mesenteric vessels (**Figures 15.2.3 and 15.2.4**).
3. **What is your sonographic diagnosis?**
   Intussusception.
4. **What additional sonographic examination or finding would help you rule in or rule out your differential diagnoses (if necessary)?**

- If unsure of the findings or if nothing is identified on POCUS evaluation, a complete abdominal ultrasound by a board-certified radiologist or another very experienced sonographer is indicated in a patient with significant abdominal pain and systemic signs of illness. Additionally, when an intussusception is identified, a more thorough GI ultrasound by these experienced sonographers may reveal additional findings related to the underlying cause of the intussusception.

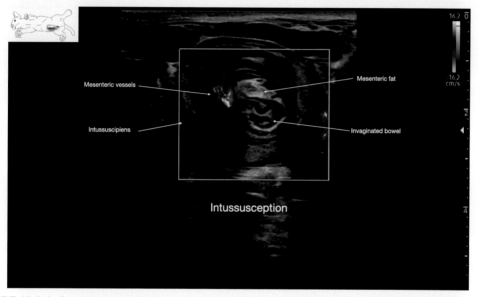

FIGURE 15.2.3 Same image as Figure 15.2.1 with labels added. Vessels identified by color Doppler.

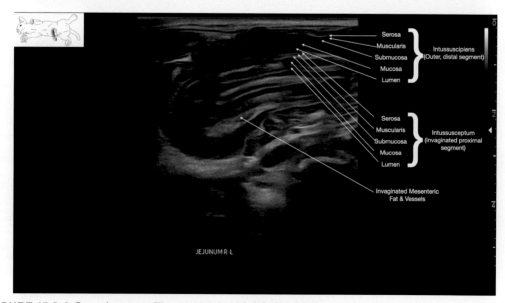

FIGURE 15.2.4 Same image as Figure 15.2.2 with labels added.

CLINICAL INTEGRATION

Suspected diagnosis based on ultrasound findings:
Intussusception secondary to intestinal parasites.

Sonographic interventions, monitoring, and outcome:
- GI assessment was performed with the patient in dorsal recumbency, which identified an intussusception.
- Three attempts to manually manipulate and non-surgically reduce the intussusception were made, and ultrasound was used to check for success/failure between attempts.[1,2]
- These attempts were unsuccessful and the patient was ultimately taken to surgery where the surgeon assessed the affected segment as questionably viable, so resection and anastomosis was performed.
- Treatment for the identified intestinal parasites was initiated 12 h post-operatively. The puppy recovered uneventfully and was discharged 48 h later.

Important concepts regarding the sonographic diagnosis of intussusception:
- Intussusception is defined as the invagination of one segment of bowel into an immediately adjacent segment. The intussusceptum refers to the proximal segment which telescopes into the intussuscipiens (the adjacent distal segment).[1-4]
- Note that when the bowel invaginates, the attached mesentery (and the associated mesenteric fat) and adjacent vessels are pulled along with the bowel and are found within the intussuscipiens along with the invaginated bowel segment.[3]
- The mesenteric fat will be seen as hyperechoic tissue within the intussuscipiens either surrounding or more often displacing the invaginated bowel off to one side. Invaginated blood vessels can be identified with color Doppler.[3]
- Intussusception is rarely a spontaneous/isolated occurrence and is usually caused by other underlying pathology. Some of these pathologies can be identified with ultrasound. For example, linear foreign bodies can often be identified by a skilled sonographer. GI parasites (such as roundworms) can also occasionally be identified.[3,4]
- The segment of small bowel leading into the intussusception is usually significantly fluid dilated due to obstruction as the intussusceptum is compressed within the intussuscipiens. Tracing these dilated bowel segments can lead to the identification of the obstruction (in this case, intussusception).[3,4]
- Patients with intussusception often have significant abdominal pain and may require optimizing their pain control in order to tolerate imaging.[1-4]
- GI assessment is usually performed with the patient in dorsal recumbency, but sometimes imaging with the patient in left and right lateral recumbency can help image around air artifact when significant GI gas is present.[3,4]

Technique and tips to identify the appropriate ultrasound images/views:
- Use a methodical approach to scanning the small bowel. Beginning at a designated location (e.g., the right cranial abdomen) and moving methodically back and forth across the abdomen in a "mowing the grass" pattern will provide an overview of the GI tract and identify areas that need more focus.[3,4]
- Tracing dilated bowel segments in both directions will help to identify obstructing pathologies including intussusception. Use these dilated segments as

"road maps" by following them until the obstructing pathology (or normal bowel) is identified.[3,4]

- Occasionally there will be minimal or no mesenteric tissue within the intussusception – these lesions look like intestine with too many layers, sometimes described as a bull's eye appearance.[3,4]

Take-home message

Identification of intussusception should be followed by an investigation for underlying causes, some of which can be identified with POCUS.

REFERENCES

1. Moores AL, Urraca CI, de Sousa RJR, Jenkins G, Anderson DM. Nonsurgical reduction of prolapsed colocolic intussusception in 2 puppies. *J Vet Emerg Crit Care (San Antonio)* 2021;31(5):656–660.
2. Patsikas MN, Papazoglou LG, Paraskevas GK. Current views in the diagnosis and treatment of intestinal intussusception. *Top Companion Anim Med* 2019;37:100360.
3. Penninck, D and d'Angjou, MA. Gastrointestinal tract. In Pennick D and d'Anjou MA, ed. *Atlas of Small Animal Ultrasonography*, 2nd ed. Singapore: Wiley Blackwell; 2015, pp. 259–308.
4. Garcia DA, Froes TR, Vilani RG, Guérios SD, Obladen A. Ultrasonography of small intestinal obstructions: A contemporary approach. *J Small Anim Pract* 2011;52(9):484–490.

CHAPTER 15 – CASE 3

PIT BULL WITH 2 DAYS OF PROGRESSIVE VOMITING, LETHARGY, AND INAPPETENCE AFTER GETTING INTO THE TRASH 72 HOURS EARLIER

Armi Pigott

HISTORY, TRIAGE, AND STABILIZATION

A 3-year-old castrated male Pit Bull presented with 2 days of progressive vomiting, lethargy, and inappetence after getting into the trash 72 h earlier.

Triage exam findings (vitals):
- Mentation: dull
- Respiratory rate: 28 breaths per minute
- Heart rate: 150 beats per minute
- Mucous membranes: pink, dry, 1 sec capillary refill time
- Femoral and dorsal pedal arterial pulses: bounding
- Temperature: 38.6°C (101.4°F)

POCUS exam(s) to perform:
- Abdominal POCUS
- Gastrointestinal (GI) assessment

Abnormal POCUS exam results:
See **Figures 15.3.1** and **15.3.2**.

Additional point-of-care diagnostics and initial management:
Initial treatments included 0.2 mg/kg methadone IV, 0.5 mg/kg ondansetron IV, and a 20 mL/kg IV fluid bolus over 20 min. Venous blood gas showed a mild hypochloremic metabolic alkalosis. Blood glucose and lactate were normal.

QUESTIONS AND ANSWERS

1. What are the differentials to rule in or out with POCUS based on history and physical exam?
2. What are the sonographic findings?
3. What is the sonographic diagnosis?
4. If necessary, what additional sonographic examination or findings would help rule in or out the differential diagnoses?

DOI: 10.1201/9781003436690-79

CHAPTER 15 – CASE 3

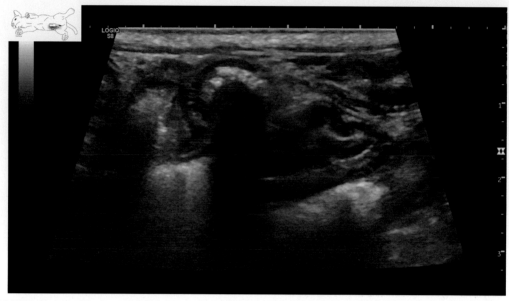

FIGURE 15.3.1 GI assessment. Patient in right lateral recumbency. Mid-abdomen cranial to umbilicus, transducer oriented parallel to spine, depth 3.5 cm, frequency 10 MHz. The transducer was passed back and forth across the abdomen in a "mow-the-lawn" pattern across the abdominal wall.

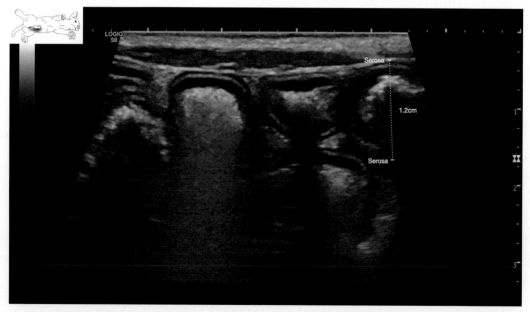

FIGURE 15.3.2 GI assessment. Patient in left lateral recumbency. Mid-abdomen near umbilicus, transducer oriented parallel to spine, depth 3.5 cm, frequency 10 MHz.

Pit Bull with 2 Days of Progressive Vomiting, Lethargy, and Inappetence

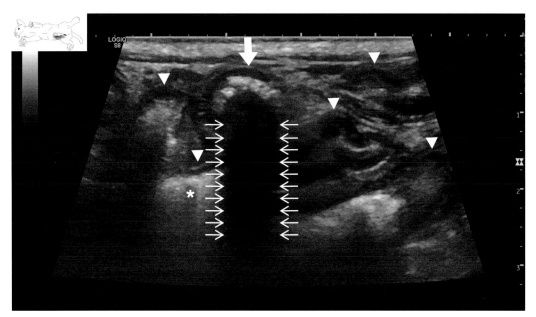

FIGURE 15.3.3 Same image as Figure 15.3.1 with annotations. Several segments of small intestine (solid individual arrowheads) are seen. One segment is gas filled, creating a gas shadow (asterisk), and another segment contains material that creates a hyperechoic interface (large thick arrow) with a strong acoustic shadow (between the stacked arrows). The bowel segment containing material causing an acoustic shadow was traced in both directions (oral and aboral) to estimate size and look for other evidence of obstruction including small bowel fluid and gas dilation upstream to the obstructive material (not shown, very mild at presentation).

1. **Differential diagnosis to rule in or rule out with POCUS:**
   - Gastrointestinal (GI) foreign body obstruction, pancreatitis, uncomplicated gastroenteritis, GI perforation/septic abdomen.
2. **Describe your sonographic findings:**
   - Material in the small bowel with strong acoustic shadow (**Figure 15.3.3**).
   - Moderate mixed gas and fluid small bowel dilation (**Figure 15.3.4**).
3. **What is your sonographic diagnosis?**
   - Hard shadowing foreign material in the small bowel, and moderate mixed gas and fluid small bowel dilation.
4. **What additional sonographic examination or finding would help you rule in or rule out your differential diagnoses (if necessary)?**
   - Pancreatic assessment to rule in/out concurrent pancreatitis.

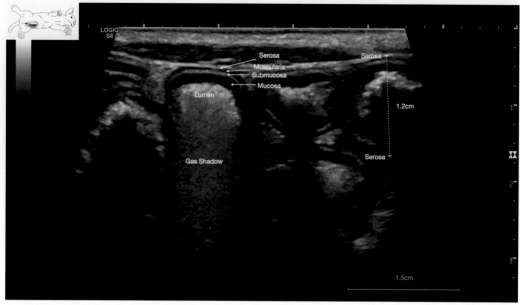

FIGURE 15.3.4 Same image as Figure 15.3.2 with annotations. Several small bowel loops with mild to moderate mixed fluid and gas dilation are shown. All segments measure less than 1.5 cm when measured from serosa to serosa.

CLINICAL INTEGRATION

Suspected diagnosis based on ultrasound findings:
- Small intestinal foreign body obstruction.

Sonographic interventions, monitoring, and outcome:
- Following IV fluid resuscitation from shock, a nasogastric tube was placed to maintain gastric decompression. The patient was continued on IV fluids for rehydration and monitored in the hospital.
- Serial GI assessments over the subsequent 12 h showed worsening bowel dilation and lack of foreign material progression through the bowel. The patient was taken to surgery, where a single enterotomy was performed to remove a mid-jejunal foreign body (a foam practice golf ball).
- The patient recovered quickly post-operatively and was discharged the following morning.

Important concepts regarding the sonographic diagnosis of small intestinal foreign bodies:
- GI assessment is usually performed with the patient in dorsal recumbency, but sometimes imaging with the patient in left and right lateral recumbency can help image around an air artifact when significant GI gas is present.[1-4]
- Typically, there are two populations of small bowel in patients with GI foreign body obstruction. The bowel is dilated with a mixture of fluid and gas proximal to the foreign body, and normal with a

nearly empty lumen distal to the obstructing object. Tracing the dilated bowel loops can help to locate the foreign body. Tracing a dilated bowel segment to a segment of bowel with material creating an artifact, and with normal-appearing bowel on the opposite side of this luminal material is diagnostic for small bowel foreign material causing an obstruction.[1-4]

- The ultrasonographic characteristics of foreign bodies will depend on the physical properties of the obstructing material (cloth, plastic, bone, hairballs, etc). Many (but not all) will be associated with ultrasonographic shadowing or other artifacts.[1-6]
- Many (but not all) GI foreign bodies will cause a distinct hyperechoic interface with a strong distal acoustic shadow. The acoustic shadow should arise from the lumen of the GI tract (not from the transducer–skin interface).[1-6]
- Linear foreign bodies often cause a very distinct bunched or plicated appearance to the small bowel. The linear foreign material in these bunched segments may be seen as a non-moving hyperechoic linear structure within the bowel lumen.[1-4, 6]
- The presence of foreign material in the GI tract does not always indicate an obstructive process requiring surgery. Imaging findings should be considered with presenting signs. In some cases, serial assessments or other imaging modalities (radiographs, CT scans, ultrasound performed by a board-certified radiologist or another very experienced sonographer) may be needed to confirm obstruction or provide enough supporting evidence to justify proceeding to surgery.
- In dogs, measuring the total diameter of the jejunum from serosa to serosa may be useful when trying to differentiate between ileus-associated bowel dilation and a mechanical obstruction. Measurements over 1.5 cm are much more likely to be caused by an obstruction and may help to support clinical decision-making when a definite foreign body cannot be identified with ultrasound. It should be noted that this measurement only applies when the jejunal enlargement is due to luminal dilation (not wall thickening) and not to the duodenum or ileum. Measurements less than 1.5 cm may be the result of many causes, including obstruction.[1]

Technique and tips to identify the appropriate ultrasound images/views:

- If the stomach is massively gas or fluid filled and interfering with imaging, pass a nasogastric tube and decompress the stomach. This makes imaging easier, improves patient comfort, and reduces the risk of regurgitation and aspiration during imaging.
- Scanning methodically for evidence of foreign body obstruction will reduce the chances of missing a bowel region. Start in a designated location (right cranial abdomen, for example) and use a methodical back-and-forth "mow-the-lawn" technique to obtain an overview and scan for areas to evaluate more extensively.[1-6]

Take-home messages

- Identifying the combination of foreign material and an obstructive bowel pattern with ultrasound is diagnostic for GI foreign body obstruction.[1-6] If definitive foreign material cannot be found, jejunal lumen dilation causing a jejunal serosa-to-serosa diameter greater than 1.5 cm is strongly suggestive of a mechanical obstruction.[1]
- Trace dilated bowel segments to look for foreign material.
- GI assessment can be used to serially monitor patients if not going directly to surgery.

REFERENCES

1. Sharma A, Thompson MS, Scrivani PV, et al. Comparison of radiography and ultrasonography for diagnosing small-intestinal mechanical obstruction in vomiting dogs. *Vet Radiol Ultrasound* 2011;52(3):248–255. doi:10.1111/j.1740-8261.2010.01791.x
2. Garcia DA, Froes TR, Vilani RG, Guérios SD, Obladen A. Ultrasonography of small intestinal obstructions: A contemporary approach. *J Small Anim Pract* 2011;52(9):484–490. doi:10.1111/j.1748-5827.2011.01104.x
3. Tidwell AS, Penninck DG. Ultrasonography of gastrointestinal foreign bodies. *Vet Rad Ultrasound* 1992;33(3):160–169. doi:10.1111/j.1740-8261.1992.tb01439.x
4. Manczur F, Vörös K, Vrabély T, Wladár S, Németh T, Fenyves B. Sonographic diagnosis of intestinal obstruction in the dog. *Acta Vet Hung* 1998;46(1):35–45.
5. Penninck, D and d'Angjou, MA. Gastrointestinal tract. In Pennink D and d'Anjou MA, eds. *Atlas of Small Animal Ultrasonography*, 2nd ed., Singapore: Wiley Blackwell; 2015, pp. 259–308.
6. Hoffmann KL. Sonographic signs of gastroduodenal linear foreign body in 3 dogs. *Vet Radiol Ultrasound* 2003;44(4):466–469. doi:10.1111/j.1740-8261.2003.tb00486.x

CHAPTER 15 – CASE 4

MIXED-BREED K9 WITH A 5-DAY HISTORY OF ANOREXIA, VOMITING, AND ABDOMINAL PAIN

Elizabeth Gribbin and Rachael Birkbeck

HISTORY, TRIAGE, AND STABILIZATION

A 12-year-old male neutered crossbreed dog with a 5-day history of anorexia, vomiting, and abdominal pain.

Triage exam findings (vitals):
- Mentation: quiet, alert, responsive
- Respiratory rate: 32 breaths per minute, no increase in respiratory effort
- Heart rate: 160 beats per minute, regular rhythm, no cardiac murmur detected
- Mucous membranes: pink, tacky, <2 sec capillary refill time
- Femoral and dorsal pedal arterial pulses: strong and regular
- Doppler blood pressure: 140 mmHg systolic
- Abdominal palpation: muscle tensing and vocalization on cranial abdominal palpation
- Temperature: 39.4°C (102.9°F)

POCUS exam(s) to perform:
- Abdominal POCUS
- Pancreatic assessment

Abnormal POCUS exam results:
See **Figures 15.4.1** and **15.4.2**.

Additional point-of-care diagnostics and initial management:
Point-of-care blood tests revealed a packed cell volume (PCV) of 55% and total solids of 82 g/L, and a venous blood gas analysis was unremarkable. Analgesia was provided with intravenous methadone 0.2 mg/kg. In addition, 5 mL/kg Hartmann's solution fluid was administered as a bolus over 15 min.

QUESTIONS AND ANSWERS

1. What are the differentials to rule in or out with POCUS based on history and physical exam?
2. What are the sonographic findings?
3. What is the sonographic diagnosis?
4. If necessary, what additional sonographic examination or findings would help rule in or out the differential diagnoses?

DOI: 10.1201/9781003436690-80

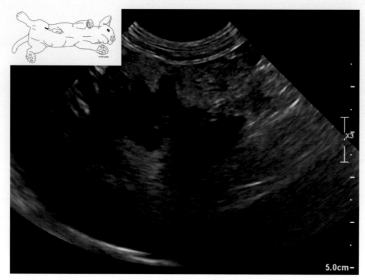

FIGURE 15.4.1 Pancreatic assessment. The patient is positioned in left lateral recumbency with the probe positioned longitudinally and the marker directed cranially, using a microconvex curvilinear probe with the depth set at 5 cm and the frequency set at 30 Hz. Image is of the body of the pancreas, visible from tracing the right limb of the pancreas and sweeping the ultrasound probe medially.

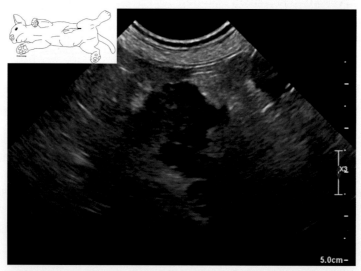

FIGURE 15.4.2 Pancreatic assessment. The patient is positioned in right lateral recumbency with the probe positioned longitudinally and the marker directed cranially, using a microconvex curvilinear probe with the depth set at 5 cm and the frequency set at 30 Hz. Image is of the left limb of the pancreas, located by tracing from the body of the pancreas toward the left lateral abdomen.

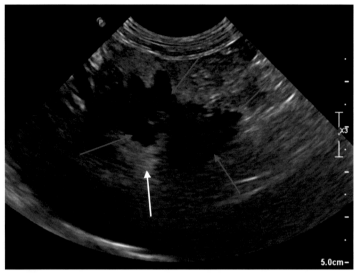

FIGURE 15.4.3 Annotated version of Figure 15.4.1, with the red arrows indicating the body of the pancreas and the white arrow indicating the location of hyperechoic fat. Note the hypoechogenicity of the inflamed pancreatic parenchyma.

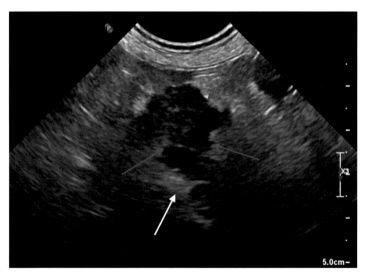

FIGURE 15.4.4 Annotated version of Figure 15.4.2, with the red arrows indicating the left limb of the pancreas and the white arrow indicating the location of hyperechoic fat. Note the hypoechogenicity of the inflamed pancreatic parenchyma.

1. **Differential diagnosis to rule in or rule out with POCUS:**
   - Gastrointestinal tract obstruction (mechanical vs functional), septic peritonitis, pancreatitis, biliary disease (cholangiohepatitis, gallbladder mucocele, bile peritonitis), prostatitis, abdominal organ volvumlus/torsion (gastrointestinal tract, spleen, retained testical, etc), neoplasia.

2. **Describe your sonographic findings:**
   - Irregular delineation and hypoechogenicity of the body and left lobe of

the pancreas (**Figures 15.4.3** and **15.4.4**).
- Hyperechoic fat surrounding the affected areas of the pancreas (**Figures 15.4.3** and **15.4.4**).
- Normal right pancreatic lobe appearance.
- No evidence of free abdominal fluid or gas.

3. **What is your sonographic diagnosis?**
   Pancreatitis with evidence of focal peritonitis.
4. **What additional sonographic examination or finding would help you rule in or rule out your differential diagnoses (if necessary)?**
   - Gallbladder assessment would be expected to show a normal wall and gallbladder contents. Gallbladder sludge is an incidental finding on abdominal ultrasonography in up to 67% of healthy dogs.[1]
   - Pancreatitis is a common cause of extrahepatic biliary tract obstruction in cats and dogs and surgery may be required. Dilation of the common bile duct can be detected ultrasonographically.[2]
   - Gastrointestinal assessment may reveal evidence of inflammation including gastric wall edema and duodenal corrugation, secondary to pancreatitis causing local inflammation.
   - Serial abdominal POCUS is used to monitor for the development of free abdominal fluid. If peritoneal effusion is present then, when possible, a sample should be obtained for fluid analysis.
   - Complete abdominal ultrasound examination to evaluate for further abdominal pathology.

CLINICAL INTEGRATION

Suspected diagnosis based on ultrasound findings:
The history and physical examination findings alongside the pancreatic changes on abdominal POCUS, and the absence of other sonographic abnormalities, raise the index of suspicion for acute pancreatitis.

Sonographic interventions, monitoring, and outcome:
- A complete abdominal ultrasound scan was performed and confirmed the sonographic diagnosis of pancreatitis.
- Serial abdominal POCUS and a pancreatic assessment were performed daily during hospitalization, and no change in pancreatic findings or the progression of peritonitis was noted. The result of a quantitative ELISA specific for canine pancreatic lipase (SpecCPLi) was 682 μg/L, with results ≥400 μg/L considered consistent with pancreatitis.[3]
- The dog received symptomatic treatment and supportive care and was discharged on Day 5 of hospitalization.

Important concepts regarding the sonographic diagnosis of pancreatitis:
- Normal appearance of the pancreas on sonographic examination does not rule out pancreatitis. Ultrasonographic signs of pancreatitis are not always present at the time of presentation and may develop later.[3]
- Alterations from the normal sonographic appearance of the pancreas include enlargement of the pancreas, heterogeneous parenchyma, and hyper- or hypoechogenicity.[4,5]
- Hypoechoic parenchyma, hyperechoic mesentery, and peripancreatic free fluid are sonographic alterations associated with acute pancreatitis.[4,5]

- Pathology in adjacent organs may be present secondary to pancreatitis and observed during ultrasound examination.
- Alterations seen in the duodenum of dogs can include poor layer definition, ileus, and corrugation.[6]
- Evidence of gastric wall edema, visualized as gastric wall thickening which may be generalized or focal, can be present in cases of acute pancreatitis.[7]
- Pancreatitis can lead to extra-hepatic bile duct obstruction, which can be visualized ultrasonographically as bile duct dilatation >3 mm in diameter.[6]

Techniques and tips to identify the appropriate ultrasound images/views:

- To evaluate the pancreas, the patient can be standing or positioned in lateral, ventral, or dorsal recumbency.[8]
- The normal pancreas in a Beagle-sized dog is approximately 1–3 cm wide and 1 cm thick. In cats, the right lobe is normally 0.3–0.6 cm thick, with the body and left lobe 0.5–0.9 cm thick.
- The echogenicity of the pancreatic parenchyma is isoechoic or slightly hypoechoic to that of the surrounding mesenteric fat; however, the capsular serosa appears hyperechoic.[9]
- The pancreaticoduodenal vein is more apparent in dogs, whereas the pancreatic duct is conspicuous in cats.
- Pancreatic location in dogs:
- The right lobe is most easily identified.[9] Place the probe in transverse orientation on the right abdominal wall and locate the right kidney. In deep-chested dogs an intercostal approach may be required. Using the right kidney as a landmark, scan ventrally to obtain a transverse view of the duodenum, which is located close to the body wall, either medial or lateral to the right kidney. Compared to the adjacent jejunal segments, the descending duodenum is larger in diameter and follows a more linear cranial-to-caudal path when viewed in the sagittal plane. In the transverse plane, the triangular right lobe of the pancreas lies dorsomedial to the descending duodenum, ventral to the right kidney, and lateral to the portal vein.[8] The pancreaticoduodenal vein and artery can be visualized in the center of the pancreas and show flow signal on Doppler examination.
- The body of the pancreas can be traced from the right lobe, caudo-dorsal to the pylorus, near the portal vein.[9]
- The left lobe lies between the stomach, transverse colon, and spleen.[9]
- Pancreatic location in cats:
- The left lobe and the body of the pancreas are most readily identified. With the probe in transverse orientation on the cranial abdomen, the portal vein is located and traced toward the caudal margin of the stomach. The body is positioned ventral to the portal vein.[9]
- The left lobe is located cranial to the transverse colon and caudal to the stomach. It follows the course of the splenic vein from the portal vein medially toward the spleen.[8] The pancreatic duct can be visualized and lacks Doppler flow.[10]
- The right lobe is situated medially to the descending duodenum, found by tracing caudally from the body.[8]

Take-home messages

- Pancreatitis is diagnosed clinically based on compatible history, physical examination, blood biochemical analysis, and diagnostic imaging findings.[3]
- Initial pancreatic sonographic examination may be normal in cases of acute pancreatitis. Repeat sonography is recommended if there is a high index of suspicion for pancreatitis.

- Pancreatitis can lead to alterations in the ultrasonographic appearance of adjacent abdominal organs.
- Patients presenting with findings compatible with pancreatitis and peritonitis should have a full abdominal imaging evaluation to rule out other causes of acute abdomen.
- If peritoneal effusion is present, then an abdominocentesis and fluid analysis/cytology should be performed to rule out surgical disease.

REFERENCES

1. Butler T, Bexfield N, Dor C, et al. A multicentre retrospective study assessing progression of biliary sludge in dogs using ultrasonography. *J Vet Intern Med* Apr 2022;36(3):976–985.
2. Wilkinson AR, DeMonaco SM, Panciera DL, et al. Bile duct obstruction associated with pancreatitis in 46 dogs. *J Vet Intern Med* 2020;34(5):1794–1800.
3. Cridge H, Twedt DC, Marlof AJ, et al. Advances in the diagnosis of acute pancreatitis in dogs. *J Vet Intern Med* Nov 2021;35(6):2572–2587.
4. Cridge H, Sullivant AM, Wills RW and Lee AM. Association between abdominal ultrasound findings, the specific canine pancreatic lipase assay, clinical severity indices, and clinical diagnosis in dogs with pancreatitis. *J Vet Intern Med* Mar 2020;34(2):636–643.
5. Gori E, Pierini A, Lippi I, et al. Evaluation of diagnostic and prognostic usefulness of abdominal ultrasonography in dogs with clinical signs of acute pancreatitis. *J Am Vet Med Assoc* Sep 2021;259(6):631–636.
6. Wilkinson AR, DeMonaco SM, Panciera DL, et al. Bile duct obstruction associated with pancreatitis in 46 dogs. *J Vet Intern Med* Aug 2020;34(5):1794–1800.
7. Murakami M, Heng HG, Lim CK, et al. Ultrasonographic features of presumed gastric wall edema in 14 dogs with pancreatitis. *J Vet Intern Med* Apr 2022;33(3):1260–1265.
8. Mattoon JS and Nyland TG. *Small Animal Diagnostic Ultrasound*, 4th ed., Elsevier Inc, 2020.
9. Hecht S and Henry G. Sonographic evaluation of the normal and abnormal pancreas. *Clin Tech Small Anim Pract* Aug 2007;22(3):115–121.
10. Etue SM, Penninck DG, Labato MA. et al. Ultrasonography of the normal feline pancreas and associated anatomic landmarks: A prospective study of 20 cats. *Vet Radiol Ultrasound* 2001;42(4):330–336.

CHAPTER 15 – CASE 5

AUSTRALIAN SHEPHERD THAT IS UNRESPONSIVE FOLLOWING A MOTOR VEHICLE ACCIDENT 30 MINUTES PRIOR TO PRESENTATION

Mark W Kim and Alexandra Nectoux

HISTORY, TRIAGE, AND STABILIZATION

A 4-year-old female spayed Australian Shepherd dog presented unresponsive following a motor vehicle accident 30 min prior to presentation.

Triage exam findings (vitals):
- Mentation: unconscious
- Respiratory rate: apnea
- Cardiovascular: no palpable femoral pulse
- Oral cavity: white mucous membranes

Cardiopulmonary resuscitation (CPR) was started immediately based on a strong suspicion of cardiopulmonary arrest.

POCUS exam(s) to perform:
- Abdominal POCUS

Abnormal POCUS exam results:
See **Figure 15.5.1** and **Video 15.5.1**.

QUESTIONS AND ANSWERS

1. What are the differentials to rule in or out with POCUS based on history and physical exam?
2. What are the sonographic findings?
3. What is the sonographic diagnosis?
4. If necessary, what additional sonographic examination or findings would help rule in or out the differential diagnoses?

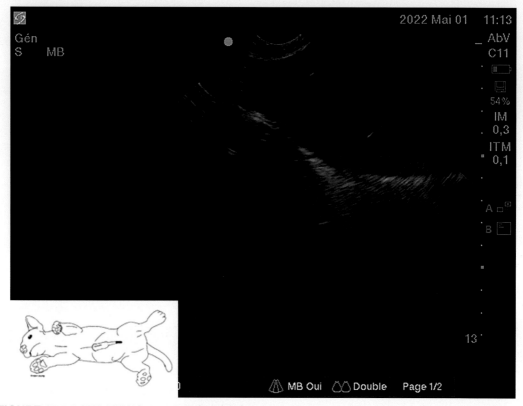

FIGURE 15.5.1 AND VIDEO 15.5.1 Subxiphoid view of the heart taken with the dog in lateral position during chest compressions. Alternatively, this can be performed during rhythm-check pauses. Image and video taken with a 7.5 MHz microconvex curvilinear probe placed longitudinally, the marker directed cranially. The fur was not clipped and gel was applied to the area.

1. **Differential diagnosis to rule in or rule out with POCUS:**
   - Cardiac abnormalities: pericardial effusion and cardiac tamponade, cardiac standstill, pulseless electrical activity.
   - Non-cardiac abnormalities: hemothorax, hemoperitoneum, pneumothorax.

2. **Describe your sonographic findings:**
   - Anechoic rim at apical margin of heart (**Figures 15.5.2** and **15.5.3**).
   - B-lines along the caudal part of the diaphragm (**Figure 15.5.3**)

3. **What is your sonographic diagnosis?**
   Pericardial effusion.

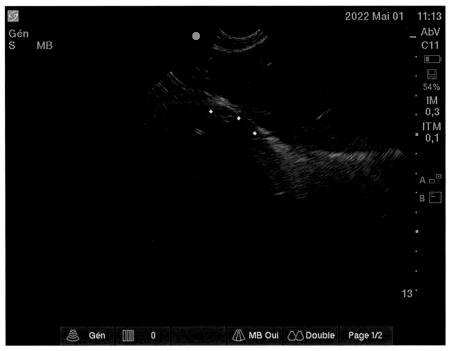

FIGURE 15.5.2 Labeled version of Figure 15.5.1. An anechoic rim (*) at the apical margin of the heart becomes evident between the 15 and 21-sec mark of the cineloop.

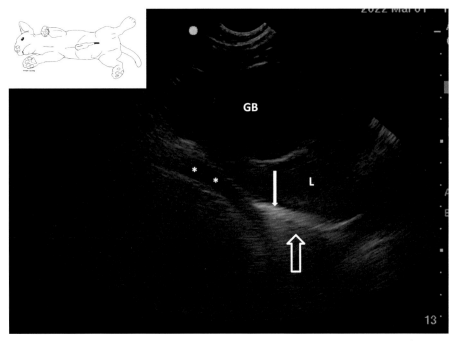

FIGURE 15.5.3 Still image from Video 15.5.1. Hyperechoic line of the diaphragm (solid arrow) separating the liver (L) and gallbladder (GB) from the thoracic cavity. Anechoic rim (*) following the curvature of the heart. B-lines along the caudal part of the diaphragm (hollow arrow).

CLINICAL INTEGRATION

Suspected diagnosis based on ultrasound findings:
Acute traumatic hemopericardium due to a motor vehicle accident.

Sonographic interventions, monitoring, and outcome:
- Emergency pericardiocentesis was performed as another team member was obtaining permission from the client for open-chest cardiopulmonary resuscitation.
- Pericardiocentesis was performed with minimal preparation to minimize the time to start effective CPR.
- Despite successful pericardiocentesis and a reduction in hemopericardium volume, CPR was unsuccessful and the clients declined open-chest CPR.

Important concepts regarding the sonographic diagnosis of pericardial effusion during CPR:
- Identifying pericardial effusions during CPR can dramatically change the prognosis and clinical outcome. It is one of the few indications for open-chest CPR, as closed chest compressions are considered ineffective. Although pericardiocentesis may theoretically restore the effectiveness of closed-chest compressions, the delay in CPR can considerably diminish the chance of the return of spontaneous circulation (ROSC). Excellent client communication is always desirable in these situations.
- Clinicians must be aware that POCUS during CPR is currently not recommended in the RECOVER guidelines. Given the absolute requirement for high-quality basic life support, the authors' experience is that POCUS can very easily disrupt CPR and distract clinicians from basic life support.[1] For this reason, we do not recommend routine POCUS during CPR until further supportive evidence is published.
- Human studies show the potential benefit in evaluating the quality of cardiac compressions by visually guiding the site of compressions toward the heart apex (ventricles) and away from the heart base (outflow tract).[2] Equally, POCUS has been used to distinguish cardiac standstill from true pulseless electrical activity in conjunction with electrocardiogram (ECG) interpretation. These studies have demonstrated significantly poorer ROSC rates in patients with cardiac standstill versus those with pulseless electrical activity that continue to show myocardial activity on POCUS.[3,4]

Techniques and tips to identify the appropriate ultrasound images/views:
- Firm pressure under the xiphoid is often needed to obtain a sufficiently shallow angle to image the heart.
- During CPR, avoid using isopropyl alcohol as a coupling agent as this may be hazardous if a defibrillator is used during CPR.

Limitations and risks:
- The POCUS techniques, performance, and utility have not been investigated in veterinary patients. The interpretation and use of POCUS images during CPR may lead to decisions that harm the patient.
- Using POCUS during CPR may interfere with and reduce the quality of basic and advanced life support, the core of evidence-based CPR.[1]
- In the authors' experience, the ability to obtain meaningful POCUS images during chest compressions is highly variable (**Video 15.5.1**). While somewhat easier during rhythm-check pauses, the

limited timeframe keeps it a challenging proposition.
- There is no evidence from human studies demonstrating an outcome benefit with POCUS during CPR.[5]

Take-home message

POCUS during CPR has the potential to improve outcomes in veterinary patients. Until further studies are performed, POCUS is not recommended as a routine procedure during CPR, and, if performed, must not interfere with the quality of basic and advanced life support.

REFERENCES

1. Huis In 't Veld MA, Allison MG, Bostick DS, et al. Ultrasound use during cardiopulmonary resuscitation is associated with delays in chest compressions. Resuscitation. 2017;119:95–98. doi:10.1016/j.resuscitation.2017.07.021
2. Hwang SO, Zhao PG, Choi HJ, et al. Compression of the left ventricular outflow tract during cardiopulmonary resuscitation. Academic Emergency Medicine. 2009;16(10):928–933. doi:10.1111/j.1553-2712.2009.00497.x
3. Flato UAP, Paiva EF, Carballo MT, Buehler AM, Marco R, Timerman A. Echocardiography for prognostication during the resuscitation of intensive care unit patients with non-shockable rhythm cardiac arrest. Resuscitation. 2015;92:1–6. doi:10.1016/j.resuscitation.2015.03.024
4. Tsou PY, Kurbedin J, Chen YS, et al. Accuracy of point-of-care focused echocardiography in predicting outcome of resuscitation in cardiac arrest patients: A systematic review and meta-analysis. Resuscitation. 2017;114:92–99. doi:10.1016/j.resuscitation.2017.02.021
5. Panchal AR, Bartos JA, Cabañas JG, et al. Part 3: Adult basic and advanced life support: 2020 American Heart Association guidelines for cardiopulmonary resuscitation and emergency cardiovascular care. Circulation. 2020;142(16_suppl_2):S366–S468. doi:10.1161/CIR.0000000000000916

CHAPTER 15 – CASE 6

GREAT DANE WITH A 3-DAY HISTORY OF LETHARGY AND PIGMENTURIA FOLLOWING A SINGLE EPISODE OF VOMITING 24 HOURS PREVIOUSLY

Dave Beeston and Stefano Cortellini

HISTORY, TRIAGE, AND STABILIZATION

A 2-year-old, male intact Great Dane presented with a 3-day history of lethargy and pigmenturia. There was a single episode of vomiting on the day of presentation.

Triage exam findings (vitals):
- Mentation: alert and responsive
- Respiratory rate: 20 breaths per minute with normal effort
- Heart rate and auscultation: 130 beats per minute with normal sinus rhythm and no audible murmur
- Mucous membranes: pale pink, 3 sec capillary refill time
- Femoral and dorsal pedal arterial pulses: hyperdynamic and synchronous
- Temperature: 37.7°C (99.9°F)
- Abdominal palpation: the abdomen was painful on palpation and a mass effect was felt in the cranial abdomen.

POCUS exam(s) to perform:
- Abdominal POCUS

Abnormal POCUS exam results:
See **Figures 15.6.1–15.6.3** and **Video 15.6.1–15.6.3**.

Additional point-of-care diagnostics and initial management:
Packed cell volume and total solids were 26% and 89 g/L, respectively, with hemolyzed serum. Emergency venous blood gas and electrolyte analysis documented hyperlactatemia (3.4 mmol/L; 0.6–2.5). Coagulation times (prothrombin time/activated partial thromboplastin time [PT/aPTT]) were off-scale high. In-house blood smear analysis documented a regenerative anemia and ghost cells with occasional schistocytes present. Platelets were within normal limits.

The patient received 0.2 mg/kg of methadone intravenously and 15 mL/kg of compound sodium lactate over 15 min prior to the initiation of a fresh frozen plasma transfusion.

DOI: 10.1201/9781003436690-82

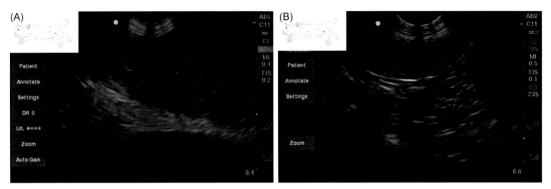

FIGURE 15.6.1 AND VIDEO 15.6.1 (A) Abdominal POCUS at the left paralumbar view using an 8.5 MHz curvilinear probe. Image taken with probe placed in longitudinal orientation with marker directed toward the head. (B) A normal splenic anatomy for comparison.

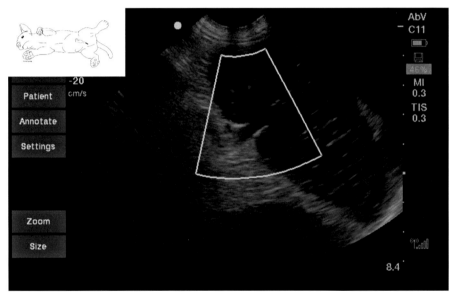

FIGURE 15.6.2 AND VIDEO 15.6.2 Abdominal POCUS at the left paralumbar view using an 8.5 MHz curvilinear probe. Image taken with probe placed in longitudinal orientation with marker directed toward the head. Color Doppler has been placed over the splenic vein.

QUESTIONS AND ANSWERS

1. What are the differentials to rule in or out with POCUS based on history and physical exam?
2. What are the sonographic findings?
3. What is the sonographic diagnosis?
4. If necessary, what additional sonographic examination or findings would help rule in or out the differential diagnoses?

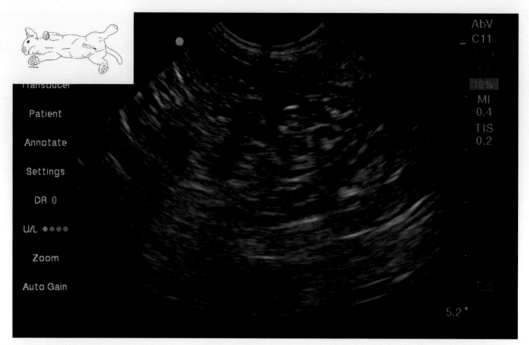

FIGURE 15.6.3 AND VIDEO 15.6.3 Abdominal POCUS at the left paralumbar view using an 8.5 MHz curvilinear probe. Image taken with probe placed in longitudinal orientation with marker directed toward the head.

1. **Differential diagnosis to rule in or rule out with POCUS:**
   Splenic torsion, hemoabdomen (secondary to splenic or hepatic mass), less likely intra-abdominal mass or abscessation and septic peritonitis.
2. **Describe your sonographic findings:**
   - Abnormally positioned, diffusely hypoechoic, enlarged spleen with a coarse "lacy" texture surrounded by hyperechoic mesentery (**Figures 15.6.1, 15.6.3,** and **15.6.4** and **Videos 15.6.1** and **15.6.3**).
   - Absent flow on Doppler ultrasonography of the splenic vein (**Figure 15.6.2** and **15.6.4** and **Video 15.6.2**).
   - Abdominal fluid score 1/4 (**Figure 15.6.3** and **Video 15.6.3**).
3. **What is your sonographic diagnosis?**
   Splenic torsion with peritoneal effusion.
4. **What additional sonographic examination or finding would help you rule in or rule out your differential diagnoses (if necessary)?**
   - Formal abdominal ultrasound and assessment of splenic vasculature for thrombotic disease and confirmation of splenic torsion.
   - Abdominocentesis may be performed to characterize the nature of the abdominal effusion and rule in or out hemoabdomen or septic peritonitis.

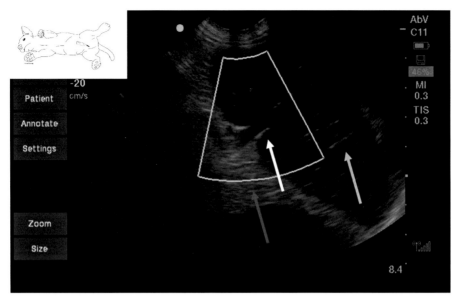

FIGURE 15.6.4 Labeled image of 15.6.2. The spleen is diffusely hypoechoic with a coarse lacy texture (blue arrow) surrounded by hyperechoic mesentery (red arrow). Color Doppler of the splenic vein shows absent flow (white arrow).

CLINICAL INTEGRATION

Suspected diagnosis based on ultrasound findings:
Splenic torsion.

Sonographic interventions, monitoring, and outcome:
- Complete abdominal ultrasound performed by an ultrasonographer confirmed splenic torsion without obvious thromboembolic disease.
- The patient received a fresh frozen plasma transfusion to correct the underlying coagulopathy, likely related to disseminated intravascular coagulopathy (DIC), prior to exploratory laparotomy and splenectomy.
- The patient made a full recovery.

Important concepts regarding the sonographic diagnosis of splenic torsion:
- Splenic torsion is a rare condition in dogs and occurs when the spleen rotates around the gastrosplenic and phrenicosplenic ligaments.[1] The underlying pathophysiology is not entirely understood; however, splenic torsion has been associated with concurrent abdominal pathology such as gastric dilation and volvulus.[2]
- Clinical signs can be vague and non-specific including lethargy, anorexia, vomiting, altered urination, and collapse. Roughly 70% of dogs present with a palpably enlarged spleen, and one-third of patients will display signs consistent with hypovolemic shock.[1]
- Although CT is considered the gold standard for the diagnosis of splenic torsion in people,[3] CT is not always readily available in veterinary primary-care practice.
- Ultrasonographic findings of splenic torsion in dogs include splenomegaly, splenic venous thrombosis, absent or minimal intra-splenic blood flow on color Doppler, diffuse hypoechogenic splenic parenchyma

with short and linear lacy echogenic separations ("starry night" pattern), and a hilar perivenous hyperechoic triangle.[1,4,5]
- Free abdominal fluid is commonly found, although not all effusion is hemorrhagic in origin, and abdominocentesis should be performed where possible to rule out septic effusions due to rupture of the abdominal viscera.[1] Interestingly, a septic peritonitis might be concurrent with splenic torsion and is associated with a higher risk of mortality in these cases.[1]
- Splenectomy is the treatment of choice for splenic torsion and the outcome is favorable if treated early in the disease course with a reported mortality rate of 8.8%.[1]

Techniques and tips to identify the appropriate ultrasound images/views:
- Splenomegaly is commonly appreciated in splenic torsion, but the spleen can be difficult to recognize due to its irregular hypoechoic parenchyma and starry night pattern.
- Use and interpretation of color Doppler ultrasonography can be challenging at first, and so finding the splenic vein prior to the initiation of Doppler can be helpful. The splenic vein can be found by placing the probe behind the last rib and fanning through the spleen from dorsal to ventral. Once the splenic vein has been identified, its long axis can be found through a combination of rocking and rotating.

Take-home messages
- Splenic torsion can be readily identified during a standard five-point abdominal POCUS exam and has a characteristic sonographic appearance.
- Use of color Doppler ultrasound can help raise suspicion of splenic torsion by identifying absent flow through splenic vasculature.
- Splenic torsion has been associated with DIC; therefore, coagulation testing prior to surgical exploration and treatment may be warranted.[6]

REFERENCES

1. DeGroot W, Giuffrida MA, Rubin J, et al. Primary splenic torsion in dogs: 102 cases (1992–2014). *J Am Vet Med Assoc* 2016; 248(6): 661–668.
2. Song KK, Goldsmid SE, Lee J, et al. Retrospective analysis of 736 cases of canine gastric dilatation volvulus. *Aust Vet J* 2020; 98(6): 232–238.
3. Bough GM, Gargan KE, Cleeve SJ, et al. Diagnosis, management and outcome of splenic torsion; a systematic review of published studies. *Surg* 2021. DOI: 10.1016/J.SURGE.2021.08.006.
4. Mai W. The hilar perivenous hyperechoic triangle as a sign of acute splenic torsion in dogs. *Vet Radiol Ultrasound* 2006; 47(5): 487–491.
5. Saunders HM, Neath PJ, Brockman DJ. B-mode and Doppler ultrasound imaging of the spleen with canine splenic torsion: A retrospective evaluation. *Vet Radiol Ultrasound* 1998; 39(4): 349–353.
6. Stoneham AE, Henderson AK, O'Toole TE. Resolution of severe thrombocytopenia in two standard Poodles with surgical correction of splenic torsion. *J Vet Emerg Crit Care* 2006; 16(2): 131–135.

APPENDIX: TABLE OF CONTENTS BY DIAGNOSIS

SECTION AND TITLE (NOTE: TO SEARCH BY DIAGNOSIS SEE INDEX AT THE BACK OF THE BOOK)	Page
Section 1: Introduction to Veterinary Point-of-Care Ultrasound	1
Chapter 1: **Basic Ultrasound Physics**	3
Chapter 2: **Binary Question Approach to Point-of-Care Ultrasound**	10
Chapter 3: **Basics of Image Acquisition for Point-of-Care Ultrasound**	13
Section 2: Pleural Space and Lung Ultrasound (PLUS) – section editor Boysen	17
Chapter 4: **Introduction to Pleural Space and Lung Ultrasound (PLUS)**	19
Chapter 5: **Pleural and Lung Ultrasound (PLUS) Cases**	35
Case 5.1: **Aspiration Pneumonia**	36
Case 5.2: **Hemothorax**	42
Case 5.3: **Pneumothorax (Absent Lung Sliding)**	48
Case 5.4: **Cardiogenic Pulmonary Edema (myxomatous mitral Valve Disease (MMVD) and Left-Sided Congestive Heart Failure (CHF))**	53
Case 5.5: **Pulmonary Metastatic Neoplasia**	59
Case 5.6: **Post-Operative Atelectasis**	64
Case 5.7: **Acute Respiratory Distress Syndrome (B-lines)**	70
Case 5.8: **Fungal Pneumonia**	75
Case 5.9: **Dilated Cardiomyopathy And Right-Sided Congestive Hart Failure**	79
Case 5.10: **Pulmonary Abscesses**	87
Case 5.11: **Pyothorax**	91
Case 5.12: **Pulmonary Contusions**	97
Case 5.13: **Negative Pressure Pulmonary Edema**	101
Case 5.14: **Pulmonary Thromboembolism**	107
Case 5.15: **Pulmonary Metastatic Neoplasia**	116
Case 5.16: **Pneumothorax (Abnormal Curtain Sign)**	122

Case 5.17: **Acute Respiratory Distress Syndrome (Lung Consolidation)**	127
Case 5.18: **Canine Idiopathic Pulmonary Fibrosis**	132
Case 5.19: **Pneumothorax (Lung Point)**	137
Case 5.20: **Atelectasis Due to Recumbency**	142
Case 5.21: **Hemothorax (Small Volume)**	147

Section 3: Veterinary Cardiac Point-of-Care Ultrasound – Section Editor Stastny — 153

Chapter 6: **Introduction to Veterinary Cardiac Point-of-Care Ultrasound**	155
Chapter 7: **Cardiac Point-of-Care Ultrasound Cases**	162
Case 7.1: **MMVD and Left-Sided CHF**	163
Case 7.2: **Idiopathic Pericardial Effusion (Without Tamponade)**	168
Case 7.3: **Left Atrial Rupture**	174
Case 7.4: **Pericardial Effusion And Cardiac Tamponade**	181
Case 7.5: **Heartworm Disease And Caval Syndrome**	187
Case 7.6: **Congestive Heart Failure Due to Hypertrophic Cardiomyopathy (Feline)**	192
Case 7.7: **Dilated Cardiomyopathy and Right-Sided CHF**	198
Case 7.8: **Pericarditis (Coccidioidomycosis)**	206
Case 7.9: **Severe Pulmonary Hypertension**	213
Case 7.10 Traumatic cardiac tamponade	220

Section 4: Vascular Applications of Veterinary Point-of-Care Ultrasound - Section Editor Stastny — 227

Chapter 8a: **Introduction to Ultrasound Guided Vascular Access**	229
Chapter 8b: **Introduction to Caudal Vena Cava Collapsibility Index**	234
Chapter 9: **Vascular Point-of-Care Cases**	238
Case 9.1: **Central Line Placement**	239
Case 9.2: **Hypovolemia**	246
Case 9.3: **Thrombophlebitis**	253
Case 9.4: **Volume Overload**	257

Section 5: Abdominal Point-of-Care Ultrasound - Section Editor Binagia — 265

Chapter 10: **Introduction to Abdominal Point-of-Care Ultrasound**	267
Chapter 11: **Abdominal Point-of-Care Ultrasound Cases**	273
Case 11.1: **Cystolithiasis**	274
Case 11.2: **Gallbladder Mucocele and Rupture**	278
Case 11.3: **Retroperitoneal Effusion Due to Pheochromocytoma**	283

Case 11.4: **Pyelonephritis**	288
Case 11.5: **Pyometra**	294
Case 11.6: **Septic Abdomen**	299
Case 11.7: **Splenic Mass and Spontaneous Hemoabdomen**	305
Case 11.8: **Dystocia**	312
Case 11.9: **Free Peritoneal Gas**	318
Case 11.10: **Gallbladder Wall Edema (Thickening)**	324
Case 11.11: **Uroabdomen Due to Bladder Rupture**	329
Case 11.12: **Ureteral Obstruction**	333
Case 11.13: **Urinary Volume Estimation**	339
Case 11.14: **Leptospirosis**	343
Case 11.15: **Gastrointestinal Ileus**	348
Case 11.16: **Traumatic Hemoabdomen**	354

Section 6: Miscellaneous Point-of-Care Ultrasound Cases- Section Editor Binagia — 361

Chapter 12: **Introduction to Miscellaneous Point-of-Care Ultrasound Cases**	363
Chapter 13: **Miscellaneous Point-of-Care Ultrasound Cases**	364
Case 13.1: **Intracranial Hypertension (Optic Nerve Sheath Diameter)**	365
Case 13.2: **Tracheal Tear (Subcutaneous Emphysema)**	370

Section 7: Advanced Multisystem Point-of-Care Ultrasound Cases - Section Editor Binagia — 375

Chapter 14: **Introduction to Systemic Point-of-Care Ultrasound Cases**	377
Chapter 15: **Systemic Point-of-Care Ultrasound Cases**	378
Case 15.1: **Anaphylaxis**	379
Case 15.2: **Intussusception Due to Intestinal Parasitism**	386
Case 15.3: **Gastrointestinal Foreign Body Obstruction**	391
Case 15.4: **Pancreatitis**	397
Case 15.5: **Cardiopulmonary Resuscitation**	403
Case 15.6: **Splenic Torsion**	408

INDEX

Acute kidney injury
    Due to Leptospirosis (Case 11.13 and 11.14), 339, 343
    Due to pyelonephritis (Case 11.4), 288
    Due to ureteral obstruction (Case 11.12), 333
Actinomyces spp, 95
Anaphylaxis (Case 15.1), 379
Atelectasis
    Post-op (Case 5.6), 64
    Recumbency (Case 5.20), 142

Bladder
    Measurements (Case 11.13), 270, 339
    Stones (Case 11.1), 274
    Rupture (Case 11.11), 329
Blastomycosis, 76, 77
Binary question, 10, 268, 269

Cardiac tamponade, 184
Cardiopulmonary resuscitation (Case 15.5), 403
Caval syndrome (Case 7.5), 187
Coccidioidomycosis (*Coccidioides immitis*), 206, 212
Congestive heart failure (CHF)
    Left-sided, Feline Hypertrophic Cardiomyopathy (HCM) (Cardiac POCUS) (Case 7.6,) 192
    Left-sided, Myxomatous mitral valve disease (MMVD) (PLUS) (Case 5.4), 53
    Left-sided, MMVD (Cardiac POCUS) (Case 7.1), 163
    Right-sided, Dilated cardiomyopathy (DCM) (PLUS) (Case 5.9), 79
    Right-sided, DCM (Cardiac POCUS) (Case 7.7), 198
    Right-sided, constrictive pericarditis (Case 7.8), 206
Central line placement (Case 9.1), 239
Cystolithiasis (see bladder stones)

Dilated cardiomyopathy (DCM) (Case 5.9 and 7.7), 79, 198
Dystocia (Case 11.8), 312

Effusion
    Pericardial effusion
        Cardiac tamponade (hemangiosarcoma, Case 7.4), 181
        Cardiopulmonary Resuscitation (Case 15.5), 403
        Idiopathic (no tamponade, Case 7.2), 168
        Left atrial rupture (and pericardial thrombus) (Case 7.3), 174
        Pericarditis (constrictive, fungal) (Case 7.8), 206
        Right heart failure (Case 5.9), 79

Peritoneal effusion
    Anaphylaxis (Case 15.1), 379
    Bile peritonitis (Case 11.2), 278
    Congestive heart failure (ascites) (Case 7.7) (Cardiac POCUS), 198
    Hemoabdomen (Case 11.7 (splenic mass) and Case 11.16 (trauma)), 305, 354
    Leptospirosis (Case 11.14), 343
    Septic abdomen (Case 11.6 and Case 11.9), 299, 318
    Splenic torsion (Case 15.6), 408
    Uroabdomen (Case 11.11), 329
Pleural effusion
    Congestive heart failure (Case 5.9, 7.6, 7.8), 79, 192, 206
    Hemothorax (penetrating injury) (Case 5.2), 42
    Hemothorax (Small volume) (Case 5.21), 147
    Left atrial rupture (Case 7.3), 174
    Pyothorax (Case 5.11), 91
    Volume overload, 263
Retroperitoneal effusion
    Perirenal effusion/fluid, 345–347
    Pheochromocytoma (Case 11.3), 283
Scant effusion, 322, 347

Fetal heart rate (cardiac activity) (Case 11.8), 312
Fluid overload
    Volume overload (Vascular POCUS) (Case 9.2), 246
    Leptospirosis (Abdominal POCUS) (Case 11.14), 343

Gallbladder
    Mucocele (Case 11.2), 278
    Wall edema/thickening
        due to anaphylaxis (Case 15.1), 379
        due to dexmedetomidine (Case 11.10), 324
        due to thromboembolic disease (Case 5.14), 107
        Pathophysiology, 411
    Pseudothickening, 327
Gastric
    Distension, 348, 349, 351–353
    Delayed emptying, 352
    Residual volume, 271, 352
    Rugal folds, 349–352

Hemoabdomen (see peritoneal effusion)
    Confirmation, 359
    Spontaneous, 381, 384
    Traumatic, 359
Heartworms, 189–191

HeLP score, 309
Hypertrophic Cardiomyopathy (HCM) (Case 7.6), 192
Hypervolemia (see fluid overload)
Hypovolemia
    Cardiac evidence – 93, 300, 303, 352, 355, 359
    Vascular evidence – 248, 250, 251, 253, 269

Ileus
    Functional (Case 11.5), 294
    Mechanical (Case 15.3), 391, 395
Intestinal
    Contraction, 352
    Dilation (distension), 348, 391
    intussusception (Case 15.2), 386
        Intussusceptum, 389
        Intussuscipiens, 387, 389
    foreign body obstruction (Case 15.3), 391
    measurements, 270, 392, 393
    motility (see ileus)
    Peristalsis, 349-352
Intracranial hypertension (Case 13.1), 365

Leptospirosis (see acute kidney injury)

Myxomatous mitral valve degeneration (MMVD) (Case 5.4, 7.1, 7.3), 53, 163, 174

Nephritis (due to Leptospirosis) (Case 11.14), 343

Optic nerve sheath diameter (Case 13.1), 365-358

Pancreas (normal), 270, 399
Pancreatitis (Case 15.4), 270, 397
Pericarditis (Case 7.8), 206
Pheochromocytoma (Case 11.3), 283
Pneumonia
    Aspiration (Case 5.1), 36
    Fungal (Case 5.8), 75
Pneumoperitoneum, 301, 321, 322
Pneumothorax
    Absent lung sliding (Case 5.3), 48
    Abnormal curtain sign (Case 5.3), 48
    Abnormal curtain sign (Case 5.16), 122
    Lung point (Case 5.19), 137
POCUS – Abdominal (Section 5), 263
    Abdominal fluid score, 187, 220, 268, 286, 301, 303, 306, 355, 359, 381, 411
    Assessment of
        Caudal vena cava, 81
        Gallbladder, 208, 280, 324, 326, 327, 379
        Gastrointestinal, 348, 386, 391, 400
        Liver, 306
        Kidneys, 333
        Pancreatic, 393, 397, 398
        Splenic, 305
        Stomach, 349-351
        Urinary bladder, 274, 294, 339, 340
        Urinary tract, 288
    Enhanced peritoneal stripe sign (EPSS), 321
    Left paralumbar site/view/window, 268, 322
    Right paralumbar site/view/window, 268, 303, 333
    Measurements, 257
    Shadow (see ultrasound acoustic shadow)
    Subxiphoid site/view/window, 32, 33, 36, 82, 85, 160, 173, 184, 201, 208, 212, 225, 251, 268, 269, 303, 327, 381, 384
    Technique, 268
    Umbilical site/view/window, 268, 269
    Urinary bladder site/view/window, 268, 276, 303, 332, 342, 359
POCUS – Cardiac (Section 6), 155
    Contractility (decreased),
    Eccentricity Index, 218
    "Four-chamber" (right parasternal long axis) view, 158, 173, 184
    Fractional shortening, 86, 203
    Heart rate measurement, 314
    Left atrial enlargement, 57-58, 74, 208, 210
    Left atrium to aorta (La:Ao) ratio
        enlarged, 104, 111, 172, 196, 263
        decreased, 248
    Left atrium to aorta (right parasternal short axis La:Ao) view, 156, 173, 250
    Interatrial septum (deviated), 110, 111
    Interventricular septum (flattening), 218
    "Mercedes" sign, 158
    Mitral regurgitation, 210
    M-mode, 207, 312
    "Mushroom" (right parasternal short axis left ventricular) view, 156, 188, 251
    Pericardial thickening, 211, 212
    Pericardial thrombus, 179
    Pulmonary artery to aorta ratio (PA:Ao), 218
    Right atrial inversion/collapse, 174, 185, 212
    Right atrial mass, 186
    Right heart enlargement, 86, 110, 201, 209, 218
    Right ventricular outflow tract (RVOT), 210
    Septal bounce, 211
    "Smoke", 196
    Subxiphoid cardiac long-axis view, 159
    Subxiphoid caudal vena cava view, 159
    Transaortic view, 210
POCUS – interventional, 11
POCUS – Miscellaneous (Section 6), 361
    Limb, 253
    Optic nerve sheath (Case 13.1), 365
    Subcutaneous tissue (Case 9.3 and 13.2), 253, 370
POCUS – Pleural space and lung ultrasound (PLUS) (Section 2), 19-34
    Air bronchogram, 76, 111, 112, 143
    A lines, 21, 22, 39, 49
    B lines, 24, 37, 43, 55, 71, 98, 104, 110, 118, 129, 133, 143, 166, 201, 260, 371, 404
    Barcode sign, 123
    BAT sign, 20, 22
    Curtain sign, 25, 49, 123
    E- lines, 320

Fluid bronchogram, 77
I-lines, 25, 57
Jellyfish sign, 93
Landmarks, 32
Lung consolidation, 21, 29, 30, 37-40, 43, 45, 62, 65-67, 71, 73, 76, 77, 93, 100, 105, 108, 112, 118-120, 129-132, 140, 142, 144, 145, 150
Lung slide (present), 76, 118
Lung slide (absent), 48, 89
Lung recruitment (ultrasound-guided), 66, 67, 144, 145, 146
modified approach, 40
Nodule sign, 60, 62, 76, 90, 119, 135
Normal findings, 23
pleural and lung borders, 26-38
pleural line
  abnormal, 60, 65, 66, 70, 75, 91, 100, 108, 115, 125, 130, 135-137, 144, 146, 168, 198, 256, 365
  normal, 21, 22, 30, 50, 51, 55, 56, 77, 103, 104, 125, 145
pseudohypertrophy, 248, 381
Sail sign (mediastinal triangle), 150, 171
Shred sign, 37, 39, 44, 46, 62, 65, 143
Technique, 29
Z- lines, 23, 49, 58
POCUS – Vascular
  Ultrasound-guided vascular access (Section 4, Chapter 8a), 229
    In-plane technique, 230, 233
    Out-of-plane technique, 229, 233
  Caudal Vena Cava (CVC)
    Collapsed/flattened, 245, 347
    Collapsibility index (CVC-CI) (Chapter 8b), 85, 230, 244, 256-257
    CVC:Ao index, 85, 244, 247
    Distended (non-compliant), 85, 173, 205, 211, 219, 245
    Central line placement (Case 9.1), 239
    Phlebitis (Case 9.2), 246
    Venous thrombus (Case 9.2), 246
POCUS – serial, 112, 269, 278, 292, 299, 330, 342, 371, 384, 386
POCUS – systemic, 10, 107, 109, 111, 114, 147, 278, 365
POCUS – triage, 10
Pulmonary
  abscess (Case 5.10), 87
  fibrosis (CIPF, Case 5.18), 132
  contusions (Case 5.2 and 5.12), 44, 100
  edema
    cardiogenic
      dilated cardiomyopathy (DCM) (PLUS, Case 5.9), 85
      feline (Cardiac POCUS, Case 7.6), 192
      myxomatous mitral valve disease (MMVD) (PLUS, Case 5.4), 53
      myxomatous mitral valve disease (MMVD) (Cardiac POCUS, Case 7.1), 164
      Volume overload (Case 9.4), 257
    non-cardiogenic
      acute respiratory distress syndrome (ARDS)
        B-lines (Case 5.7), 76
        Lung consolidation (Case 5.17), 127
        negative pressure (PLUS, Case 5.13), 103
      hypertension (Severe PH, Case 7.9), 213
      metastatic neoplasia (Case 5.5 and 5.15), 61, 119
      thromboembolism (Case 5.14), 110
Pyelonephritis (Case 11.4), 288
Pyometra (Case 11.5), 294

Renal
  Cortex, 270, 337
  Corticomedullary distinction, 332, 336
  Crest, 334
  Medulla, 270, 336
  pyelectasia (hydronephrosis), 270, 290, 291
  pelvis dilation, 265, 289, 292, 323
  pelvic width, 271, 335
Renomegaly, 270, 290, 292, 334, 337

Subcutaneous emphysema (Case 13.2), 370

Thromboembolism (see pulmonary thromboembolism)
Thrombophlebitis (Case 9.3), 253
Tracheal tear (see subcutaneous emphysema)

Ultrasound
  artifacts, 5
  acoustic shadow, 5, 275, 277, 294, 341, 393
  color doppler, 410
  "twinkle" artifact, 277
  image acquisition, 13
  image optimization, 7
  physics, 3
  probe (transducer) choice 6-7, 20, 164, 263
  probe maneuvers, 13-14
  reverberation, 373
Spleen (normal), 268, 301-302
Splenic
  mass (Case 11.7), 305
  thromboembolism (Case 5.14), 111, 114
  torsion (Case 15.6), 408
Subcutaneous emphysema (tracheal tear) (Case 13.2), 370
Ureter(al)
  dilation (hydroureter), 269, 289, 292, 327
  obstruction (Case 11.4 and 11.12), 288, 333
  pyelogram, 325
Uterine wall, 296
Uterus (distended), 271, 296
Urinary volume estimation (Case 11.13), 339

Volume overload (see fluid overload)